SCHAUM'S OUTLINE OF

THEORY AND PROBLEMS

OF

FINITE ELEMENT ANALYSIS

•

GEORGE R. BUCHANAN

Department of Civil Engineering
Tennessee Technological University
Cookeville, Tennessee

SCHAUM'S OUTLINE SERIES

McGRAW-HILL

New York   San Francisco   Washington, D.C.   Auckland   Bogotá   Caracas
Lisbon   London   Madrid   Mexico City   Milan   Montreal
New Delhi   San Juan   Singapore
Sydney   Tokyo   Toronto

GEORGE R. BUCHANAN is Professor of Civil Engineering at the Tennessee Technological University, Cookeville. He received B.S. and M.S. degrees in Civil Engineering from the University of Kentucky and the Ph.D. from the Virginia Polytechnic Institute. He began his university career in the Department of Engineering Science and Mechanics at Tennessee Tech, where he served for 16 years including four years as chairperson. After a year as a visiting scientist at the Los Alamos National Laboratory he returned to Tennessee Tech with the Department of Civil Engineering. He is the author of a textbook, *Mechanics of Materials* (Saunders). He has served as a consultant with the U.S. Army Missile Command, Tennessee Valley Authority, and Los Alamos National Laboratory.

Schaum's Outline of Theory and Problems of
FINITE ELEMENT ANALYSIS

8 9 0 VLP VLP 0 5 4 3

ISBN 0-07-008714-8

Sponsoring Editor: Arthur Biderman
Production Supervisor: Kathy Porzio
Editing Supervisor: Patty Andrews

Library of Congress Cataloging-in-Publication Data

Buchanan, George R.
    Schaum's outline of theory and problems of finite element analysis
/ George R. Buchanan.
      p.   cm. – – (Schaum's outline series)
    Includes index.
    ISBN 0-07-008714-8
    1. Finite element method. I. Title. II. Title: Finite element
analysis. III. Series.
TA347.F5B83   1994
620'.001'51535 – – dc20   94-11362
                                                              CIP

McGraw-Hill

A Division of The McGraw-Hill Companies

To Chris,

my trusted friend and beloved wife,

who made me complete

Preface

During my academic career the topic of finite element analysis has literally grown from a mere concept into one of the most powerful methods of numerical analysis that exist today. My first significant work with finite element modeling and computer coding of those models was in the early 1970s, and I was not convinced then that the results that could be obtained were worth the effort that went into using the method. Isoparametric finite elements, Galerkin methods, and the use of the more powerful numerical integration techniques were new and emerging. At that time I was not yet aware of the impact these ideas were having on the use and development of the finite element method. As time passed, I became a serious student of the method and developed some capability for successful application of the method.

The Outline is based upon notes that I developed, over a period of several years, for a first course in finite element analysis. My own introduction to finite element methods was somewhat naive, and I think that is often the case for someone with prior knowledge of matrix analysis of structures. I must emphasize that the finite element analysis is not an extension of either the stiffness method or matrix theory of structures. Note that the method is not used to solve engineering problems but is used to solve differential equations, a subtle but significant difference.

The Outline is written with emphasis on applied techniques rather than theoretical justification of the techniques and methods. Each chapter of the Outline, especially Chapters 2 through 6, serves a specific purpose. Chapter 1 contains a brief review of specific mathematical topics. Chapter 2 begins with the Rayleigh-Ritz method and a variational statement of a standard second-order, one-dimensional differential equation that appears in numerous applications in applied physics. It is illustrated that the finite element method is an organized application of the Rayleigh-Ritz method of numerical analysis. Chapter 3 is an extension to two dimensions. The problems are formulated in the standard cartesian coordinate system with emphasis on the formulation of the element, area integration, and subsequent formulation of the global finite element model. Chapters 2 and 3 are academic, but very necessary since they serve as an introduction to the more powerful modern applications of the finite element method. Chapter 4 is intended to show the connection between finite element analysis and matrix analysis of structures and can be omitted by the reader who is not interested in beam and column structures. Chapter 5 is important for the reader who intends to use the finite element method for solving problems that involve coupled partial differential equations. There is an overview of the underlying mathematics that supports the use of the variational functions that were introduced in Chapters 2 and 3. The very powerful Galerkin method of numerical analysis is introduced in Chapter 5 and used to derive finite element models of partial differential equations that govern several different physical phenomena. Chapter 6 is devoted to isoparametric finite elements and the coordinate transformations and numerical integrations that pertain to that topic. Chapter 7 is a collection of several applied topics. A computer code is included as an Appendix for readers desiring some connection between theory and computer application. An index of solved problems is included that will assist the reader who is searching for a particular application.

Several people should be acknowledged for their assistance with the preparation of this book. Professor John Peddieson, Jr., Tennessee Technological University, for many discussions, over the years, concerning applied mathematics; Jeffery Abston for the analytical solution given in Problem 7.23; Mean-Fun Cheng for the numerical results given in Problem 7.11; and Satya P. Narimetla for the numerical solutions given in Problem 7.18. I wish to thank Ms. Yvette Clark for her very gracious assistance with computers, software, and printers. The draft copy was meticulously reviewed by Abraham J. Rokach, Hypermedia Systems Inc., Chicago. I also wish to thank the staff of editors of the Schaum's Division of McGraw-Hill, John Aliano, David Beckwith, and Arthur Biderman for their patience and encouragement.

George R. Buchanan
Cookeville, Tennessee

Contents

Chapter 1

Mathematical Background

1.1. INTRODUCTION

The mathematics required for the study of finite element analysis can vary from elementary to sophisticated. Fortunately, most concepts can be mastered with a reasonable knowledge of vector analysis, matrix theory, and differential equations. Pertinent mathematical concepts will be reviewed in this chapter, and the reader who needs more information may consult the references listed in the bibliography. The review of vector analysis ranges from elementary definitions to more advanced integral theorems. The matrix theory that is covered consists of elementary definitions, matrix manipulations, and the solution of simultaneous equations. A brief treatment of differential equations is also included. Differential equations are required for solving boundary-value problems that can be used as a check on numerical solutions obtained using the finite element method. Finally, a discussion of tensor analysis is included but is limited to cartesian tensor notation. The formulation of problems using cartesian tensor (subscript) notation occurs in the literature of finite elements and usually streamlines the mathematical presentation.

1.2. VECTOR ANALYSIS

A vector is defined as a physical quantity that can be described by a single magnitude and a direction that is related to a coordinate reference frame. A fundamental concept, which justifies the use of vector analysis, is that physical quantities that are arbitrarily directed in space can be resolved into orthogonal components corresponding to the reference frame. Once the components are found, they can be manipulated using standard algebraic operations. Several specialized vector operations required throughout this text will be reviewed in this chapter. Vectors, in this chapter, will be written using boldface lowercase letters. The vector of Fig. 1-1 is

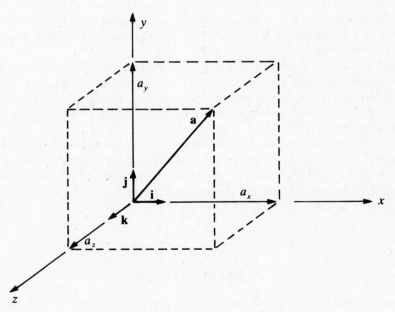

Fig. 1-1

1

$$\mathbf{a} = a_x\mathbf{i} + a_y\mathbf{j} + a_z\mathbf{k} \tag{1.1}$$

where $\mathbf{i}$, $\mathbf{j}$, and $\mathbf{k}$ are unit vectors directed along the x, y, and z axes, respectively.

The *vector differential operator del* ∇ is defined as

$$\nabla \equiv \frac{\partial}{\partial x}\mathbf{i} + \frac{\partial}{\partial y}\mathbf{j} + \frac{\partial}{\partial z}\mathbf{k} \tag{1.2}$$

This operator, by definition, has vector properties and is used to define three fundamental vector operations, the *gradient*, the *divergence*, and the *curl*. These vector operations are useful when defining integral vector theorems such as the *divergence theorem* and *Green's theorem*, which is sometimes called the *Green-Gauss theorem*.

1.3. MATRIX THEORY

Matrices

A rectangular array of numbers with a definite number of rows and columns is a *matrix*. Once an array has been defined as a matrix, it has certain mathematical properties that can be classified within the context of matrix theory. A comprehensive knowledge of matrix theory is not required for finite element analysis; however, certain fundamental concepts are necessary for the study of finite element theory and for its subsequent application.

The array of numbers can be written in the abstract as

$$[A] = \begin{bmatrix} a_{11} & a_{12} & a_{13} & \cdots & a_{1n} \\ a_{21} & a_{22} & a_{23} & \cdots & a_{2n} \\ a_{31} & a_{32} & a_{33} & \cdots & a_{3n} \\ \cdots & \cdots & \cdots & \cdots & \cdots \\ a_{m1} & a_{m2} & a_{m3} & \cdots & a_{mn} \end{bmatrix} \tag{1.3}$$

The notation $[A]$ will be used in this text to indicate a matrix. The terms within the matrix are called elements, and when an element or a group of elements is referred to, subscript notation will be used, such as a_{ij}, where i indicates a row number and j indicates a column number. The matrix of Eq. (*1.3*) is called an m by n matrix or simply an $m \times n$ matrix, and $m \times n$ is referred to as the *order* of the matrix. A row matrix is defined as a $1 \times m$ matrix and, similarly, a column matrix is defined as an $m \times 1$ matrix. The column matrix is often written $\{A\}$.

A matrix $[A]$ and a matrix $[B]$ can be added or subtracted, element by element, as long as both are $m \times n$ matrices. Proper addition and subtraction are not defined for matrices of unequal order. Matrix multiplication is the process of multiplying one matrix by a second matrix and is written $[A][B]$; in general, $[A][B] \neq [B][A]$. In matrix multiplication $[A][B]$, $[B]$ is said to be premultiplied by $[A]$ or $[A]$ is said to be postmultiplied by $[B]$.

The division, element by element, of one matrix by a second matrix is not defined. However, the *inverse* of a matrix, written as $[A]^{-1}$, serves a similar purpose and will be discussed later.

The *transpose* of a matrix is obtained by interchanging its rows and columns. The transpose of $[A]$ is written $[A]^T$, and in subscript notation the element interchange is

$$[a_{ij}]^T = [a_{ji}] \tag{1.4}$$

It follows that

$$[[A][B]]^T = [B]^T[A]^T \tag{1.5}$$

A *symmetric* matrix is defined as a square matrix with property $a_{ij} = a_{ji}$ for $i \neq j$. A *diagonal* matrix has all elements of a square matrix equal to zero except those on the *principal* diagonal, which is the diagonal from upper left to lower right. A *unit* matrix is a special case of a diagonal matrix; all diagonal elements are equal to 1, and all off-diagonal elements are equal to 0.

Determinants

An understanding of selected topics from the theory of determinants is necessary for successful solution of the simultaneous equations that result in finite element analysis. The determinant used throughout this text is of course a square matrix and as a square matrix has certain mathematical properties an ordinary $m \times n$ matrix does not have. The determinant is used in Prob. 1.4 to define the vector product. Determinants of order 2 or 3 can generally be used to illustrate all the concepts required for understanding the manipulation of determinants. A determinant is usually symbolized by enclosing the array of numbers within vertical lines rather than brackets. For a matrix denoted [A], the notation for the determinant might be $|A|$, det [A], or $|\det A|$ and indicates the determinant of the matrix [A].

Every determinant has a *determinantal equation*. For higher-order determinants it can be quite formidable from a computational standpoint to obtain that equation. The determinantal equation of a determinant of order 3 is obtained as follows:

$$|\det A| = \begin{vmatrix} a_{11} & a_{12} & a_{13} \\ a_{21} & a_{22} & a_{23} \\ a_{31} & a_{32} & a_{33} \end{vmatrix}$$

$$= a_{11}a_{22}a_{33} + a_{12}a_{23}a_{31} + a_{13}a_{32}a_{21} - a_{11}a_{23}a_{32} - a_{21}a_{12}a_{33} - a_{31}a_{22}a_{13} \qquad (1.6)$$

and can be described as the product of the principal diagonal terms minus the product of the secondary diagonal terms. This elementary concept can be applied to determinants of order 2 or 3 but fails for higher-order determinants.

The *minor* of a determinant is the determinant that remains after a row and a column are removed from the original determinant. The minor can be referenced to a particular element of the determinant using the notation a_{ij}. The minor of $|a_{22}|$ of Eq. (1.6) is

$$|M_{22}| = \begin{vmatrix} a_{11} & a_{13} \\ a_{31} & a_{33} \end{vmatrix}$$

a determinant of order 2. The *cofactor* of an element of a determinant is defined as $C_{ij} = (-1)^{i+j}|M_{ij}|$, where $|M_{ij}|$ is the minor of the element a_{ij}. It is now possible to define a *cofactor matrix* as the square matrix constructed by replacing each element of a square matrix by the cofactor of the determinant corresponding to the original square matrix. The *adjoint matrix* is defined as the transpose of the cofactor matrix. The adjoint matrix is used to compute the inverse of a matrix in Prob. 1.14.

Simultaneous Equations

Numerous methods have been proposed for the solution of a set of simultaneous equations, and two of these procedures will be emphasized in this text. A set of simultaneous equations can be written in matrix form as

$$[A]\{x\} = \{f\} \qquad (1.7)$$

The matrix [A] represents the matrix of coefficients that are multiplied by the unknown quantities $\{x\}$. The column matrix on the right-hand side contains the known quantities f. Multiplying by the inverse of [A] gives

$$[A]^{-1}[A]\{x\} = [I]\{x\} = \{x\} = [A]^{-1}\{f\} \qquad (1.8)$$

The use of the inverse for solving a set of simultaneous equations is inefficient for large sets of equations.

A method that is sometimes called *gaussian elimination* is faster and hence more efficient than the inverse method. Gaussian elimination is an organized method of substituting each equation into the previous equation until the last equation contains only one unknown. The unknowns are determined sequentially, starting with the last equation and proceeding upward. The method is sometimes referred to as upper trianglization and is best illustrated by example, as in Prob. 1.16.

1.4. DIFFERENTIAL EQUATIONS

Finite element analysis is a method for the numerical solution of a differential equation. It follows that without differential equations there would not be a finite element method. Many practicing engineers and scientists learned the finite element method as an application of structural analysis for civil engineering or aircraft structures. The state of the art of finite element analysis several decades ago was responsible for that situation. The classical stiffness method of structural analysis, as discussed in Chap. 4, can be derived without mention of the governing differential equations. That is, the fundamental relationships for deriving the stiffness method are based upon solutions of differential equations, but the user can easily lose sight of the origin of the analysis. Finite element analysis for beam and frame structures can be based upon energy theorems without considering differential equations. Again, the fault is not with the engineer or scientist, but historically the connection between energy methods in structural analysis and the governing differential equation has not been emphasized.

Differential equations are emphasized in this text. Beginning in Chap. 2 the differential equation is associated with the corresponding variational function (energy theorem). The finite element method can be derived in a variety of ways, but regardless of the derivation, the method is the *numerical solution of a differential equation*. The differential equations in this text are for the most part elementary. A very basic differential equation is considered in Chap. 2, where it is shown that the same equation governs numerous physical theories. The most elementary equation will probably allow the reader to become acquainted with the connection between finite element theory and differential equations, and for this reason Chap. 2 is a very important chapter. From a practical viewpoint finite element analysis would not be used to solve a one-dimensional second-order differential equation, however, Chap. 2 is an absolute necessity for understanding the more complicated analysis problems.

The analytical solution of the differential equation is important as a check on the numerical solution obtained using the finite element method. How else will the user know if the computer code that generates a numerical solution is correct? The fundamental differential equations of Chap. 2 are of the form

$$\frac{d}{dx}\,\alpha(x)A(x)\,\frac{d\phi(x)}{dx} + C(x)A(x) = 0 \qquad\qquad (1.9)$$

where $\alpha(x)$ is a material parameter that can be a function of x, $C(x)$ is an external source, and $A(x)$ is the cross-sectional area. If material parameters, external source terms, and area are functions of x, they are allowed to change from element to element. In other words, the finite element is not expected to model a variable area or material parameter within the element since they can be modeled from element to element. The functional form is usually disregarded, and Eq. (1.9) is written in a more elementary form as

$$\alpha\,\frac{d^2\phi}{dx^2} + C = 0 \qquad\qquad (1.10)$$

The analytical solution is elementary. Nevertheless, such elementary solutions are invaluable for successful application of the finite element method. The most general form of Eq. (1.10) is given as Eq. (2.19) and appears as

$$\alpha\,\frac{d^2\phi(x)}{dx^2} - \beta\,\frac{d\phi}{dx}(x) - \gamma\phi(x) + C = 0 \qquad\qquad (1.11)$$

Solutions for one-dimensional differential equations are given in Probs. 2.1, 2.3, 2.17, 2.18, and 2.28. Problem 2.18 provides a solution for an equation of the general form of Eq. (*1.11*), and Prob. 2.28 represents a solution for a differential equation in one-dimensional cylindrical coordinates with a change in material properties. The one-dimensional counterpart of Eq. (*1.11*) in cylindrical coordinates is

$$\frac{\alpha}{r}\frac{d}{dr}\left[r\frac{d\phi(r)}{dr}\right] - \beta\frac{d\phi}{dr}(r) - \gamma\phi(r) + C = 0 \qquad (1.12)$$

The two-dimensional counterpart of Eq. (*1.11*) is a partial differential equation and is discussed in Chap. 3 and modeled primarily in x, y coordinates. [See Eqs. (*3.1*) and (*3.2*).] The analytical solutions become more complicated and are obtained using the separation of variables technique. Steady-state temperature distribution for a rectangular plate is discussed in Prob. 3.6 using both a classical approach and finite element analysis. Additionally, the classical approach is discussed in Prob. 1.17. An analytical solution for an equation similar to the mass transport, Eq. (*3.2*), is given by Prob. 3.36. Partial differential equations occur when both spatial and time coordinates are included in the same problem. This type of partial differential equation is introduced in Chap. 5 and discussed in some detail in Chap. 7.

The finite element method is quite powerful for solving coupled ordinary or partial differential equations. Analytical solutions for coupled partial differential equations can become a challenge. The equations of elasticity were one of the early topics to be studied using the finite element method and in two or more dimensions are always a set of coupled equations. These equations are given in Chap. 3 for both cartesian and cylindrical coordinates. Coupled partial differential equations are discussed in Chaps. 3 and 5–7.

Homogeneous differential equations occur in mathematical physics and offer a somewhat different challenge for the analyst than the nonhomogeneous equation. The resulting analysis is referred to as an *eigenvalue problem*. Methods for solving the differential equation eigenvalue problem and the algebraic eigenvalue problem can be found in numerous textbooks. Several important problems are formulated in Chap. 7, and elementary finite element solutions are given.

1.5.   CARTESIAN TENSORS

Cartesian tensor notation is the simplified version of tensor notation that can be referenced to any curvilinear coordinate system. Cartesian tensor notation is often referred to as *indicial notation* or *subscript tensor notation* since only subscripts are required for proper representation of physical equations, but the reader must keep in mind that the notation is valid only for the classical x, y, z coordinate system. Cartesian tensor notation is used in this text for convenience when writing the governing equations for a problem to be modeled using the finite element method. Also, once the analyst has become comfortable with finite element modeling, it is possible to directly visualize a matrix finite element equation based upon the tensor equation.

Visualize the standard x, y, z coordinate system. Rather than use x, y, z let the coordinates be called x_1, x_2, x_3, where $x_1 \Rightarrow x$, $x_2 \Rightarrow y$, and $x_3 \Rightarrow z$ (the symbol $\Rightarrow$ has the meaning "is the same as"). Any vector can be written as $\mathbf{f} = f_x\mathbf{i} + f_y\mathbf{j} + f_z\mathbf{k}$ in the x, y, z system. A similar vector equation in the x_1, x_2, x_3 system would be written $\mathbf{f} = f_1\mathbf{i} + f_2\mathbf{j} + f_3\mathbf{k}$. The essence of subscript notation is to write the vector equation

$$\mathbf{f} \Rightarrow f_1\mathbf{i} + f_2\mathbf{j} + f_3\mathbf{k} \Rightarrow f_i \qquad (1.13)$$

where i has the *range* 1, 2, 3. Cartesian tensor notation is a shorthand version of vector notation. A vector, based upon its mathematical definition, can be classified as a *first-order tensor*. Higher-order tensors can be defined, but there are no corresponding quantities in vector analysis; however, all vector operations are defined for tensors of any order.

There are two basic types of subscripts, *range* and *summation*. The range subscript corresponds to the coordinate direction as illustrated by Eq. (*1.13*), and a single unrepeated subscript implies three quantities, one corresponding to each coordinate direction. A repeated subscript implies a summation and is illustrated in Prob. 1.18. Partial differentiation with respect to the space coordinates can be represented using a comma following the variable that is to be differentiated and is illustrated in Prob. 1.19. Time differentiation is indicated by placing a dot above the variable.

Coordinate transformation plays a significant role in the mathematical theory of tensor analysis. The elementary definition of a vector is based upon the idea that any physical quantity that can be described with a magnitude and a direction qualities as a vector. Besides the elementary definition, a vector quantity must exist in all coordinate systems and there must be a valid transformation relationship between the coordinate systems in order for it to quality as a vector. Coordinate transformations are discussed in Chap. 3 using a matrix to represent the transformation. The coordinate transformation matrix is not a tensor because it does not have any properties that allow for the transformation between coordinate systems. Additional discussion of this concept is given in Prob. 1.20.

Solved Problems

1.1. A position vector emanates from point P in space defined by coordinate locations $(10, 15, 5)$ and terminates at point Q defined by $(-2, 5, 3)$ as shown in Fig. 1-2. Use the concept of addition and subtraction of vectors to determine the components of the position vector and compute the magnitude of the position vector.

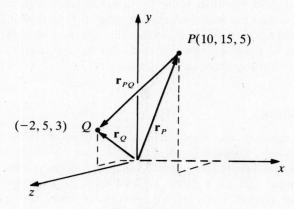

Fig. 1-2

Define position vectors $\mathbf{r}_P = 10\mathbf{i} + 15\mathbf{j} + 5\mathbf{k}$ and $\mathbf{r}_Q = -2\mathbf{i} + 5\mathbf{j} + 3\mathbf{k}$. Use Fig. 1-2 to write the vector equation

$$\mathbf{r}_Q = \mathbf{r}_P + \mathbf{r}_{PQ} \qquad \text{or} \qquad \mathbf{r}_{PQ} = \mathbf{r}_Q - \mathbf{r}_P \qquad (a)$$

Substituting and adding components gives the required result:

$$\mathbf{r}_{PQ} = -12\mathbf{i} - 10\mathbf{j} - 2\mathbf{k} \qquad (b)$$

The magnitude, in this case, is the actual length of the position vector:

$$|\mathbf{r}_{PQ}| = [(-12)^2 + (-10)^2 + (-2)^2]^{1/2} = 15.7 \qquad (c)$$

1.2. Define a unit vector and compute the components of a unit vector directed from point P to point Q of Fig. 1-2.

A unit vector **u** is a vector with a magnitude of unity. The components of the unit vector must satisfy the relation

$$[u_x^2 + u_y^2 + u_z^2]^{1/2} = 1 \qquad (a)$$

where u_x, u_y, and u_z are the x, y, z components of the unit vector, respectively. The components of the unit vector are computed using the position vector of Fig. 1-2. The position vector $\mathbf{r}_{PQ}$ is defined in terms of a local coordinate system in Fig. 1-3 using the angles θ_x, θ_y, and θ_z. The direction cosines are defined as follows with $\mathbf{r}_{PQ}$ as the magnitude of the vector:

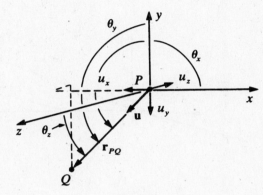

Fig. 1-3

$$\cos \theta_x = \frac{r_{PQx}}{r_{PQ}} = \frac{-12}{15.7} = -0.762 \qquad r_{PQx} = r_{PQ} \cos \theta_x$$

$$\cos \theta_y = \frac{r_{PQy}}{r_{PQ}} = \frac{-10}{15.7} = -0.635 \qquad r_{PQy} = r_{PQ} \cos \theta_y \qquad (b)$$

$$\cos \theta_z = \frac{r_{PQz}}{r_{PQ}} = \frac{-2}{15.7} = -0.127 \qquad r_{PQz} = r_{PQ} \cos \theta_z$$

Write the vector in component form, as in Eq. (*1.1*), and substitute Eq. (*b*):

$$\mathbf{r}_{PQ} = r_{PQx}\mathbf{i} + r_{PQy}\mathbf{j} + r_{PQz}\mathbf{k} = r_{PQ}(\cos \theta_x \mathbf{i} + \cos \theta_{yj} + \cos \theta_z \mathbf{k}) \qquad (c)$$

Recognize that Eq. (*c*) is

$$\mathbf{r}_{PQ} = r_{PQ}(u_x \mathbf{i} + u_y \mathbf{j} + u_z \mathbf{k}) = r_{PQ}\mathbf{u} \qquad (d)$$

and the direction cosines are the components of the unit vector, then

$$\mathbf{u} = -0.762\mathbf{i} - 0.635\mathbf{j} - 0.127\mathbf{k} \qquad (e)$$

It is easily verified that Eq. (*d*) satisfies Eq. (*a*).

1.3. Define the *scalar* product of two vectors and use that definition to compute the component of the vector **f** (**f** = $-20\mathbf{i} + 5\mathbf{j} + 12\mathbf{k}$) in Fig. 1-4(*b*) that passes through point *B*. Write the result in vector format.

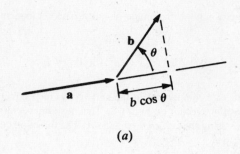

(a)

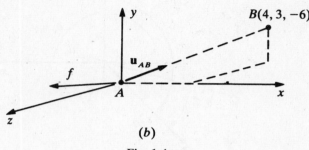

(b)

Fig. 1-4

The scalar product or *dot* product is illustrated in Fig. 1-4(a) and by definition is a scalar quantity given by

$$\mathbf{a} \cdot \mathbf{b} = a(b \cos \theta) = \mathbf{b} \cdot \mathbf{a} = b(a \cos \theta) \tag{a}$$

The component of $\mathbf{f}$ that passes through point B is computed as $\mathbf{f} \cdot \mathbf{u}_{AB}$, where $\mathbf{u}_{AB}$ is the unit vector from A to B. Refer to Prob. 1.2 for computation of the unit vector:

$$\mathbf{f} \cdot \mathbf{u}_{AB} = (-20\mathbf{i} + 5\mathbf{j} + 12\mathbf{k}) \cdot \frac{4\mathbf{i} + 3\mathbf{j} - 6\mathbf{k}}{7.81}$$

$$= [(-20)(4)\mathbf{i} \cdot \mathbf{i} + (5)(3)\mathbf{j} \cdot \mathbf{j} + (12)(-6)\mathbf{k} \cdot \mathbf{k}] \div 7.81 = -17.5$$

Note that $\mathbf{i} \cdot \mathbf{i} = (1)(1) \cos 0 = 1$; similarly, $\mathbf{i} \cdot \mathbf{j} = (1)(1) \cos (\pi/2) = 0$; and so forth.

1.4. Define the *vector* product of two vectors and use that definition to compute the vector product $\mathbf{a} \times \mathbf{b}$ of the vectors $\mathbf{a} = 2\mathbf{i} + 3\mathbf{j} + \mathbf{k}$ and $\mathbf{b} = -\mathbf{i} + 2\mathbf{j} - 4\mathbf{k}$.

The vector product or *cross* product is illustrated in Fig. 1-5 and by definition is a vector result given by

$$\mathbf{a} \times \mathbf{b} = a(b \sin \theta) \tag{a}$$

The resulting vector is directed perpendicular to the plane formed by $\mathbf{a}$ and $b \sin \theta$. The same numerical result is obtained for $\mathbf{b} \times \mathbf{a}$, but the resulting vector is opposite in sign. The vector product can be computed using the determinant

$$\mathbf{a} \times \mathbf{b} = \begin{vmatrix} \mathbf{i} & \mathbf{j} & \mathbf{k} \\ 2 & 3 & 1 \\ -1 & 2 & -4 \end{vmatrix} = (-12 - 2)\mathbf{i} + (-1 + 8)\mathbf{j} + (4 + 3)\mathbf{k} = -14\mathbf{i} + 7\mathbf{j} + 7\mathbf{k}$$

It follows that $\mathbf{i} \times \mathbf{i} = (1)(1) \sin 0 = 0$; similarly, $\mathbf{i} \times \mathbf{j} = (1)(1) \sin (\pi/2) = \mathbf{k}$, a vector normal to the x, y plane of unit magnitude. Also, $\mathbf{j} \times \mathbf{i} = -\mathbf{i} \times \mathbf{j} = -\mathbf{k}$.

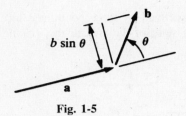

Fig. 1-5

1.5. Compute the components of a unit vector directed normal to the line segment defined by x_1, y_1 and x_2, y_2 as shown in Fig. 1-6.

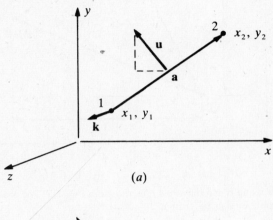

(a)

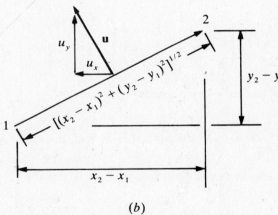

(b)

Fig. 1-6

Given the definition of the vector product, $\mathbf{k} \times \mathbf{a}$ is a vector perpendicular to the plane of $\mathbf{k}$ and $\mathbf{a}$, where $\mathbf{k}$ is the unit vector in the z-coordinate direction as shown in Fig. 1-6.

$$\mathbf{k} \times \mathbf{a} = \begin{vmatrix} \mathbf{i} & \mathbf{j} & \mathbf{k} \\ 0 & 0 & 1 \\ x_2 - x_1 & y_2 - y_1 & 0 \end{vmatrix} = -(y_2 - y_1)\mathbf{i} + (x_2 - x_1)\mathbf{j}$$

The unit vector parallel to $\mathbf{k} \times \mathbf{a}$ is

$$\mathbf{u} = \frac{-(y_2 - y_1)\mathbf{i} + (x_2 - x_1)\mathbf{j}}{[(y_2 - y_1)^2 + (x_2 - x_1)^2]^{1/2}}$$

The result can be interpreted as shown in Fig. 1-6(b).

1.6.   A plane triangular area is defined by three coordinate locations as shown in Fig. 1-7. Use vector analysis to derive an expression for the area of the triangle in terms of the coordinate locations.

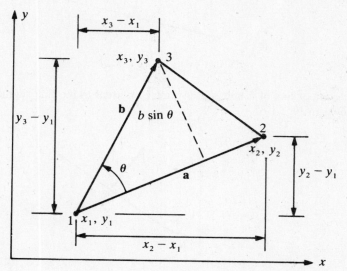

Fig. 1-7

Define the vectors **a** and **b** and the angle θ as shown in Fig. 1-7. The area of the triangle is $\frac{1}{2}ab \sin \theta$, and by Eq. (*a*) of Prob. 1.4 the area can be written

$$\mathbf{A} = \frac{1}{2}\mathbf{a} \times \mathbf{b} = \frac{1}{2}\begin{vmatrix} \mathbf{i} & \mathbf{j} & \mathbf{k} \\ x_2 - x_1 & y_2 - y_1 & 0 \\ x_3 - x_1 & y_3 - y_1 & 0 \end{vmatrix} = \frac{1}{2}[(x_2 - x_1)(y_3 - y_1) - (x_3 - x_1)(y_2 - y_1)]\mathbf{k} \qquad (a)$$

The result shows that area has vector properties.

1.7.   Define the gradient of a scalar function. Let ϕ and ψ be two scalar functions and show that $\nabla(\phi\psi) = \phi \, \nabla\psi + \psi \, \nabla\phi$.

The gradient of a scalar function ϕ is defined as

$$\nabla\phi = \left(\frac{\partial}{\partial x}\mathbf{i} + \frac{\partial}{\partial y}\mathbf{j} + \frac{\partial}{\partial z}\mathbf{k} \right)\phi = \frac{\partial \phi}{\partial x}\mathbf{i} + \frac{\partial \phi}{\partial y}\mathbf{j} + \frac{\partial \phi}{\partial z}\mathbf{k} \qquad (a)$$

Similarly,

$$\nabla(\phi\psi) = \frac{\partial}{\partial x}(\phi\psi)\mathbf{i} + \frac{\partial}{\partial y}(\phi\psi)\mathbf{j} + \frac{\partial}{\partial z}(\phi\psi)\mathbf{k}$$

$$= \phi\left(\frac{\partial \psi}{\partial x}\mathbf{i} + \frac{\partial \psi}{\partial y}\mathbf{j} + \frac{\partial \psi}{\partial z}\mathbf{k} \right) + \psi\left(\frac{\partial \phi}{\partial x}\mathbf{i} + \frac{\partial \phi}{\partial y}\mathbf{j} + \frac{\partial \phi}{\partial z}\mathbf{k} \right) = \phi \, \nabla\psi + \psi \, \nabla\phi \qquad (b)$$

1.8.   The divergence of vector function is defined as $\nabla \cdot \mathbf{a}$. Discuss this vector operation in terms of the definition of a scalar product.

It follows from Eqs. (*1.1*) and (*1.2*) that

$$\nabla \cdot \mathbf{a} = \left(\frac{\partial}{\partial x} \mathbf{i} + \frac{\partial}{\partial y} \mathbf{j} + \frac{\partial}{\partial z} \mathbf{k} \right) \cdot (a_x \mathbf{i} + a_y \mathbf{j} + a_z \mathbf{k}) \tag{a}$$

$$= \frac{\partial a_x}{\partial x} + \frac{\partial a_y}{\partial y} + \frac{\partial a_z}{\partial z} \tag{b}$$

It follows that $\nabla \cdot \mathbf{a} \neq \mathbf{a} \cdot \nabla$ because the ∇ operator should act (operate) on $\mathbf{a}$.

1.9. The *divergence theorem*, sometimes called *Gauss' divergence theorem*, can be written in vector notation (Spiegel, 1959) as

$$\int_V \nabla \cdot \mathbf{a} \, dV = \int_S \mathbf{a} \cdot \mathbf{n} \, dS \tag{a}$$

where $\mathbf{n}$ is a unit outward normal vector acting on the surface (boundary) of the volume (region) described by V. Equation (a) simply states that the change in the quantity $\mathbf{a}$ within the region is equal to the quantity flowing into or out of the region through the boundary. Discuss the derivation of the *Green-Gauss theorem* using the divergence theorem and the application to finite element concepts.

The Green-Gauss theorem is derived in numerous textbooks. The application that is of interest pertains to Galerkin's method for deriving finite element models (see Chap. 5). Assume a one-dimensional case and consider the derivative of the following function, where k can be considered a constant:

$$\frac{d}{dx} \left(k \frac{d\phi}{dx} \psi \right) = k \frac{d^2\phi}{dx^2} \psi + k \frac{d\phi}{dx} \frac{d\psi}{dx} \tag{b}$$

Integrate over an interval a to b:

$$k \int_a^b \frac{d}{dx} \left(\frac{d\phi}{dx} \psi \right) dx = k \int_a^b \frac{d^2\phi}{dx^2} \psi \, dx + k \int_a^b \frac{d\phi}{dx} \frac{d\psi}{dx} \, dx \tag{c}$$

Consider the term on the left-hand side a perfect differential:

$$k \int_a^b \frac{d}{dx} \left(\frac{d\phi}{dx} \psi \right) dx = k \int_a^b d\left(\frac{d\phi}{dx} \psi \right) = k \left. \frac{d\phi}{dx} \psi \right|_a^b \tag{d}$$

Substitute Eq. (d) into Eq. (c) and rearrange terms:

$$k \int_a^b \frac{d^2\phi}{dx^2} \psi \, dx = k \left. \frac{d\phi}{dx} \psi \right|_a^b - k \int_a^b \frac{d\phi}{dx} \frac{d\psi}{dx} \, dx \tag{e}$$

In one dimension Eq. (e) could be obtained using integration by parts. However, the development above could be extended to two or three dimensions, and the divergence theorem is useful for that purpose.

Let $\mathbf{a}$ of Eq. (a) be the product of a scalar β and a vector $\mathbf{b}$ and substitute into Eq. (a):

$$\int_V \nabla \cdot (\beta \mathbf{b}) \, dV = \int_S \beta \mathbf{b} \cdot \mathbf{n} \, dS \tag{f}$$

The vector identity $\nabla \cdot (\beta \mathbf{b}) = \beta \nabla \cdot \mathbf{b} + \nabla \beta \cdot \mathbf{b}$ is substituted into Eq. (f) to obtain the desired result:

$$\int_V \beta \nabla \cdot \mathbf{b} \, dV = \int_S \beta \mathbf{b} \cdot \mathbf{n} \, dS - \int_V \nabla \beta \cdot \mathbf{b} \, dV \tag{g}$$

Equation (g) is the classical result using vector analysis but can be extended to additional situations other than a scalar and vector. In finite element theory Eq. (e) is an extension of Eq. (g), where ϕ and ψ represent matrices of interpolation functions.

1.10. Given the following matrices, where [A] is a 3×2 matrix and [B] is a 2×2 matrix, discuss the process of performing matrix multiplication.

Matrix multiplication is defined for [A][B] where the number of columns in [A] is equal to the number of rows in [B]. The first row of [A] is multiplied, term by term, by the first column of [B] as shown below to give element 1, 1 of the product matrix. The first row of [A] is then multiplied by the second column of [B], and that becomes element 1, 2 of the product matrix. The result is shown below, and it is sometimes convenient to write the matrices with [A] to the left and below [B] so that a horizontal line through a row of [A] intersects a vertical line through a column of [B] to locate the corresponding element in the product matrix. This procedure is helpful when the multiplication is being done by hand.

$$[B] = \begin{bmatrix} b_{11} & b_{12} \\ b_{21} & b_{22} \end{bmatrix}$$

$$[A] = \begin{bmatrix} a_{11} & a_{12} \\ a_{21} & a_{22} \\ a_{31} & a_{32} \end{bmatrix} \quad \begin{bmatrix} a_{11}b_{11} + a_{12}a_{21} & a_{11}b_{12} + a_{12}b_{22} \\ a_{21}b_{11} + a_{22}b_{21} & a_{21}b_{12} + a_{22}b_{22} \\ a_{31}b_{11} + a_{32}b_{21} & a_{31}b_{12} + a_{32}b_{22} \end{bmatrix}$$

The product of an $m \times n$ matrix and an $n \times p$ matrix is an $m \times p$ matrix. It should be obvious that [B][A] cannot exist for the matrices above and remain within the definition of a proper matrix multiplication.

1.11. Given a scalar function ϕ defined as

$$\phi = N_1\phi_1 + N_2\phi_2 + N_3\phi_3 = \sum_{i=1}^{3} N_i\phi_i \qquad (a)$$

write ϕ in matrix format.

Define $N_i \Rightarrow [N] = [N_1 \quad N_2 \quad N_3]$ and $\phi_i \Rightarrow [\phi] = [\phi_1 \quad \phi_2 \quad \phi_3]$. To obtain the result of Eq. (a) the matrix equation should be constructed in one of the equivalent forms:

$$\phi = [N]\{\phi\} = [N][\phi]^T = [N_1 \quad N_2 \quad N_3]\begin{Bmatrix} \phi_1 \\ \phi_2 \\ \phi_3 \end{Bmatrix} = [N_1 \quad N_2 \quad N_3][\phi_1 \quad \phi_2 \quad \phi_3]^T$$

1.12. The determinantal equation of a determinant can be obtained using the cofactors of the determinant. Given the 3×3 determinant

$$\begin{vmatrix} a_{11} & a_{12} & a_{13} \\ a_{21} & a_{22} & a_{23} \\ a_{31} & a_{32} & a_{33} \end{vmatrix} \qquad (a)$$

the determinantal equation can be written in terms of cofactors

$$a_{11}|a_{11}| - a_{12}|a_{12}| + a_{13}|a_{13}| \qquad (b)$$

where $|a_{ij}|$ are the cofactors and a_{ij} are the corresponding elements of the determinant. Expand Eq. (b) and show that it is equivalent to Eq. (1.6).

Equation (b) is expanded using the definition of the cofactor in terms of the minor, or

$$a_{11}\begin{vmatrix} a_{22} & a_{23} \\ a_{32} & a_{33} \end{vmatrix} - a_{12}\begin{vmatrix} a_{21} & a_{23} \\ a_{31} & a_{33} \end{vmatrix} + a_{13}\begin{vmatrix} a_{21} & a_{22} \\ a_{31} & a_{32} \end{vmatrix}$$

$$= a_{11}(a_{22}a_{33} - a_{32}a_{23}) - a_{12}(a_{21}a_{33} - a_{31}a_{23}) + a_{13}(a_{21}a_{32} - a_{31}a_{22})$$

and is equivalent to Eq. (1.6).

1.13. Given a function $y = ax^2$, where a is a constant. Then, $dy/dx = 2ax$. This mathematical operation occurs in finite element theory, except that the function is represented as a matrix equation. Given that [A] is a symmetric $n \times n$ matrix and {X} is an $n \times 1$ matrix, a matrix equation that is equivalent to the function y above is $y = [X]^T[A]\{X\}$. Show that $dy/d\{X\} = 2[A]\{X\}$.

Assume [A] is 2×2 and {X} is 2×1. Then,

$$y = \begin{bmatrix} x_1 & x_2 \end{bmatrix} \begin{bmatrix} a_{11} & a_{12} \\ a_{21} & a_{22} \end{bmatrix} \begin{Bmatrix} x_1 \\ x_2 \end{Bmatrix} = a_{11}x_1^2 + a_{12}x_2x_1 + a_{21}x_1x_2 + a_{22}x_2^2$$

$$\frac{\partial y}{\partial x_1} = 2a_{11}x_1 + a_{12}x_2 + a_{21}x_2$$

$$\frac{\partial y}{\partial x_2} = a_{12}x_1 + a_{21}x_1 + 2a_{22}x_2$$

Because [A] is symmetric, $a_{12} = a_{21}$. Then,

$$2\begin{bmatrix} a_{11} & a_{12} \\ a_{21} & a_{22} \end{bmatrix} \begin{Bmatrix} x_1 \\ x_2 \end{Bmatrix} = 2[A]\{X\}$$

a result that is valid for matrices of any order.

1.14. Given a matrix

$$[A] = \begin{bmatrix} 3 & 1 & 4 \\ -1 & 4 & 2 \\ -2 & 2 & -2 \end{bmatrix}$$

Define and compute, where appropriate, (a) the minors of the matrix (determinant) [A], M_{ij}, (b) the cofactors of the matrix (determinant) [A], C_{ij}, (c) the adjoint of the matrix (determinant) [A], (d) the value of the determinant |A| of the matrix [A], and (e) the inverse of the matrix [A].

(a) The minor of the term a_{11} is the 2×2 matrix remaining after deleting row 1 and column 1:

$$M_{11} = \begin{bmatrix} 4 & 2 \\ 2 & -2 \end{bmatrix} \quad \text{similarly} \quad M_{12} = \begin{bmatrix} -1 & 2 \\ -2 & -2 \end{bmatrix} \quad M_{13} = \begin{bmatrix} -1 & 4 \\ -2 & 2 \end{bmatrix}$$

$$M_{21} = \begin{bmatrix} 1 & 4 \\ 2 & -2 \end{bmatrix} \quad M_{22} = \begin{bmatrix} 3 & 4 \\ -2 & -2 \end{bmatrix} \quad M_{23} = \begin{bmatrix} 3 & 1 \\ -2 & 2 \end{bmatrix}$$

$$M_{31} = \begin{bmatrix} 1 & 4 \\ 4 & 2 \end{bmatrix} \quad M_{32} = \begin{bmatrix} 3 & 4 \\ -1 & 2 \end{bmatrix} \quad M_{33} = \begin{bmatrix} 3 & 1 \\ -1 & 4 \end{bmatrix}$$

(b) The cofactor of the term a_{11} is obtained formally as $(-1)^{i+j}|M_{11}|$, or

$$C_{11} = (-1)^2(-8-4) = -12 \qquad C_{12} = (-1)^3(2+4) = -6 \qquad C_{13} = (-1)^4(-2+8) = 6$$

$$C_{21} = (-1)^3(-2-8) = 10 \qquad C_{22} = (-1)^4(-6+8) = 2 \qquad C_{23} = (-1)^5(6+2) = -8$$

$$C_{31} = (-1)^4(2-16) = -14 \qquad C_{32} = (-1)^5(6+4) = -10 \qquad C_{33} = (-1)^5(12+1) = 13$$

The determinant of the cofactors is a 3×3 matrix.

(c) The adjoint matrix is the transpose of the cofactor matrix, or

$$[C]^T = \begin{bmatrix} -12 & 10 & -14 \\ -6 & 2 & -10 \\ 6 & -8 & 13 \end{bmatrix}$$

(*d*) The determinant can be evaluated using the minors of part (*a*) and the cofactors of part (*b*). For instance, along the top row,

$$|\det A| = (3)(-12) + (1)(-6) + (4)(6) = -18$$

(*e*) The inverse of [A] can be computed as

$$[A]^{-1} = \frac{[C]^T}{|\det A|}$$

or
$$[A]^{-1} = \begin{bmatrix} -12 & 10 & -14 \\ -6 & 2 & -10 \\ 6 & -8 & 13 \end{bmatrix} \div (-18)$$

The reader can prove the result $[A]^{-1}[A] = [I]$, the unit matrix.

1.15. The inverse of a square matrix can be used in the computation for simultaneous equations. The method of computing the inverse discussed in Prob. 1.14 is limited in application, and another method is preferable for computer applications. The inverse is defined as the matrix $[A]^{-1}$ that can be multiplied by [A] to give a unit matrix, or

$$[A]^{-1}[A] = [I] \tag{a}$$

The inverse can be computed by writing the matrix [A] augmented by a unit matrix

$$[A\,|\,I] \tag{b}$$

Row and column substitutions are applied to Eq. (*b*) that will convert [A] to a unit matrix, and the same operations are applied to the unit matrix that will give the inverse matrix, or

$$[I\,|\,A^{-1}] \tag{c}$$

Compute $[A]^{-1}$ for the matrix of Prob. 1.14 using the method described above.

Write the matrix in the form

$$\begin{bmatrix} 3 & 1 & 4 & \cdot & 1 & 0 & 0 \\ -1 & 4 & 2 & \cdot & 0 & 1 & 0 \\ -2 & 2 & -2 & \cdot & 0 & 0 & 1 \end{bmatrix}$$

Divide row 1 by 3, and the first diagonal term becomes 1:

$$\begin{bmatrix} 1 & 0.3333 & 1.3333 & \cdot & 0.3333 & 0 & 0 \\ -1 & 4 & 2 & \cdot & 0 & 1 & 0 \\ -2 & 2 & -2 & \cdot & 0 & 0 & 1 \end{bmatrix}$$

Multiply the new row 1 by -1 and subtract it from row 2, and the a_{21} term becomes 0:

$$\begin{bmatrix} 1 & 0.3333 & 1.3333 & \cdot & 0.3333 & 0 & 0 \\ 0 & 4.3333 & 3.3333 & \cdot & 0.3333 & 1 & 0 \\ -2 & 2 & -2 & \cdot & 0 & 0 & 1 \end{bmatrix}$$

Multiply row 1 by -2 and subtract it from row 3:

$$\begin{bmatrix} 1 & 0.3333 & 1.3333 & \cdot & 0.3333 & 0 & 0 \\ 0 & 1.3333 & 3.3333 & \cdot & 0.3333 & 1 & 0 \\ 0 & 2.6667 & 0.6667 & \cdot & 0.6667 & 0 & 1 \end{bmatrix}$$

Operate on the second column using the second row. Reduce the diagonal term to 1 by dividing by 4.3333:

$$\begin{bmatrix} 1 & 0.3333 & 1.3333 & \cdot & 0.3333 & 0 & 0 \\ 0 & 1 & 0.7692 & \cdot & 0.0769 & 0.2308 & 0 \\ 0 & 2.6667 & 0.6667 & \cdot & 0.6667 & 0 & 1 \end{bmatrix}$$

Multiply the new row 2 by 0.3333 and subtract it from row 1. Also, multiply row 2 by 2.6667 and subtract it from row 3:

$$\begin{bmatrix} 1 & 0 & 1.0769 & \cdot & 0.3077 & -0.0769 & 0 \\ 0 & 1 & 0.7692 & \cdot & 0.0769 & 0.2308 & 0 \\ 0 & 0 & -1.3843 & \cdot & 0.4616 & -0.6155 & 1 \end{bmatrix}$$

Divide row 3 by -1.3843:

$$\begin{bmatrix} 1 & 0 & 1.0769 & \cdot & 0.3077 & -0.0769 & 0 \\ 0 & 1 & 0.7692 & \cdot & 0.0769 & 0.2308 & 0 \\ 0 & 0 & 1 & \cdot & -0.3334 & 0.4446 & -0.7224 \end{bmatrix}$$

Multiply the new row 3 by 1.0769 and subtract it from row 1. Also, multiply row 3 by 0.7692 and subtract it from row 2:

$$\begin{bmatrix} 1 & 0 & 0 & \cdot & 0.6668 & -0.5557 & 0.7780 \\ 0 & 1 & 0 & \cdot & 0.3334 & -0.1112 & 0.5557 \\ 0 & 0 & 1 & \cdot & -0.3334 & 0.4446 & -0.7224 \end{bmatrix} \qquad (d)$$

The matrix of (d) is now in the form of that given by (c). The inverse of $[A]$ is

$$[A]^{-1} = \begin{bmatrix} 0.6668 & -0.5557 & 0.7780 \\ 0.3334 & -0.1112 & 0.5557 \\ -0.3334 & 0.4446 & -0.7224 \end{bmatrix} \qquad (e)$$

and can be verified by a comparison with the results of Prob. 1.14 or by formulating the matrix multiplication $[A]^{-1}[A] = [I]$.

1.16. Use the gaussian elimination method to solve the simultaneous equations

$$4x_1 + 2x_2 - 2x_3 - 8x_4 = 4$$
$$x_1 + 2x_2 + x_3 \qquad = 2$$
$$0.5x_1 - x_2 + 4x_3 + 4x_4 = 10$$
$$-4x_1 - 2x_2 \qquad - x_4 = 0$$

The equations can be written as the matrix equation

$$\begin{bmatrix} 4 & 2 & -2 & -8 \\ 1 & 2 & 1 & 0 \\ 0.5 & -1 & 4 & 4 \\ -4 & -2 & 0 & 1 \end{bmatrix} \begin{Bmatrix} x_1 \\ x_2 \\ x_3 \\ x_4 \end{Bmatrix} = \begin{Bmatrix} 4 \\ 2 \\ 10 \\ 0 \end{Bmatrix} \qquad (a)$$

Divide row 1 by 4. Subtract the new row 1 from row 2. Multiply the new row 1 by 0.5 and subtract it from row 3. Multiply row 1 by -4 and subtract it from row 4. The result is

$$\begin{bmatrix} 1 & 0.5 & -0.5 & -2 \\ 0 & 1.5 & 1.5 & 2 \\ 0 & -1.25 & 4.25 & 5 \\ 0 & 0 & -2 & -7 \end{bmatrix} \begin{Bmatrix} x_1 \\ x_2 \\ x_3 \\ x_4 \end{Bmatrix} = \begin{Bmatrix} 1 \\ 1 \\ 9.5 \\ 4 \end{Bmatrix}$$

Divide row 2 by 1.5. Multiply the new row 2 by -1.25 and subtract it from row 3. A zero already appears in row 4, and no modification is required. The result is

$$\begin{bmatrix} 1 & 0.5 & -0.5 & -2 \\ 0 & 1 & 1 & 1.3333 \\ 0 & 0 & 5.5 & 6.6667 \\ 0 & 0 & -2 & -7 \end{bmatrix} \begin{Bmatrix} x_1 \\ x_2 \\ x_3 \\ x_4 \end{Bmatrix} = \begin{Bmatrix} 1 \\ 0.6667 \\ 10.3333 \\ 4 \end{Bmatrix}$$

Divide row 3 by 5.5. Multiply the new row 3 by -2 and subtract it from row 4:

$$\begin{bmatrix} 1 & 0.5 & -0.5 & -2 \\ 0 & 1 & 1 & 1.3333 \\ 0 & 0 & 1 & 1.2121 \\ 0 & 0 & 0 & -4.5758 \end{bmatrix} \begin{Bmatrix} x_1 \\ x_2 \\ x_3 \\ x_4 \end{Bmatrix} = \begin{Bmatrix} 1 \\ 0.6667 \\ 1.8788 \\ 7.7576 \end{Bmatrix}$$

Divide row 4 by -4.5758 and solve for the unknowns by substitution:

$$x_1 = 0.0794 \qquad x_2 = -1.0066 \qquad x_3 = 3.9338 \qquad x_4 = -1.6954$$

1.17. Use the separation of variables technique to solve the problem of steady-state temperature distribution defined by Prob. 3.6. The separation of variables method is defined in numerous textbooks on partial differential equations; see for instance, Hildebrand (1962).

The problem is described in Fig. 3-6 and is defined by Laplace's equation

$$\frac{\partial^2 T}{\partial x^2} + \frac{\partial^2 T}{\partial y^2} = 0 \qquad (a)$$

with $T(0, y) = T(x, 0) = T(L, y) = 0$ and $T(x, W) = T_0$. Assume a solution that separates the dependent variables:

$$T(x, y) = X(x)Y(y) \qquad (b)$$

and substitute into Eq. (a):

$$\frac{d^2 X}{dx^2} Y + X \frac{d^2 Y}{dy^2} = 0$$

Group the functions and equate to a constant:

$$-\frac{1}{X} \frac{d^2 X}{dx^2} = \frac{1}{Y} \frac{d^2 Y}{dy^2} = k^2 \qquad (c)$$

This results in the two ordinary differential equations

$$\frac{d^2X}{dx^2} + k^2 X = 0 \quad \text{and} \quad \frac{d^2Y}{dy^2} - k^2 Y = 0 \qquad (d)$$

Both equations can be solved as homogeneous linear equations with constant coefficients. Assume $X = Ce^{mx}$ for the first equation. Substitute into the equation and solve for m. It follows that the characteristic equation is $m = \pm ki$. The general solution for X is

$$X = \sum A_n \sin\left(\frac{n\pi x}{L}\right) + B_n \cos\left(\frac{n\pi x}{L}\right) \quad \text{where } k_n = \frac{n\pi}{L} \qquad (e)$$

The boundary conditions $X(0) = 0$ and $X(L) = 0$ give $B_n = 0$ and

$$0 = A_n \sin\left(\frac{n\pi L}{L}\right)$$

A similar assumption, $Y = Ce^{my}$, gives a solution for the second equation. The characteristic equation is $m^2 - k^2 = 0$, and the general solution is

$$Y = \sum C_n \sinh\left(\frac{n\pi y}{L}\right) + D_n \cosh\left(\frac{n\pi y}{L}\right) \qquad (f)$$

where k_n was previously defined as $n\pi/L$. The boundary condition $Y(0) = 0$ gives $D_n = 0$, and Eq. (b) becomes

$$T = \sum_{n=1}^{\infty} E_n \sin\left(\frac{n\pi x}{L}\right) \sinh\left(\frac{n\pi y}{L}\right) \quad \text{where } E_n = A_n C_n \qquad (g)$$

The last boundary condition $T(x, W) = T_0$ is substituted into Eq. (g), and the orthogonality of the trigonometric functions is employed to determine E_n. Multiply both sides of the equation by $\sin(m\pi x/L)$ and integrate from 0 to L.

$$\int_0^L T_0 \sin\left(\frac{m\pi x}{L}\right) dx = \int_0^L \sum_{n=1} E_n \sinh\left(\frac{n\pi W}{L}\right) \sin\left(\frac{n\pi x}{L}\right) \sin\left(\frac{m\pi x}{L}\right) dx \qquad (h)$$

It follows that

$$\int_0^L \sin\left(\frac{m\pi x}{L}\right) \sin\left(\frac{n\pi x}{L}\right) dx = \begin{cases} 0 & \text{for } m \neq n \\ L/2 & \text{for } m = n \end{cases}$$

The constant E_n can be obtained from Eq. (h):

$$E_n \sinh\left(\frac{n\pi W}{L}\right) = \frac{2}{L} \int_0^L T_0 \sin\left(\frac{n\pi x}{L}\right) dx \qquad (i)$$

Evaluating E_n and substituting into Eq. (g) gives the analytical result for Prob. 3.6.

1.18. Given a vector **a** and a vector **b**, use the scalar product to illustrate the summation subscript.

The scalar product is often called an *inner* product and if there is an inner product, there should be an *outer* product, which there is but it is not used in vector analysis since it leads to defining a higher-order tensor. The scalar product for the vectors **a** and **b** is formally written

$$\mathbf{a} \cdot \mathbf{b} = (a_x \mathbf{i} + a_y \mathbf{j} + a_z \mathbf{k}) \cdot (b_x \mathbf{i} + b_y \mathbf{j} + b_z \mathbf{k}) = a_x b_x + a_y b_y + a_z b_z \qquad (a)$$

The equivalent tensor statement is written

$$a_i b_i = \sum_{i=1}^{3} a_i b_i = a_1 b_1 + a_2 b_2 + a_3 b_3 \qquad (b)$$

Equations (a) and (b) are equivalent. The repeated subscript implies the summation even if the summation symbol is omitted, which it usually is. The outer product of the two vectors would be written $a_i b_j$, where i ranges from 1 to 3 and j ranges from 1 to 3.

1.19.   The divergence of a vector function is illustrated in Prob. 1.8. Write the same function using cartesian tensor notation.

Differentiation is denoted using a comma. Partial differentiation of a scalar quantity A is written

$$\nabla \cdot A = \frac{\partial A}{\partial x}\mathbf{i} + \frac{\partial A}{\partial y}\mathbf{j} + \frac{\partial A}{\partial z}\mathbf{k} \tag{a}$$

and has vector properties. The corresponding statement in subscript tensor notation is $A_{,i}$, where i is a range subscript and implies three separate quantities, one for each coordinate direction:

$$A_{,i} \Rightarrow \frac{\partial A}{\partial x_i} \Rightarrow \left(\frac{\partial A}{\partial x_1}, \frac{\partial A}{\partial x_2}, \frac{\partial A}{\partial x_3} \right) \tag{b}$$

The divergence of a vector is written $\nabla \cdot \mathbf{a} \Rightarrow a_{i,i}$, where the repeated subscript implies the summation:

$$a_{i,i} \Rightarrow \sum_{i=1}^{3} \frac{\partial a_i}{\partial x_i} = \frac{\partial a_1}{\partial x_1} + \frac{\partial a_2}{\partial x_2} + \frac{\partial a_3}{\partial x_3} \tag{c}$$

Equation (c) can be compared with Eq. (b) of Prob. 1.8.

1.20.   Coordinate transformation for a vector in two-dimensional space is defined in Chap. 3, Eqs. (3.14) and (3.15), and is defined and used again in Chap. 4. Refer to Fig. 3.2 and discuss the vector transformation using subscript tensor notation.

Refer to Eq. (3.14) and Fig. 3-2 and extend the transformation to three dimensions by visualizing the rotation of the coordinate system to be about the z axis. The z axis is common to both the x, y, z system and the ξ, η, z system. It is common practice when using tensor notation to use primed (new system) and unprimed (old system) coordinates. The vector transformation for a vector $\mathbf{f}$ from the old system (x_1, x_2, x_3) to the new system (x_1', x_2', x_3') is

$$\begin{Bmatrix} f_1' \\ f_2' \\ f_3' \end{Bmatrix} = \begin{bmatrix} \cos\theta & \sin\theta & 0 \\ -\sin\theta & \cos\theta & 0 \\ 0 & 0 & 1 \end{bmatrix} \begin{Bmatrix} f_1 \\ f_2 \\ f_3 \end{Bmatrix} \tag{a}$$

The corresponding subscript tensor statement of Eq. (a) is

$$f_i' = a_{ij} f_j \tag{b}$$

The transformation a_{ij} as it is written in Eq. (a) can be thought of as a matrix since matrix multiplication is used to compute f' in terms of f. When tensor notation is used, the same transformation is merely referred to as the transformation. The reason for this is that the transformation can be extended to higher-order tensors, but the transformation for higher-order tensors cannot be computed using matrix multiplication. Refer to Eq. (b) and compute the transformation as follows.

Let $i = 1$ and expand the equation for f_1' by summing over j:

$$f_1' = \sum_{j=1}^{3} a_{1j} f_j = a_{11} f_1 + a_{12} f_2 + a_{13} f_3$$

Let $i = 2$ and expand the equation for f_2' by summing over j:

$$f_2' = \sum_{j=1}^{3} a_{2j} f_j = a_{21} f_1 + a_{22} f_2 + a_{23} f_3$$

Let $i = 3$ and expand the equation for f_3' by summing over j:

$$f_3' = \sum_{j=1}^{3} a_{3j}f_j = a_{31}f_1 + a_{32}f_2 + a_{33}f_3$$

The transformation for second- and higher-order tensors is computed in a similar way. The inverse transformation that corresponds to transforming from the primed coordinate system to the original system is the reverse of Eq. (b):

$$f_k = a_{jk}f_j' \qquad (c)$$

and can be expanded and verified.

1.21. The governing equation for three-dimensional heat transfer can be written

$$k_x \frac{\partial^2 T}{\partial x^2} + k_y \frac{\partial^2 T}{\partial y^2} + k_z \frac{\partial^2 T}{\partial z^2} = Q \qquad (a)$$

and for the case $k_x = k_y = k_z = k$ can be written

$$\nabla^2 T = \frac{Q}{k} \qquad (b)$$

(a)   Write Eq. (a) using subscript tensor notation.
(b)   Write Eq. (b) using subscript tensor notation.

(a)   The material constant k can be defined as a second-order tensor in a cartesian coordinate system as

$$k_{kj} = \begin{pmatrix} k_{11} & 0 & 0 \\ 0 & k_{22} & 0 \\ 0 & 0 & k_{33} \end{pmatrix} \qquad k_{kj} = 0 \qquad \text{for } k \neq j \qquad (c)$$

Equation (a) becomes

$$k_{kj}T_{,kj} = Q \qquad (d)$$

A double summation is implied by Eq. (d), and there is no range index, which means there is only one term. In view of Eq. (c) the summation becomes

$$k_{11}T_{,11} + k_{22}T_{,22} + k_{33}T_{,33} = Q$$

(b) $T_{,kk} = Q/k$ implies $T_{,11} + T_{,22} + T_{,33} = Q/k$.

1.22. The *Kronecker delta* is defined as

$$\delta_{ij} = \begin{cases} 1 & \text{if } i = j \\ 0 & \text{if } i \neq j \end{cases} \qquad (a)$$

It follows that a first-order tensor can be written as

$$f_i = \delta_{ij}f_j \qquad (b)$$

and can be verified by expanding the equation. Show that the Kronecker delta is related to the transformation a_{ij} of Prob. 1.20 and defines the orthonormal conditions that a_{ij} must satisfy.

Begin with Eq. (b) above and transform the left-hand side using Eq. (c) of Prob. 1.20:

$$f_i = a_{ki}f_k' = \delta_{ij}f_j$$

Transform f'_k using Eq. (b) of Prob. 1.20, $f'_k = a_{kj}f_j$,

$$a_{ki}a_{kj}f_j = \delta_{ij}f_j$$

$$(a_{ki}a_{kj} - \delta_{ij})f_j = 0 \qquad \text{or} \qquad a_{ki}a_{kj} = \delta_{ij} \tag{c}$$

Any valid coordinate transformation must satisfy Eq. (c).

1.23. The vector product of Prob. 1.4 is written in cartesian tensor notation using the permutation symbol defined as

$$\epsilon_{ijk} = \begin{cases} 0 & \text{if the values of } i,\,j,\,k \text{ do not form a permutation of 1, 2, 3} \\ +1 & \text{if the values of } i,\,j,\,k \text{ form an even permutation of 1, 2, 3} \\ -1 & \text{if the values of } i,\,j,\,k \text{ form an odd permutation of 1, 2, 3} \end{cases}$$

The vector product is written

$$\mathbf{c}_i = \epsilon_{ijk}\mathbf{a}_j\mathbf{b}_k \tag{a}$$

Use Eq. (a) to write the curl of a vector in subscript notation.

  The curl of a vector is written

$$\text{curl } \mathbf{f} = \nabla \times \mathbf{f} = \begin{vmatrix} \mathbf{i} & \mathbf{j} & \mathbf{k} \\ \partial/\partial x & \partial/\partial y & \partial/\partial z \\ f_x & f_y & f_z \end{vmatrix} \tag{b}$$

The corresponding statement in subscript notation is

$$\epsilon_{ijk}f_{k,\,j} = [(f_{3,\,2} - f_{2,\,3}),\,(f_{1,\,3} - f_{3,\,1}),\,(f_{2,\,1} - f_{1,\,2})] \tag{c}$$

Equation (b) can be expanded to verify Eq. (c).

Supplementary Problems

1.24. Show that the area of the triangular area defined in Fig. 1-7 is given by

$$A = \frac{1}{2}\det\begin{bmatrix} 1 & x_1 & y_1 \\ 1 & x_2 & y_2 \\ 1 & x_3 & y_3 \end{bmatrix}$$

1.25. A triangular finite element is shown in Fig. 1-8. Compute the unit vector normal to side 2-3.

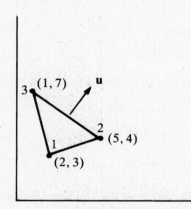

Fig. 1-8

1.26. A three-dimensional finite element can appear as a four-sided tetrahedron. Compute the components of a unit vector normal to the plane defined by points 1, 2, 4 for the three-dimensional element shown in Fig. 1-9.

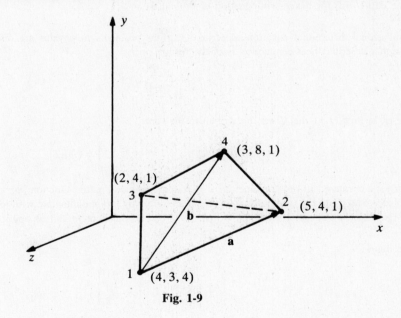

Fig. 1-9

1.27. Show that

$$[[A][B]]^T = [B]^T[A]^T$$

1.28. Use the results of Prob. 1.11 to write the divergence of a vector **A** in matrix format.

1.29. A set of coupled partial differential equations is

$$C_{11} \frac{\partial^2 u}{\partial x^2} + C_{12} \frac{\partial^2 v}{\partial x\, \partial y} = A$$

$$C_{12} = C_{21} \qquad\qquad (a)$$

$$C_{21} \frac{\partial^2 u}{\partial x\, \partial y} + C_{22} \frac{\partial^2 v}{\partial y^2} = B$$

where u and v are defined as

$$u = [N(x, y)]\{u\} \qquad \text{and} \qquad v = [N(x, y)]\{v\}$$

or

$$u = N_1 u_1 + N_2 u_2 + N_3 u_3 \qquad \text{and} \qquad v = N_1 v_1 + N_2 v_2 + N_3 v_3$$

It is desirable to define $\{u\}$ and $\{v\}$ in one array as

$$[U] = [u_1 \quad u_2 \quad u_3 \quad v_1 \quad v_2 \quad v_3]^T$$

Construct the matrix equation that represents Eq. (a).

1.30. Refer to Prob. 1.12 and show that $a_{11}|a_{11}| - a_{12}|a_{12}| + a_{13}|a_{13}|$ is equivalent to $a_{11}|a_{11}| - a_{21}|a_{21}| + a_{31}|a_{31}|$. The determinant can be expanded by any column or row.

1.31. Given a scalar function $J(u) = k(Bu)^2$, where k, B, and u are defined as follows: $[k]$ is an $n \times n$ matrix, $[B]$ is an $n \times m$ matrix, and $[u]$ is an $m \times p$ matrix. Describe $J(u)$ using a matrix equation.

1.32. Compute $\partial J(u)/\partial[u]$ for the matrix equation of Prob. 1.31.

1.33. One-dimensional differential equations describing various problems in physics are discussed in Chap. 2. An equation that describes heat conduction is of the form

$$k \frac{d^2 T}{dx^2} + Q = 0 \tag{a}$$

Assume as in Prob. 1.11 that T can be written in the form

$$T = \phi_1 T_1 + \phi_2 T_2 = [\phi_1 \quad \phi_2]\begin{Bmatrix} T_1 \\ T_2 \end{Bmatrix} = [\phi]\{T\} \tag{b}$$

where $\{T\}$ is a constant and $[\phi]$ is a function of x. Assume a second matrix function $[\psi] = [\psi_1 \quad \psi_2]$ that is also a function of x. Substitute Eq. (b) into Eq. (a), then premultiply by $[\psi]$ and integrate with respect to x with limits of a to b. Use Eq. (e) of Prob. 1.9 and show that the Green-Gauss theorem can be applied to a matrix equation.

1.34. Given a matrix

$$[A] = \begin{bmatrix} 4 & -2 & -1 \\ 1 & 1 & -4 \\ -1 & 2 & 2 \end{bmatrix}$$

Define and compute, where appropriate, the (a) minors of the matrix (determinant) [A], M_{ij}, (b) cofactors of the matrix (determinant) [A], C_{ij}, (c) adjoint of the matrix (determinant) [A], (d) value of the determinant $|A|$ of the matrix [A], (e) inverse of the matrix [A].

1.35. Use the method of Prob. 1.15 to compute the inverse of the matrix given in Prob. 1.34.

1.36. Use the gaussian elimination method to solve the simultaneous equations

$$2a + b + 2c - 3d = 0$$
$$2a - 2b + c - 4d = 5$$
$$a + 2c - 3d = -4$$
$$4a + 4b - 4c + d = -6$$

1.37. Write Eqs. (f) and (g) of Prob. 1.9 using cartesian tensor notation.

1.38. The two-dimensional equations of equilibrium are written in terms of stress in Chap. 3, Eqs. (*3.4*) and (*3.5*). In three dimensions there are nine components of stress, and the complete set can be written as a second-order tensor that is referred to as the stress tensor. In cartesian coordinates the stresses are subscripted using *x*, *y*, *z*. In cartesian tensor notation they are subscripted using 1, 2, 3, and the following analogy is used:

$$\begin{pmatrix} \sigma_{xx} & \sigma_{xy} & \sigma_{xz} \\ \sigma_{yx} & \sigma_{yy} & \sigma_{yz} \\ \sigma_{zx} & \sigma_{zy} & \sigma_{zz} \end{pmatrix} \Rightarrow \sigma_{ij} = \begin{pmatrix} \sigma_{11} & \sigma_{12} & \sigma_{13} \\ \sigma_{21} & \sigma_{22} & \sigma_{23} \\ \sigma_{31} & \sigma_{32} & \sigma_{33} \end{pmatrix} \qquad (a)$$

The stress tensor is symmetric, and that is indicated as $\sigma_{ij} = \sigma_{ji}$ for $i \neq j$. Refer to Eqs. (*3.4*) and (*3.5*) and deduce the three-dimensional equations of equilibrium and then write the equations using subscript tensor notation.

1.39. The coordinate transformation for a second-order tensor is an extension of the vector transformation discussed in Prob. 1.20 and is written

$$\sigma'_{rs} = a_{ri}a_{sj}\sigma_{ij} \qquad (a)$$

Show that the stress transformation equations for two-dimensional elasticity that are derived in elementary mechanics of materials are given by Eq. (*a*).

1.40. The gradient of a scalar function is defined in Prob. 1.7. Rewrite Eq. (*b*) of Prob. 1.7 using cartesian tensor notation.

1.41. Write Eqs. (*f*) and (*g*) and the vector identity $\nabla \cdot (\beta \mathbf{b}) = \beta \nabla \cdot \mathbf{b} + \nabla \beta \cdot \mathbf{b}$ of Prob. 1.11 using subscript tensor notation.

Answers to Supplementary Problems

1.24. Expand the determinant and expand Eq. (*a*) of Prob. 1.6 to show that they are equivalent.

1.25. Refer to Prob. 1.5. $\mathbf{u} = 0.6\mathbf{i} + 0.8\mathbf{j}$.

1.26. Refer to Prob. 1.5 and Fig. 1-9. $\mathbf{a} \times \mathbf{b} = 12\mathbf{i} + 6\mathbf{j} + 6\mathbf{k}$.

$$\mathbf{u} = \frac{\mathbf{a} \times \mathbf{b}}{|\mathbf{a} \times \mathbf{b}|} = \frac{(12\mathbf{i} + 6\mathbf{j} + 6\mathbf{k})}{(216)^{1/2}}$$

1.27. Assume a matrix [A] and a matrix [B]. Expand the matrices to prove the result.

1.28. The divergence of a vector is $\nabla \cdot \mathbf{A}$. Define $[\nabla] = [\partial/\partial x \quad \partial/\partial y \quad \partial/\partial z]$ and $[A] = [A_x \quad A_y \quad A_z]$, and the result is $[\nabla][A]^T$.

1.29. Define the following matrices:

$$[C] = \begin{bmatrix} C_{11} & C_{12} \\ C_{21} & C_{22} \end{bmatrix} \quad [L] = \begin{bmatrix} \partial/\partial x & 0 \\ 0 & \partial/\partial y \end{bmatrix} \quad [N] = \begin{bmatrix} N_1 & N_2 & N_3 & 0 & 0 & 0 \\ 0 & 0 & 0 & N_1 & N_2 & N_3 \end{bmatrix} \quad \{A\} = [A \quad B]^t$$

Then, $[L][C][L][N]\{U\} = \{A\}$.

1.30. Expand the cofactors and compare terms.

1.31.
$$J(u) = \underset{p \times m}{[u]^T} \underset{m \times n}{[B]^T} \underset{n \times n}{[k]} \underset{n \times m}{[B]} \underset{m \times p}{\{u\}}$$

1.32. Use the results of Prob. 1.31 and note that $[B]^T[k][B]$ is symmetric. Refer to Prob. 1.13 and

$$\frac{\partial J(u)}{\partial [u]} = 2[B]^T[k][B]\{u\}$$

1.33. Assume k is a constant and Eq. (a) can be written

$$k\frac{d^2}{dx^2}[\phi]\{T\} + Q = 0 \tag{c}$$

or define additional matrices $k \Rightarrow [k]$ and $d^2/dx^2 \Rightarrow [d^2/dx^2]$ and Eq. (c) can be written

$$[k]\left[\frac{d^2}{dx^2}\right][\phi]\{T\} + Q = 0 \tag{d}$$

Multiply Eq. (d) by $[\psi]^T$:

$$\begin{Bmatrix} \psi_1 \\ \psi_2 \end{Bmatrix}_{2\times1} [k]_{1\times1} \left[\frac{d^2}{dx^2}\right]_{1\times1} [\phi_1 \quad \phi_2]_{1\times2} \begin{Bmatrix} T_1 \\ T_2 \end{Bmatrix}_{2\times1} + \begin{Bmatrix} \psi_1 \\ \psi_2 \end{Bmatrix}_{2\times1} Q = 0 \tag{e}$$

Note that the result given by Eq. (d) is a 1×1 matrix and that the result given by Eq. (e) is a 2×1 matrix, indicating that there are two equations. The two equations, after matrix multiplication, are

$$\psi_1 k \left|\frac{d^2\phi_1}{dx^2}\right| T_1 + \psi_1 k \left|\frac{d^2\phi_2}{dx^2}\right| T_2 + \psi_1 Q = 0$$
$$\psi_2 k \left|\frac{d^2\phi_1}{dx^2}\right| T_1 + \psi_2 k \left|\frac{d^2\phi_2}{dx^2}\right| T_2 + \psi_2 Q = 0 \tag{f}$$

Integrate both equations and use the Green-Gauss theorem [Eq. (e) of Prob. 1.9]:

$$-\int_a^b \frac{d\psi_1}{dx} k \frac{d\phi_1}{dx} T_1\, dx - \int_a^b \frac{d\psi_1}{dx} k \frac{d\phi_2}{dx} T_2\, dx + \psi_1 k \frac{d\phi_1}{dx} T_1 \Big|_a^b + \psi_1 k \frac{d\phi_2}{dx} T_2 \Big|_a^b + \int_a^b \psi_1 Q\, dx = 0$$
$$-\int_a^b \frac{d\psi_2}{dx} k \frac{d\phi_1}{dx} T_1\, dx - \int_a^b \frac{d\psi_2}{dx} k \frac{d\phi_2}{dx} T_2\, dx + \psi_2 k \frac{d\phi_1}{dx} T_1 \Big|_a^b + \psi_2 k \frac{d\phi_2}{dx} T_2 \Big|_a^b + \int_a^b \psi_2 Q\, dx = 0$$

or the result can be written in matrix form as

$$\int_a^b \begin{bmatrix} \dfrac{d\psi_1}{dx} k \dfrac{d\phi_1}{dx} & \dfrac{d\psi_1}{dx} k \dfrac{d\phi_2}{dx} \\ \dfrac{d\psi_2}{dx} k \dfrac{d\phi_1}{dx} & \dfrac{d\psi_2}{dx} k \dfrac{d\phi_2}{dx} \end{bmatrix} \begin{Bmatrix} T_1 \\ T_2 \end{Bmatrix} dx = \begin{bmatrix} \psi_1 k \dfrac{d\phi_1}{dx} & \psi_1 k \dfrac{d\phi_2}{dx} \\ \psi_2 k \dfrac{d\phi_1}{dx} & \psi_2 k \dfrac{d\phi_2}{dx} \end{bmatrix} \begin{Bmatrix} T_1 \\ T_2 \end{Bmatrix}_a^b + \int_a^b \begin{Bmatrix} \psi_1 \\ \psi_2 \end{Bmatrix} [Q]\, dx \tag{g}$$

Equation (g) is the result after applying the Green-Gauss theorem to Eq. (e).

1.34. (a)
$$M_{11} = \begin{bmatrix} 1 & -4 \\ 2 & 2 \end{bmatrix} \quad M_{12} = \begin{bmatrix} 1 & -4 \\ -2 & 2 \end{bmatrix} \quad M_{13} = \begin{bmatrix} 1 & 4 \\ -1 & 2 \end{bmatrix}$$
$$M_{21} = \begin{bmatrix} -2 & -1 \\ 2 & 2 \end{bmatrix} \quad M_{22} = \begin{bmatrix} 4 & -1 \\ -1 & 2 \end{bmatrix} \quad M_{23} = \begin{bmatrix} 4 & -2 \\ -1 & 2 \end{bmatrix}$$
$$M_{31} = \begin{bmatrix} -2 & -1 \\ 1 & -4 \end{bmatrix} \quad M_{32} = \begin{bmatrix} 4 & -1 \\ 1 & -4 \end{bmatrix} \quad M_{33} = \begin{bmatrix} 4 & -2 \\ 1 & 1 \end{bmatrix}$$

(b)
$$C_{11} = 10 \qquad C_{12} = 2 \qquad C_{13} = 3$$
$$C_{21} = 2 \qquad C_{22} = 7 \qquad C_{23} = -6$$
$$C_{31} = 9 \qquad C_{32} = 15 \qquad C_{33} = 6$$

(c)   The adjoint matrix is the transpose of the cofactor matrix, or

$$[C]^T = \begin{bmatrix} 10 & 2 & 9 \\ 2 & 7 & 15 \\ 3 & -6 & 6 \end{bmatrix}$$

(d)   The determinant can be evaluated using the minors of part (a) and the cofactors of part (b).

$$|\det A| = 33$$

(e)   The inverse of [A] can be computed as

$$[A]^{-1} = \frac{[C]^T}{|\det A|}$$

or
$$[A]^{-1} = \begin{bmatrix} 10 & 2 & 9 \\ 2 & 7 & 15 \\ 3 & -6 & 6 \end{bmatrix} \div 33$$

1.36.   The matrix is

$$\begin{bmatrix} 2 & 1 & 2 & -3 \\ 2 & -2 & 1 & -4 \\ 1 & 0 & 2 & -3 \\ 4 & 4 & -4 & 1 \end{bmatrix} \begin{Bmatrix} a \\ b \\ c \\ d \end{Bmatrix} = \begin{Bmatrix} 0 \\ 5 \\ -4 \\ -6 \end{Bmatrix}$$

The upper triangular matrix is

$$\begin{bmatrix} 1 & 0.5 & 1 & -3 \\ 0 & 1 & 0.3333 & 0.3333 \\ 0 & 0 & 1 & -1.1429 \\ 0 & 0 & 0 & 1 \end{bmatrix} \begin{Bmatrix} a \\ b \\ c \\ d \end{Bmatrix} = \begin{Bmatrix} 0 \\ -1.6667 \\ -4.1429 \\ 10.80 \end{Bmatrix}$$

Solving gives $a = 12.0$, $b = -8.0$, $c = -8.2$, $d = 10.2$.

1.37.
$$\int_V (\beta b_i)_{,i}\, dV = \int_S \beta b_i \mathbf{n}_i\, dS$$

$$\int_V \beta b_{i,i}\, dV = \int_S \beta b_i \mathbf{n}_i\, dS - \int_V \beta_{,i} b_i\, dS$$

1.38.
$$\frac{\partial \sigma_{xx}}{\partial x} + \frac{\partial \sigma_{xy}}{\partial y} + \frac{\partial \sigma_{xz}}{\partial y} + f_x = 0$$

$$\frac{\partial \sigma_{xy}}{\partial x} + \frac{\partial \sigma_{yy}}{\partial y} + \frac{\partial \sigma_{yz}}{\partial y} + f_y = 0 \qquad\qquad (a)$$

$$\frac{\partial \sigma_{xz}}{\partial x} + \frac{\partial \sigma_{yz}}{\partial y} + \frac{\partial \sigma_{zz}}{\partial y} + f_z = 0$$

In cartesian tensor notation Eq. (a) becomes

$$\sigma_{kl,l} + f_k = 0 \qquad \sigma_{kl} = \sigma_{lk} \qquad\qquad (b)$$

The first Eq. (b) is called the *balance of linear momentum*, and the second Eq. (b) is called the *balance of angular momentum*. Note that Eq. (b), when expanded, becomes

$$\sigma_{11,,1} + \sigma_{12,,2} + \sigma_{13,,3} + f_1 = 0$$

$$\sigma_{12,,1} + \sigma_{22,,2} + \sigma_{23,,3} + f_2 = 0 \qquad\qquad (c)$$

$$\sigma_{13,,1} + \sigma_{23,,2} + \sigma_{33,,3} + f_3 = 0$$

1.39.   The two-dimensional stress tensor can be written

$$\begin{pmatrix} \sigma_{xx} & \sigma_{xy} \\ \sigma_{xy} & \sigma_{yy} \end{pmatrix} \Rightarrow \begin{pmatrix} \sigma_{11} & \sigma_{12} \\ \sigma_{21} & \sigma_{22} \end{pmatrix}$$

and the transformation is

$$\begin{vmatrix} a_{11} & a_{12} \\ a_{21} & a_{22} \end{vmatrix} \Rightarrow \begin{vmatrix} \cos\theta & \sin\theta \\ -\sin\theta & \cos\theta \end{vmatrix} \qquad\qquad (b)$$

Expand Eq. (a) with $r = s = 1$ and note that i and j are summation indices and are summed from 1 to 2:

$$\sigma'_{11} = a_{11}a_{11}\sigma_{11} + a_{11}a_{12}\sigma_{12} + a_{12}a_{11}\sigma_{21} + a_{12}a_{12}\sigma_{22}$$

Expand Eq. (a) with $r = 1$ and $s = 2$:

$$a'_{12} = a_{11}a_{21}\sigma_{11} + a_{11}a_{22}\sigma_{12} + a_{12}a_{21}\sigma_{21} + a_{12}a_{22}\sigma_{22}$$

Expand Eq. (a) with $r = 2$ and $s = 1$:

$$\sigma'_{21} = a_{21}a_{11}\sigma_{11} + a_{21}a_{12}\sigma_{12} + a_{22}a_{11}\sigma_{21} + a_{22}a_{12}\sigma_{22}$$

Expand Eq. (a) with $r = s = 2$:

$$\sigma'_{22} = a_{21}a_{21}\sigma_{11} + a_{21}a_{22}\sigma_{12} + a_{22}a_{21}\sigma_{21} + a_{22}a_{22}\sigma_{22}$$

Note that $\sigma'_{12} = \sigma'_{21}$. Substitute the values of a_{ij} from Eq. (b):

$$\sigma'_{11} = \sigma_{11}\cos^2\theta + \sigma_{22}\sin^2\theta + 2\sigma_{12}\sin\theta\cos\theta$$

$$\sigma'_{12} = (\sigma_{22} - \sigma_{11})\sin\theta\cos\theta + \sigma_{12}(\cos^2\theta - \sin^2\theta) = \sigma'_{21}$$

$$\sigma'_{22} = \sigma_{11}\sin^2\theta + \sigma_{22}\cos^2\theta - 2\sigma_{12}\sin\theta\cos\theta$$

1.40. $$\nabla(\phi\psi) \Rightarrow (\phi\psi)_{,k} = \phi\psi_{,k} + \psi\phi_{,k}$$

1.41.   Equation (f) of Prob. 1.11 is

$$\int_V (\beta b_j)_{,j}\, dV = \int_S \beta b_j n_j\, dS \qquad\qquad (a)$$

The vector identity $\nabla \cdot (\beta\mathbf{b}) = \beta\nabla \cdot \mathbf{b} + \nabla\beta \cdot \mathbf{b}$ is

$$(\beta b_j)_{,j} = \beta b_{j,j} + \beta_{,j}b_j$$

Equation (g) of Prob. 1.11 is

$$\int_V \beta b_{j,j}\, dV = \int_S \beta b_j n_j\, dS - \int_V \beta_{,j}b_j\, dV \qquad\qquad (b)$$

Chapter 2

One-Dimensional Finite Elements

2.1. INTRODUCTION

In this chapter the fundamental methodology of finite element analysis will be introduced. Several engineering problems will be defined in terms of governing equations written in one dimension. The corresponding finite elements will be derived for one dimension and consequently will be valid for only one dimension. However, many pertinent aspects of finite element analysis can be defined and illustrated in an elementary manner.

The finite element represents an approximate numerical solution of a boundary-value problem described by a differential equation. In this chapter the differential equation is solved by solving the corresponding variational statement of the differential equation. Variational statements of physical problems usually include some statement concerning boundary conditions since boundary condition formulation is a natural result of variational formulation. In this chapter boundary conditions are not included in the variational functions in order to simplify the concepts. Modeling boundary conditions is introduced using physical reasoning.

All the fundamental concepts required for the numerical solution of differential equations are discussed in this chapter. Methods for solving physical problems defined by differential equations that do not have a corresponding variational function will be introduced in a later chapter.

2.2. MATHEMATICAL EQUATIONS OF ENGINEERING

Equations governing engineering phenomena are usually derived from a balance equation and a constitutive equation. In this section one-dimensional equations for several different physical problems will be set forth, and the form of the equations is similar for every case. The notation conforms to that commonly used in that field of engineering. The equations can be derived from a variational principle, and that will be the basis for deriving the corresponding finite element model for the physical problem. Units for the various quantities are given in terms of mass (M), force (F), length (L), time (t), temperature (T), and energy (E).

Elasticity

A one-dimensional problem in elasticity is given by the balance of forces in an elastic rod in terms of normal stress σ, area A, and axial body force f. The force in the rod is $\sigma(x)A(x)$, and the change in the force is balanced by the external body force:

$$\frac{d[\sigma(x)A(x)]}{dx} + f(x)A(x) = 0 \tag{2.1}$$

The constitutive equation, referred to as Hooke's law, relates stress to strain ϵ using the material constant $E(x)$, Young's modulus, and strain is related to axial displacement u:

$$\sigma(x) = E(x)\epsilon(x) \quad \text{and} \quad \epsilon(x) = \frac{du(x)}{dx} \tag{2.2}$$

or

$$\sigma(x) = E(x)\frac{du(x)}{dx} \tag{2.3}$$

Combining Eqs. (2.1) and (2.3) gives a second-order differential equation in terms of displacement:

$$\frac{d}{dx}\left[E(x)A(x)\frac{du(x)}{dx}\right]+f(x)A(x)=0 \tag{2.4}$$

Boundary conditions are of two types, sometimes classified as natural and geometric or essential. In this theory a boundary condition on u is essential, and a boundary condition on σ is natural.

Units: $\sigma(x)$ in F/L^2, $A(x)$ in L^2, $f(x)$ in F/L^3, ϵ in L/L, $E(x)$ in F/L^2, $u(x)$ in L.

Small deflection of a cable that is acted upon by an elastic foundation can be considered a problem in elasticity since the balance of force is combined with a geometric constitutive equation that defines the slope of the cable. The tension T in the cable is assumed to be constant for small deflection theory, and the summation of forces gives the following equation (see Prob. 2.1):

$$T\frac{d\theta(x)}{dx}-k(x)v(x)=-f(x) \tag{2.5}$$

where $\theta(x)$ is the slope of the cable, $v(x)$ is the vertical deflection of the cable, $k(x)$ is the modulus for the elastic foundation, and $f(x)$ is the vertical load acting on the cable. The geometric constitutive equation is derived from the assumption of small deflection theory that θ is small and can be approximated as

$$\theta=\frac{dv}{dx} \tag{2.6}$$

It follows from Eq. (2.6) that the vertical component of the force in the cable is related to the tension in the cable as

$$F_y=T\frac{dv}{dx} \tag{2.7}$$

Combining Eqs. (2.5) and (2.6) gives the governing equation

$$T\frac{d^2v(x)}{dx^2}-k(x)v(x)=-f(x) \tag{2.8}$$

Boundary conditions on $v(x)$ are essential, and a boundary on F_y of Eq. (2.7) would be natural, however, that would be equivalent to specifying the slope θ.

Units: T in F, $\theta(x)$ in L/L, $v(x)$ in L, $f(x)$ in F/L, $k(x)$ in F/L^2.

Heat Conduction

Equations that describe one-dimensional steady-state heat conduction are derived from the balance of energy and a constitutive equation. The balance of energy states that the change in the heat flux q is balanced by an external heat source Q:

$$\frac{d[q(x)A(x)]}{dx}=Q(x)A(x) \tag{2.9}$$

where $A(x)$ is the area and a negative value of Q implies that heat is being removed from the system. The constitutive equation is known as Fourier's law and is stated as

$$q(x)=-k(x)\frac{dT(x)}{dx} \tag{2.10}$$

where T is the temperature and k is the thermal conductivity. Combining Eqs. (2.9) and (2.10) gives the governing second-order differential equation

$$\frac{d}{dx}\left[k(x)A(x)\frac{dT(x)}{dx}\right]+Q(x)A(x)=0 \tag{2.11}$$

Boundary conditions on T are essential, and natural boundary conditions can be specified for $q(x)$.
Units: $q(x)$ in E/tL^2, $A(x)$ in L^2, $Q(x)$ in E/tL^3, $T(x)$ in T, $k(x)$ in E/tLT.

Potential Flow

Potential flow is a special area of fluid mechanics that can be applied to problems in groundwater movement. In this application the assumption of steady incompressible flow is applicable, and the problem is completely described by the continuity equation or balance of mass. A potential function is postulated and in one dimension, assuming constant area, is

$$\phi(x) = -K(x)h(x) = -K(x)\left(\frac{z+p}{\gamma}\right) \tag{2.12}$$

and

$$u(x) = \frac{d\phi}{dx} \tag{2.13}$$

where u is the fluid velocity, h is the piezometric head, z is an elevation head, γ is the specific weight of water for groundwater problems, p is the pressure, and K is the coefficient of permeability or hydraulic conductivity. The constitutive equation is Darcy's law given as

$$u(x) = -K\frac{dh(x)}{dx} \tag{2.14}$$

and it follows that Darcy's law is related to the definition of the potential of Eq. (*2.12*). The governing equation is obtained by combining Eqs. (*2.12*)–(*2.14*), and for steady incompressible flow in one dimension, $du/dx = 0$ and it follows that

$$\frac{d^2\phi}{dx^2} = 0 \tag{2.15}$$

The solution of Eq. (2.15) is a linear function, and in one dimension the velocity will be a constant. However, the two-dimensional counterpart of Eq. (*2.15*) offers more of a challenge and will be studied in a later chapter. The essential boundary condition would be on ϕ, and the natural boundary condition would be to specify the velocity.
Units: $\phi(x)$ in L^2/t, $h(x)$ in L, $K(x)$ in L/t, $u(x)$ in L/t.

Mass Transport

Diffusion, within the most elementary steady-state assumptions, results in a governing equation similar to Eq. (*2.15*) for potential flow. In this instance the balance of mass equation will be written for what is termed a dilute mixture. The theory is applicable for a variety of physical problems. In particular, when groundwater flow can be assumed to be defined as potential flow, a species within the mixture can be assumed to flow with the groundwater and diffuse into its host medium at the same time. Hence the theory of potential flow can be combined with the theory of mass diffusion, and a more complete description of a physically significant problem will result. Assuming a constant area, the balance of mass for a dilute mixture can be written as

$$u(x)\frac{dC(x)}{dx} + \frac{dj(x)}{dx} + K_rC(x) = m \tag{2.16}$$

where $u(x)$ is the velocity of the mixture, $C(x)$ is the concentration of dilute species, $j(x)$ is the flux of the species, K_r is a reaction rate that accounts for a reaction between the dilute species and its surroundings, such as a chemical reaction, and m is an external source of mass. The constitutive equation is known as Fick's law and is given as

$$j(x) = -D(x)\frac{dC(x)}{dx} \tag{2.17}$$

where $D(x)$ is the diffusivity. Combining Eqs. (2.16) and (2.17) gives the governing equation

$$u(x)\frac{dC(x)}{dx} - \frac{d}{dx}\left[D(x)\frac{dC(x)}{dx}\right] + K_r C(x) = m \tag{2.18}$$

The velocity $u(x)$ is assumed to be known. Essential boundary conditions are on $C(x)$, and a natural boundary condition would specify the flux $j(x)$.

Units: $C(x)$ in M/L^3, $D(x)$ in L^2/t, $u(x)$ in L/t, K_r in t^{-1}, $j(x)$ in $M/L^2 t$.

Electricity

The equations governing electrostatics are similar to those for heat conduction. In this case the balance of charge gives a relationship between the electric displacement $D(x)$ and the charge density $\rho(x)$ as

$$\frac{d[A(x)D(x)]}{dx} = \rho(x)A(x) \tag{2.19}$$

where $A(x)$ is the cross-sectional area perpendicular to the x axis. The electric field $E(x)$ is related to the electric potential $\phi(x)$ as

$$E(x) = -\frac{d\phi(x)}{dx} \tag{2.20}$$

The constitutive equation is

$$D(x) = \epsilon(x)E(x) = -\epsilon(x)\frac{d\phi(x)}{dx} \tag{2.21}$$

where $\epsilon(x)$ is the permittivity of the material. Combining Eqs. (2.19) and (2.21) gives the governing equation

$$\frac{d}{dx}\left[\epsilon(x)A(x)\frac{d\phi(x)}{dx}\right] + \rho(x)A(x) = 0 \tag{2.22}$$

The essential boundary condition is on ϕ, and the natural boundary would be to specify D.

Units: $D(x)$ in Q/L^2, $A(x)$ in L^2, $\rho(x)$ in Q/L^3, $E(x)$ in V/L, $\phi(x)$ in V, $\epsilon(x)$ in C/L.

Equations (2.4), (2.11), (2.15), and (2.22) are similar. Equations (2.8) and (2.18) have an additional term that contains the dependent variable. Equation (2.18) has the unknown variable, the first derivative, and the second derivative.

2.3.   VARIATIONAL FUNCTIONS

The calculus of variations is a part of mathematics that involves finding stationary values of functionals. The functional is an integral that has a specific value for each function substituted into the functional. The fundamental problem in the calculus of variations is to obtain a function $f(x)$ such that small variations in the function $\delta f(x)$ will not change the original functional. The study of the calculus of variations, as it is applied to finite element theory, can involve linear algebra, functional analysis, and topics from topology. A minimal amount of theory of the calculus of variations will be presented in this chapter. The intent is to show how the variational functional can be used to formulate the finite element model. The use of variational functions for the equations of Sec. 2.2 is analogous to the strain energy and minimum potential energy methods used in elasticity and theory of structures.

The variational function that corresponds to the governing equations set forth in Sec. 2.2 [with the exception of Eq. (2.18) since it contains a first-order derivative] can be written in a general form. However, terms such as area and material constants will be assumed as constants since they will later be assumed to be constant for an individual finite element. Let $f = f(x)$:

$$J_1(f) = \int_V \frac{1}{2} \left[\alpha \left(\frac{df}{dx} \right)^2 + \beta f^2 - 2\gamma f \right] dV \tag{2.23}$$

Equations containing a first-order derivative, such as Eq. (2.18), may not have a corresponding variational function. However, for the purpose of deriving a finite element model there may be a pseudovariational function or a quasi-variational function that can be used to represent the governing differential equation. A pseudovariational function corresponding to Eq. (2.18), in the notation of that equation with $C = C(x)$, is

$$J_2(C) = \int_V \frac{1}{2} \left[D \left(\frac{dC}{dx} \right)^2 + Cu \frac{dC}{dx} + K_r C^2 - 2mC \right] dV \tag{2.24}$$

It will be shown (see Prob. 2.2) that Eq. (2.23) is a proper variational function that when varied gives the governing differential equations of this chapter. A similar analysis (see Prob. 2.15) using Eq. (2.24) will not give Eq. (2.18) even though Eq. (2.24) can be used to derive a finite element model for Eq. (2.18). That procedure is illustrated in Prob. 2.16.

In general, variational formulations include boundary conditions, and Eq. (2.23) is not complete in this respect. However, as a variational function the function corresponds to the differential equations of this chapter and will suffice for an introductory study of the finite element method. Additional discussion of variational principles is given in Chap. 5.

The classical method for obtaining a governing equation using the variational function is illustrated in Prob. 2.2. In addition, the use of a variational function to obtain an approximate solution is illustrated in Prob. 2.3. These two examples are based upon Eq. (2.23), and similar computations using Eq. (2.24) will give incorrect results. Hence Eq. (2.24) has been called a pseudovariational function. An approximate solution can be obtained for a differential equation using the *Rayleigh-Ritz* method whereby an approximating function is substituted into the variational function. The method is illustrated in Prob. 2.3. The approximating function must satisfy the boundary conditions for the problem being studied.

2.4. INTERPOLATION FUNCTION

The fundamental concept of the finite element method is that a continuous function can be approximated using a discrete model. The discrete model is composed of one or more *interpolation polynomials*, and the continuous function is divided into finite pieces called *elements*. Each element is defined using an interpolation function to describe its behavior between its end points. The end points of the finite element are called *nodes*.

2.5. SHAPE FUNCTIONS

The *shape function* is usually denoted by the letter N and is usually the coefficient that appears in the interpolation polynomial. A shape function is written for each individual node of a finite element and has the property that its magnitude is 1 at that node and 0 for all other nodes in that element. The terminology is often interchanged between interpolation polynomial and shape function. A clear distinction is made between the two in Chap. 6.

2.6. STIFFNESS MATRIX

The term *stiffness matrix* originates from structural analysis. Early applications of the finite element method were similar to matrix analysis of structures, and the term was used to describe the matrix relation between force and displacement. The term is now used regardless of the application. The matrix relation between temperature and heat flux is called the stiffness matrix.

Finite element terminology defines two stiffness matrices. The *local* stiffness matrix corresponds to an individual element. The *global* stiffness matrix is the assembledge of all local stiffness matrices and defines the stiffness of the entire system.

2.7. CONNECTIVITY

Connectivity relates to the manner in which one element in a finite element model is connected with an adjacent element. In this chapter the local finite element derivations pertain primarily to a one-dimensional linear two-node element with one unknown at each node, and the nodes are numbered 1 for the left-hand node and 2 for the right-hand node. Obviously, all nodes in a global finite element model cannot be named 1 or 2. The global model is related to the local model using a *connectivity array*. The connectivity array would have dimensions of $N_{el} \times N_{node}$, where N_{el} is the number of elements for the global model and N_{node} is the number of nodes per element. Assume a five-element model as illustrated in Fig. 2-1 where the elements are identified with a Roman numeral. The local model is related to the global model using a 5×2 connectivity array as shown in Table 2.1.

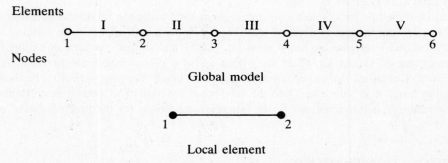

Fig. 2-1 Global and local finite elements.

Table 2.1   Connectivity Array for Fig. 2-1

Global element	Local element	
	Node 1	Node 2
I	1	2
II	2	3
III	3	4
IV	4	5
V	5	6

Connectivity is illustrated in Prob. 2.10 where it is shown graphically how the elements are connected in the global sitffness matrix. Connectivity becomes more complicated with two- and three-dimensional elements with more than one degree of freedom (unknown) per node.

2.8. BOUNDARY CONDITIONS

In Sec. 2.2 boundary conditions were classified as essential and natural. A second-order equation, such as those introduced in this chapter, can be solved analytically, and two constants of integration will result. There must be two boundary conditions in order to solve for the constants of integration, and usually they are specified on the surface at each end of the one-dimensional domain of the problem. Boundary conditions are classified mathematically according to the manner in which the unknown variable is specified.

In mathematical terminology essential boundary conditions are called *Dirichlet boundary conditions*. The *Dirichlet problem* for one-dimensional steady-state heat conduction for a rod of length L and constant area is, using Eq. (*2.11*),

$$\frac{d^2T}{dx^2} + Q = 0 \tag{2.25}$$

$$T(0) = T_0 \qquad \text{and} \qquad T(L) = T_L \tag{2.26}$$

where both boundary conditions specify the temperature. An application of this type of boundary condition is given in Probs. 2.10 and 2.12.

Neumann boundary conditions correspond to the problem where both boundary conditions specify conditions for the first derivative, and the problem is termed a *Neumann problem*. In heat conduction this would be conditions on the flux, such as Eq. (*2.25*) combined with [see Eq. (*2.10*)]

$$k_0 \frac{dT(0)}{dx} = q_0 \qquad \text{and} \qquad k_L \frac{dT(L)}{dx} = q_L \tag{2.27}$$

These boundary conditions present some difficulties for either an analytical or a numerical analysis. The solution is complete only to within an unknown constant. Finite element analysis for boundary conditions, as given by Eq. (*2.27*), will not be discussed.

The third type of boundary condition is called *mixed*. It corresponds to a combination of Eqs. (*2.26*) and (*2.27*) and is probably the most common type of boundary condition. There are actually two types of mixed boundary conditions. The first occurs when an essential condition is specified on one boundary and a natural condition is specified on the second boundary. Problem 2.13 is an example of this type of mixed boundary condition. The second type of mixed boundary condition occurs, for example, in heat conduction as

$$k \frac{dT}{dx} + h(T - T^\infty) = 0 \tag{2.28}$$

where h is a convection coefficient and T^∞ is the temperature of the medium surrounding the boundary surface. This boundary condition says that the boundary flux combined with the boundary temperature is equal to a known temperature. The second boundary condition may be essential, natural, or of the same type as Eq. (*2.28*). This boundary condition requires some special attention and is illustrated in Prob. 2.14.

2.9. PROBLEMS IN CYLINDRICAL COORDINATES

Axisymmetric formulation of problems such as heat conduction and electrostatics results in a one-dimensional differential equation that is similar to Eqs. (2.11) and (2.22). The counterpart of Eq. (2.22) for the electric potential in axisymmetric cylindrical coordinates is

$$\epsilon \frac{d^2\phi}{dr^2} + \frac{\epsilon}{r}\frac{d\phi}{dr} + \rho = 0 \tag{2.29}$$

where the area is constant because it corresponds to the circumference of the cylindrical boundary of the problem. Equation (2.29) can be written in a more concise form as

$$\frac{\epsilon}{r}\frac{d}{dr}\left(r\frac{d\phi}{dr}\right) + \rho = 0 \tag{2.30}$$

The corresponding variational function is

$$J(\phi) = \int_{r1}^{r2}\left[\pi r\epsilon\left(\frac{d\phi}{dr}\right)^2 - 2\pi r\rho\phi\right]dr \tag{2.31}$$

where dV has been replaced with $2\pi r\, dr$.

2.10. THE DIRECT METHOD

The term *direct method* is sometimes used to describe the development of a finite element using concepts from matrix analysis of structures. The fundamental idea is to use solutions for rod or beam problems that have been derived in *mechanics of materials*. For instance, in mechanics of materials axial stress in a rod subject to a constant force P is computed as $\sigma = P/A$. That definition combined with Eq. (2.3) leads to the equation for the deformation of a rod subject to an axial force:

$$u = \frac{PL}{AE} \tag{2.32}$$

where u is the total deformation for a rod of length L with uniform axial force P applied at its ends. The stiffness matrix can be derived using Eq. (2.32). Derivations based upon Eq. (2.32) are usually included in studies of trusses where each truss member is considered a finite element. The force P is normally an external joint loading. See Prob. 2.21.

Solved Problems

2.1.   Derive Eq. (2.8) for cable deflection and obtain the analytical solution for a cable with fixed ends, external upward load f, foundation modulus k, and length L.

The cable is shown in Fig. 2-2 and is shown deflecting upward in the positive y direction. It follows that the loading $f(x)$ and the deflection $v(x)$ are assumed positive. Small deflection theory implies that the cable tension T is constant. Summing forces in the y direction gives

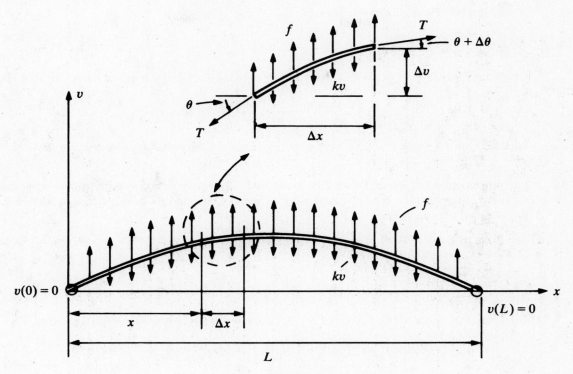

Fig. 2-2

$$-T \sin \theta + T \sin(\theta + \Delta\theta) + f(x)\,\Delta x - k(x)v(x)\,\Delta x = 0 \qquad (a)$$

Small deflection theory implies that $\sin \theta \approx \theta$ and $\sin(\theta + \Delta\theta) \approx \theta + \Delta\theta$. Canceling terms and dividing by Δx in Eq. (a) gives

$$T\,\frac{\Delta\theta}{\Delta x} - k(x)v(x) = -f(x) \qquad (b)$$

For the first term in Eq. (b),

$$\lim_{\Delta x \to 0} T\,\frac{\Delta\theta}{\Delta x} = T\,\frac{d\theta}{dx} \qquad (c)$$

Using the same limiting process and Fig. 2-2 gives

$$\theta = \frac{dv(x)}{dx} \qquad (d)$$

Substituting Eqs. (c) and (d) into Eq. (b) gives the governing differential equation

$$T\,\frac{d^2 v(x)}{dx^2} - k(x)v(x) = -f(x) \qquad (e)$$

The analytical solution for Eq. (e) is obtained by making $k(x)$ a constant and $f(x)$ a constant. The equation can be classified as linear with constant coefficients and has a solution in terms of exponential functions. The homogeneous solution is

$$v(x)_H = C_1 e^{\alpha x} + c_2 e^{-\alpha x}$$

or

$$v(x)_H = A \sinh(\alpha x) + B \cosh(\alpha x) \qquad (f)$$

where $\alpha^2 = k/T$. The particular solution is

$$v(x)_P = \frac{f}{k} \tag{g}$$

Combining Eqs. (f) and (g) and using the boundary conditions $v(0) = 0$ and $v(L) = 0$ gives the final result

$$v(x) = \frac{f[\cosh(\alpha L) - 1]\sinh(\alpha x)}{k\sinh(\alpha L)} - \frac{f}{k}\cosh(\alpha x) + \frac{f}{k} \tag{h}$$

Equation (h) will be used in a subsequent problem to make a comparison with a numerical solution.

2.2.   Use the variational function, Eq. (*2.23*), in the form of a heat conduction problem and show that the variation of the function gives the governing differential equation for heat conduction when k and Q are assumed to be constants.

Refer to Eq. (*2.23*) and make the following analogy: $f \equiv T$, $\alpha \equiv k$, $\beta \equiv 0$, and $\gamma \equiv Q$. The variational function for heat conduction becomes

$$J(T) = \int_V \left[\frac{1}{2} k \left(\frac{dT}{dx} \right)^2 - QT \right] dV \tag{a}$$

The variation is written using an operator $\delta(\ \ )$. The general form for the variation of Eq. (a) is

$$\delta J(T) = \int_V \frac{1}{2} \delta \left[k \left(\frac{dT}{dx} \right)^2 - 2QT \right] dV \tag{b}$$

Consider the first term inside the brackets:

$$\delta \left(\frac{dT}{dx} \right)^2 = \delta \left[\left(\frac{dT}{dx} \right) \left(\frac{dT}{dx} \right) \right] = \left[\delta \left(\frac{dT}{dx} \right) \right] \frac{dT}{dx} + \left(\frac{dT}{dx} \right) \delta \left(\frac{dT}{dx} \right)$$

$$= 2 \left(\frac{dT}{dx} \right) \delta \left(\frac{dT}{dx} \right)$$

It can be shown that the processes of differentiation and variation can be interchanged, or

$$\delta \left(\frac{dT}{dx} \right) = \frac{d}{dx} (\delta T)$$

Also, let the volume integral be written in terms of an intergral over the area and an integral over the length $(0, L)$. It follows that $\int_A dA = A$. The variation of Eq. (b) can be written as

$$\delta J(T) = \int_0^L \left[k \frac{dT}{dx} \frac{d(\delta T)}{dx} - Q\,\delta T \right] A\,dx \tag{c}$$

Integrate the first term by parts:

$$\delta J(T) = k \frac{dT}{dx} \delta T \bigg|_0^L - \int_0^L \left(k \frac{d^2T}{dx^2} \delta T + Q\,\delta T \right) A\,dx \tag{d}$$

The functional is to be zero for small variations in T. This is true only if $T(0) = T(L) = $ constant. Therefore it is concluded that $\delta T(0) = \delta T(L) = 0$, or $k\,dT(0)/dx = k\,dT(L)/dx = 0$, or combinations are equal to zero that make the first term zero. This implies the essential and natural boundary conditions for the problem. In the remaining integral δT is arbitrary (not necessarily zero), therefore, the remaining differential equation is set equal to zero and the result is Eq. (*2.11*).

$$k \frac{d^2T}{dx^2} + Q = 0 \tag{e}$$

2.3. Use the Rayleigh-Ritz method to obtain an approximate solution for the one-dimensional heat conduction problem discussed in Prob. 2.2. Assume a rod of length L, constant area, and constant heat source along the length of the rod. Assume boundary conditions of $T(0) = T(L) = 0$. Compare the approximate solution with the exact solution.

The exact solution is obtained by solving Eq. (e) of Prob. 2.2:

$$\frac{d^2 T}{dx^2} = -\frac{Q}{k} \tag{a}$$

The equation can be integrated twice to give

$$T = -\frac{Qx^2}{2k} + C_1 x + C_2 \tag{b}$$

Substituting the boundary conditions and rearranging gives

$$T = \frac{Q(Lx - x^2)}{2k} \tag{c}$$

The Rayleigh-Ritz method was proposed independently by Lord Rayleigh (1842–1919) and Walter Ritz (1878–1909). Stated simply, a function is assumed in terms of some unknown coefficients that would represent a solution of Eq. (a). The function is substituted into Eq. (a) of Prob. 2.2, the resulting equation is integrated, and the result is minimized with respect to the unknown coefficients. The result is a system of algebraic equations that can be solved for the unknown coefficients. Assume

$$T = \sum_{i=1}^{N} c_i x^{i-1} \tag{d}$$

where c_i are the unknown coefficients and N is the number of terms in the series. Based upon the exact solution, Eq. (c), assume a quadratic function

$$T = c_1 + c_2 x + c_3 x^2 \tag{e}$$

Substituting the boundary conditions, $T(0) = 0$ gives $c_1 = 0$, $T(L) = 0$ gives $c_2 = -c_3 L$, and Eq. (e) becomes

$$T = c_3(x^2 - Lx) \tag{f}$$

$$\frac{dT}{dx} = c_3(2x - L) \tag{g}$$

Substituting Eqs. (f) and (g) into Eq. (a) of Prob. 2.2 and replacing $\int_V dV$ with $A \int_0^L dx$ gives

$$J = \int_0^L \frac{1}{2} [kc_3^2(4x^2 - 4Lx + L^2) - 2Qc_3(x^2 - Lx)]A\, dx$$

or, integrating and substituting limits,

$$J = \frac{Akc_3^2 L^3}{6} + \frac{Ac_3 QL^3}{6} \tag{h}$$

The value of c_3 that makes J a minimum is obtained as follows:

$$\frac{\partial L}{\partial c_3} = \frac{2Akc_3 L^3}{6} + \frac{AQL^3}{6} = 0 \quad \text{and} \quad c_3 = -\frac{Q}{2k} \tag{i}$$

Substituting Eq. (i) into Eq. (f) gives the final result

$$T = \frac{Q(Lx - x^2)}{2k}$$

The approximate solution is identical to the exact solution in this application.

2.4.   Derive a one-dimensional linear interpolation formula for a function $u = u(x)$ that is valid in the range u_1 through u_2 as shown in Fig. 2-3.

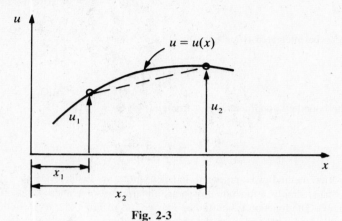

Fig. 2-3

The function $u(x)$ is shown in Fig. 2-3. A linear equation that would approximate $u(x)$ between u_1 and u_2 is assumed as

$$u = A + Bx \qquad (a)$$

where A and B are constants. Substituting the boundary conditions $u(x_1) = u_1$ and $u(x_2) = u_2$ gives two equations that can be solved for A and B:

$$u_1 = A + Bx_1$$

$$u_2 = A + Bx_2$$

Solving for A and B and substituting into Eq. (a) gives the interpolating polynomial:

$$A = \frac{u_1 x_2 - u_2 x_1}{x_2 - x_1} \qquad B = \frac{u_2 - u_1}{x_2 - x_1}$$

$$u = u_1 \frac{x_2 - x}{x_2 - x_1} + u_2 \frac{x - x_1}{x_2 - x_1} \qquad (b)$$

2.5.   Derive the shape functions for a one-dimensional linear finite element.

The results of Prob. 2.4 can be used to derive the shape function at the node located at x_1 in Fig. 2-3. The shape function at node 1 is the coefficient of u_1 in (b) of Prob. 2.4, or $N_1 = (x_2 - x)/(x_2 - x_1)$. Similarly, the shape function for node 2 is $N_2 = (x - x_1)/(x_2 - x_1)$. Note that $N_1 = 1$ for $x = x_1$ and $N_1 = 0$ for $x = x_2$. N_2 is zero at node 1 and 1 (unity) at node 2.

2.6.   Derive a linear one-dimensional interpolation formula in terms of shape functions and nodal point variables and write the result as a matrix equation.

The interpolation formula is derived in Prob. 2.4, and the corresponding shape functions are defined in Prob. 2.5. Using the notation of Fig. 2-3, the result can be written

$$u = N_1 u_1 + N_2 u_2 \qquad (a)$$

Define $[N] = [N_1 \quad N_2]$ and $\{u\} = [u_1 \quad u_2]^T$ and Eq. (a) can be written $u = [N]\{u\}$, where the notation $\{u\}$ indicates a column matrix and $[\quad]^T$ is the transpose of $[\quad]$.

2.7. Use the Rayleigh-Ritz method and two linear interpolation polynomials to solve the heat conduction problem defined in Prob. 2.3.

The problem is shown in Fig. 2-4, and the interpolation formulas of Prob. 2.4 can be used to define the approximate solution. The length is divided into two elements of length $L/2$ with nodes located at $x_1 = 0$, $x_2 = L/2$, and $x_3 = L$. Use Eq. (b) of Prob. 2.4 as a model and write the interpolation formula for each element. For the element on the left side,

$$T = T_1 \frac{x_2 - x}{L/2} + T_2 \frac{x - x_1}{L/2}$$

Insert the boundary condition $T_1 = 0$ and recognize that $x_1 = 0$:

$$T = \frac{2T_2 x}{L} \qquad 0 \le x \le \frac{L}{2} \tag{a}$$

For the element on the right side,

$$T = T_2 \frac{x_3 - x}{L/2} + T_3 \frac{x - x_2}{L/2}$$

Insert the boundary condition $T_3 = 0$ and $x_3 = L$:

$$T = 2T_2 \frac{L - x}{L} \qquad \frac{L}{2} \le x \le L \tag{b}$$

The variational equation is written

$$J(T) = \int_0^L \left[\frac{1}{2} k \left(\frac{dT}{dx} \right)^2 - QT \right] A \, dx \tag{c}$$

Substituting Eqs. (a) and (b) into Eq. (c) gives

$$J(T) = \int_0^{L/2} \left(k \frac{2T_2^2}{L^2} - 2QT_2 \frac{x}{L} \right) A \, dx + \int_{L/2}^L \left(k \frac{2T_2^2}{L^2} - 2QT_2 \frac{L - x}{L} \right) A \, dx \tag{d}$$

Integrating and collecting terms gives

$$J(T) = \left(-\frac{2kT_2^2}{L} + \frac{QT_2 L}{2} \right) A \tag{e}$$

$$\frac{\partial J(T)}{\partial T_2} = \frac{4kT_2}{L} - \frac{QL}{2} = 0 \quad \text{and} \quad T_2 = \frac{QL^2}{8k} \tag{f}$$

Substituting into Eqs. (a) and (b) gives the final result:

$$T = \frac{QLx}{4k} \qquad 0 \le x \le \frac{L}{2}$$

$$T = \frac{QL(L - x)}{4k} \qquad \frac{L}{2} \le x \le L$$

Note that at $x = L/2$ the solution is exact; $T(L/2) = QL^2/8k$. However, at $x = L/4$ the approximate solution is $T(L/4) = QL^2/16k$, and the exact solution is $T(L/4) = 3QL^2/32k$.

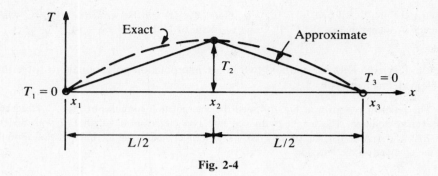

Fig. 2-4

2.8. Assume a rod of elastic material fixed at both ends with constant cross-sectional area and length of $3L$ with uniform body force loading f. Use three linear elements of length L and formulate the Rayleigh-Ritz solution using shape functions rather than interpolation formulas.

The solution process is similar to that for Prob. 2.7. Refer to Fig. 2-5 where the elements are defined using a Roman numeral and the nodes are defined with an Arabic number. For element I, using Eq. (a) of Prob. 2.6,

$$u_I = N_{I1}u_1 + N_{I2}u_2 \qquad (a)$$

and

$$\frac{du_I}{dx} = \frac{dN_{I1}}{dx}u_1 + \frac{dN_{I2}}{dx}u_2 \qquad (b)$$

where N_{I1} is the linear shape function for node 1 of element I and N_{I2} is the shape function for node 2 of element I as illustrated in Fig. 2-5(b). Here u_1 and u_2 are the nodal point values of the displacement and are constants. The shape functions in Eq. (a) are functions of x, and the derivatives in Eq. (b) affect only the shape functions. Use the results of Prob. 2.5 and let the length of all elements be L and

$$N_{I1} = \frac{L - x}{L} \qquad \frac{dN_{I1}}{dx} = \frac{-1}{L}$$

$$N_{I2} = \frac{x}{L} \qquad \frac{dN_{I2}}{dx} = \frac{1}{L}$$

Similarly, for elements II and III,

$$u_{II} = N_{II2}u_2 + N_{II3}u_3$$

$$u_{III} = N_{III3}u_3 + N_{III4}u_4$$

with

$$N_{II2} = \frac{2L - x}{L} \qquad \frac{dN_{II2}}{dx} = \frac{-1}{L}$$

$$N_{II3} = \frac{x - L}{L} \qquad \frac{dN_{II3}}{dx} = \frac{1}{L}$$

$$N_{III3} = \frac{3L - x}{L} \qquad \frac{dN_{III3}}{dx} = \frac{-1}{L}$$

$$N_{III4} = \frac{x - 2L}{L} \qquad \frac{dN_{III4}}{dx} = \frac{1}{L}$$

The shape functions are shown in Fig. 2-5(b)–(d). Figure 2-5(e) shows the final form of the solution after the shape functions have been combined and the boundary conditions applied. A significant conclusion thus far is that the first derivatives of linear shape functions are dependent only upon the length of the element. The derivative of the shape function that defines the left-hand node for a single element is always $-1/L$, and the derivative of the shape function that defines the right-hand node of the same element is always $1/L$ for linear shape functions. All the elements in an individual problem are not required to have the same length, but in this example it simplifies the writing of the shape functions.

Substituting into Eq. (2.23) and using the boundary conditions $u_1 = u_4 = 0$ gives

$$J(u) = \int_0^L \left(E\frac{u_2^2}{2L^2} - f\frac{u_2 x}{L} \right) A\,dx + \int_L^{2L} \left\{ \left(E\frac{u_2^2 - 2u_2 u_3 + u_3^2}{2L^2} \right) \right.$$
$$\left. - f\left[\frac{u_2(2L-x)}{L} + \frac{u_3(x-L)}{L} \right] \right\} A\,dx + \int_{2L}^{3L} \left(E\frac{u_3^2}{2L^2} - \frac{u_3(3L-x)}{L} \right) A\,dx$$

Integrating, substituting integration limits, and collecting terms give

$$J(u) = EA\frac{u_2^2 - u_2 u_3 + u_3}{L} - AfL(u_2 + u_3) \tag{c}$$

Minimize the function of Eq. (c) with respect to the unknown nodal point displacements:

$$\frac{\partial J}{\partial u_2} = EA\frac{2u_2 - u_3}{L} - fAL = 0 \tag{d}$$

$$\frac{\partial J}{\partial u_3} = EA\frac{-u_2 + 2u_3}{L} - fAL = 0 \tag{e}$$

Equations (d) and (e) can be written in matrix form as

$$\frac{EA}{L}\begin{bmatrix} 2 & -1 \\ -1 & 2 \end{bmatrix}\begin{bmatrix} u_2 \\ u_3 \end{bmatrix} = fAL\begin{bmatrix} 1 \\ 1 \end{bmatrix} \tag{f}$$

Solving Eq. (f) gives $u_2 = fL^2/E$ and $u_3 = fL^2/E$. The final result is

$$u_\mathrm{I} = \frac{fLx}{E} \qquad\qquad 0 \le x \le L$$

$$u_\mathrm{II} = \frac{fL^2}{E} \qquad\qquad L \le x \le 2L$$

$$u_\mathrm{III} = fL(3L - x) \qquad 2L \le x \le 3L$$

The exact solution is obtained as in Prob. 2.3 for a length of $3L$:

$$u = f\frac{3Lx - x^2}{2E}$$

The approximate solution is exact for $x = L$ and $x = 2L$. At the center of the rod, $x = 3L/2$, the exact solution is $\frac{9}{8}(fL^2/E)$, and the error is approximately 11 percent.

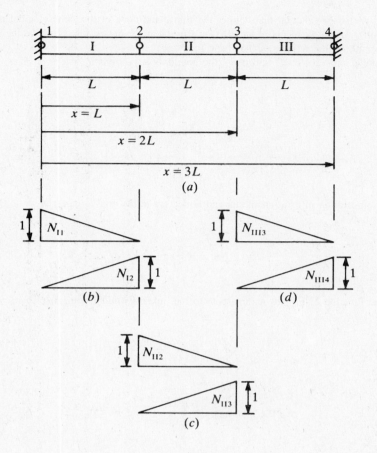

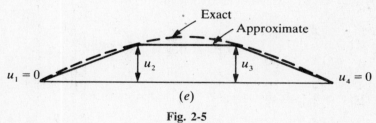

Fig. 2-5

2.9. For a rod similar to the one described in Prob. 2.8 write the variational function in terms of linear shape functions. Derive a general model or a *local stiffness matrix* for the rod using the variational function.

The variational function can be written as follows for one element of length L:

$$J(u) = \frac{1}{2} \int_0^L \frac{du}{dx} E \frac{du}{dx} A \, dx - \int_0^L ufA \, dx \qquad (a)$$

Use the matrix definition of the linear shape function that was established in Prob. 2.6 and rewrite Eq. (a):

$$J(u) = \frac{A}{2} \int_0^L \{u\}^T \left[\frac{dN}{dx} \right]^T [E] \left[\frac{dN}{dx} \right] \{u\} \, dx - A \int_0^L \{u\}^T [N]^T f \, dx \qquad (b)$$

Rewrite Eq. (b) to illustrate the matrix multiplications:

$$J(u) = \frac{A}{2} \int_0^L [u_1 \quad u_2] \begin{bmatrix} dN_1/dx \\ dN_2/dx \end{bmatrix} [E][dN_1/dx \quad dN_2/dx] \begin{bmatrix} u_1 \\ u_2 \end{bmatrix} dx - A \int_0^L [u_1 \quad u_2] \begin{bmatrix} N_1 \\ N_2 \end{bmatrix} f \, dx \qquad (c)$$

At this point the shape functions and their derivatives could be substituted into Eq. (c), the various matrix multiplications be carried out, the integration completed, and the result would be similar to the procedure developed in Prob. 2.8. However, Eq. (c) merely illustrates the matrix equation. The minimization will be carried out using Eq. (b). Refer to Chap. 1 for concepts pertaining to the derivative of a matrix equation.

$$\frac{\partial J(u)}{\partial \{u\}} = A \int_0^L \left[\frac{dN}{dx} \right]^T [E] \left[\frac{dN}{dx} \right] \{u\} \, dx - A \int_0^L [N]^T f \, dx = 0 \qquad (d)$$

The matrices are evaluated using the results of Prob. 2.8:

$$A \int_0^L \begin{Bmatrix} -1/L \\ 1/L \end{Bmatrix} [E][-1/L \quad 1/L] \begin{Bmatrix} u_1 \\ u_2 \end{Bmatrix} dx - A \int_0^L \begin{Bmatrix} (L-x)/L \\ x/L \end{Bmatrix} f \, dx = 0$$

Perform the indicated matrix multiplications:

$$A \int_0^L \begin{bmatrix} E/L^2 & -E/L^2 \\ -E/L^2 & E/L^2 \end{bmatrix} \begin{Bmatrix} u_1 \\ u_2 \end{Bmatrix} dx = A \int_0^L \begin{Bmatrix} (L-x)/L \\ x/L \end{Bmatrix} f \, dx$$

After integration the final result appears as

$$\begin{bmatrix} AE/L & -AE/L \\ -AE/L & AE/L \end{bmatrix} \begin{Bmatrix} u_1 \\ u_2 \end{Bmatrix} = \begin{Bmatrix} AfL/2 \\ AfL/2 \end{Bmatrix} \qquad (e)$$

Note that the area term is not factored out of the equation since area may vary from one finite element to the next. The term on the right of the equal sign is interpreted to mean that one-half of the body force is distributed to each node. Equation (e) is the local stiffness matrix (stiffness matrix for an individual element) for an axially loaded rod. The stiffness matrix for any similar differential equation would be the same with different material parameters. Finally, Eq. (e) is written in terms of matrices as

$$[K]\{u\} = \{f\} \qquad \text{or sometimes as} \qquad [K^e]\{u\} = \{f^e\} \qquad (f)$$

where the superscript e indicates the *element* stiffness matrix and *element* force matrix.

2.10. Formulate the elastic rod analysis of Prob. 2.8 using the local finite element developed in Prob. 2.9 and construct the global stiffness matrix.

There are three elements, each with the same length, area, and modulus of elasticity. The local stiffness matrix is given by Eq. (e) of Prob. 2.9. Elements I and II share a common node, and elements II and III share a node. This introduces the concept of *connectivity*. Elements I and II are connected at node 2. All element matrices will be identical and will be added or connected to form the global stiffness matrix as follows:

$$\begin{bmatrix} AE/L & -AE/L & 0 & 0 \\ -AE/L & AE/L & -AE/L & 0 \\ 0 & -AE/L & AE/L & -AE/L \\ 0 & 0 & -AE/L & AE/L \end{bmatrix} \begin{bmatrix} u_1 \\ u_2 \\ u_3 \\ u_4 \end{bmatrix} = \begin{bmatrix} AfL/2 \\ AfL/2 + AfL/2 \\ AfL/2 + AfL/2 \\ AfL/2 \end{bmatrix}$$

The global stiffness matrix is a 4×4 matrix before substituting boundary conditions, and the rows and columns are defined by the single lines. The three element (local) stiffness matrices are shown within the boxes. Note that for row 2, column 2, there is a contribution from two elements. The same is true for row 3, column 3. All remaining spaces in the global stiffness matrix have a contribution from only one element or they are zero. The matrix to the right of the equal sign illustrates the distribution of the uniform body force loading. One-half of the force on each element is distributed to each node. The element stiffness matrix, Eq. (e) of Prob. 2.9, has been used three times and connected at the element nodes. The boundary conditions $u_1 = u_4 = 0$ can be incorporated by merely striking out the first and fourth rows and columns. The final result is a 2×2 matrix equation identical to Eq. (f) of Prob. 2.8 and can be solved to give the same results.

In finite element analysis, boundary conditions are not dealt with by eliminating rows and columns. In this example that was done in order to make a comparison with Prob. 2.8. The standard procedure would be to set the diagonal term equal to unity and all terms in the corresponding row and column to zero and modify the matrix on the right-hand side accordingly. For the zero displacement boundary conditions of this problem, the result would be

$$\frac{EA}{L} \begin{bmatrix} 1 & 0 & 0 & 0 \\ 0 & 2 & -1 & 0 \\ 0 & -1 & 2 & 0 \\ 0 & 0 & 0 & 1 \end{bmatrix} \begin{bmatrix} u_1 \\ u_2 \\ u_3 \\ u_4 \end{bmatrix} = fAL \begin{bmatrix} 0 \\ 1 \\ 1 \\ 0 \end{bmatrix}$$

The solution of this set of equations merely gives the zero displacement boundary conditions as part of the final results. This method of analysis is preferred for computer computations. A discussion of nonzero essential boundary conditions will be included in a solved problem later in this chapter.

2.11. Do the following for the problem of the small deflection of a cable subject to a uniform load and acted upon by an elastic foundation.

(a) Write the variational function.

(b) Assume linear shape functions and derive the local stiffness matrix for an element of length L following the method used in Prob. 2.9.

(a) Equation (2.8) is the governing differential equation and is derived in Prob. 2.1. The variational function is obtained using Eq. (2.23) with $f(x) = v(x)$, $\alpha = T$, $\beta = k$, and $\gamma = f$:

$$J(v) = \int_V \frac{1}{2} \left\{ T \left[\frac{dv(x)}{dx} \right]^2 + k[v(x)]^2 - 2fv(x) \right\} dV \tag{a}$$

(b) Equation (a) is written as a matrix equation in terms of shape functions for an element of length L with constant area as

$$J(v) = \frac{A}{2} \int_0^L \{v\}^T \left[\frac{dN}{dx} \right]^T [T] \left[\frac{dN}{dx} \right] \{v\} \, dx$$

$$+ \frac{A}{2} \int_0^L \{v\}^T [N]^T [k][N]\{v\} \, dx - A \int_0^L \{v\}^T [N]^T f \, dx \tag{b}$$

Minimize the function of Eq. (b) with respect to $\{v\}$:

$$\frac{\partial J(v)}{\partial \{v\}} = \int_0^L \left[\frac{dN}{dx} \right]^T [T] \left[\frac{dN}{dx} \right] \{v\} \, dx$$

$$+ \int_0^L [N]^T [k][N]\{v\} \, dx - \int_0^L [N]^T f \, dx = 0 \tag{c}$$

The area term has been divided out since a cable would not normally have a variable area. The final element results for the first and third terms of Eq. (c) are similar to Eq. (e) of Prob. 2.9 with AE replaced by T. The second term involves the foundation modulus and the shape functions, and the matrices can be written as

$$\int_0^L \begin{Bmatrix} (L-x)/L \\ x/L \end{Bmatrix} [k][(L-x)/L \quad x/L] \begin{Bmatrix} v_1 \\ v_2 \end{Bmatrix} dx$$

or

$$\int_0^L k \begin{bmatrix} (L-x)^2/L^2 & x(L-x)/L^2 \\ x(L-x)/L^2 & x^2/L^2 \end{bmatrix} \begin{Bmatrix} v_1 \\ v_2 \end{Bmatrix} dx$$

Integrating and combining terms gives the local stiffness matrix

$$\begin{bmatrix} kL/3 & kL/6 \\ kL/6 & kL/3 \end{bmatrix} \begin{Bmatrix} v_1 \\ v_2 \end{Bmatrix}$$

The complete local stiffness matrix for the small deflection of the cable is

$$\begin{bmatrix} T/L & -T/L \\ -T/L & T/L \end{bmatrix} \begin{Bmatrix} v_1 \\ v_2 \end{Bmatrix} + \begin{bmatrix} kL/3 & kL/6 \\ kL/6 & kL/3 \end{bmatrix} \begin{Bmatrix} v_1 \\ v_2 \end{Bmatrix} = \begin{Bmatrix} fL/2 \\ fL/2 \end{Bmatrix} \qquad (d)$$

2.12. Solve the cable problem using the local element of Prob. 2.11. Assume a five-element analysis and compare the results with the exact solution given in Prob. 2.1. Assume the cable is fixed at both ends ($v_1 = v_6 = 0$) with $T = 600$ lb, $L = 120$ in, $k = 0.5$ lb/in^2, and $f = 2$ lb/in.

Substitute into Eq. (d) of Prob. 2.11 to formulate the local element stiffness matrix

$$\begin{bmatrix} 25 & -25 \\ -25 & 25 \end{bmatrix} \begin{Bmatrix} v_1 \\ v_2 \end{Bmatrix} + \begin{bmatrix} 4 & 2 \\ 2 & 4 \end{bmatrix} \begin{Bmatrix} v_1 \\ v_2 \end{Bmatrix} = \begin{Bmatrix} 24 \\ 24 \end{Bmatrix}$$

or

$$\begin{bmatrix} 29 & -23 \\ -23 & 29 \end{bmatrix} \begin{Bmatrix} v_1 \\ v_2 \end{Bmatrix} = \begin{Bmatrix} 24 \\ 24 \end{Bmatrix} \qquad (a)$$

The five elements are connected at their common nodes using Eq. (a) as the local stiffness matrix to give a 6×6 global stiffness matrix:

$$\begin{bmatrix} 29 & -23 & 0 & 0 & 0 & 0 \\ -23 & 58 & -23 & 0 & 0 & 0 \\ 0 & -23 & 58 & -23 & 0 & 0 \\ 0 & 0 & -23 & 58 & -23 & 0 \\ 0 & 0 & 0 & -23 & 58 & -23 \\ 0 & 0 & 0 & 0 & -23 & 29 \end{bmatrix} \begin{Bmatrix} v_1 \\ v_2 \\ v_3 \\ v_4 \\ v_5 \\ v_6 \end{Bmatrix} = \begin{Bmatrix} 24 \\ 48 \\ 48 \\ 48 \\ 48 \\ 24 \end{Bmatrix}$$

The boundary conditions are included by deleting the first and sixth rows and columns, or as in Prob. 2.10, the first and sixth diagonal terms are set equal to 1 with all remaining terms in these rows and columns set equal to zero. Similarly, the first and sixth terms in the right-hand column matrix are set to zero. The final system of simultaneous equations is solved for the cable deflection at the node points. The exact solution for the cable problem is given by Eq. (h) of Prob. 2.1 and is compared with the finite element solution in Table 2.2.

Table 2.2 Comparison of Finite Element and Exact Solutions for $v(x)$ (in)

Node	Finite element	Exact
1	0.0	0.0
2	1.8548	1.8173
3	2.5903	2.5444
4	2.5903	2.5444
5	1.8548	1.8173
6	0.0	0.0

2.13. Assume a rod of length 0.2 m with the temperature at $x = 0$ maintained at 100°C. Assume a constant heat source of $Q = 3(10^6)$ W/m^3 along the length of the rod, a boundary heat flux of $q = 3(10^6)$ W/m^3 along the length of the rod, a boundary heat flux of $q = 1.8(10^6)$ W/m^2 removing heat at $x = 0.2$ m, and thermal conductivity of $k = 6000$ W/(m·K). Assume the area of the rod is $0.4(10^{-3})$ m^2. Use five linear elements of equal length and compute the temperature and flux distribution at each element node.

Three new concepts are introduced in this problem: (1) an essential boundary condition other than zero by specifying the temperature as 100°C at $x = 0$, (2) a natural boundary condition by specifying the flux at $x = 0.2$ m, and (3) the computation of the flux at each element node.

The local element can be obtained using Eq. (e) of Prob. 2.9 as a model with AE/L replaced by Ak/L and f replaced by Q. Then, each element length is 0.2 m/5 = 0.04 m and $Ak/L = 0.4(10^{-3})(6000)/0.04 = 60$ W/K and $AQL = 0.400(10^{-3})(3)(10^6)(0.04) = 48$ W. The local stiffness matrix is

$$60 \begin{bmatrix} 1 & -1 \\ -1 & 1 \end{bmatrix} \begin{Bmatrix} T_1 \\ T_2 \end{Bmatrix} = 48 \begin{Bmatrix} \frac{1}{2} \\ \frac{1}{2} \end{Bmatrix} \tag{a}$$

The global stiffness matrix is

$$\begin{bmatrix} 60 & -60 & 0 & 0 & 0 & 0 \\ -60 & 120 & -60 & 0 & 0 & 0 \\ 0 & -60 & 120 & -60 & 0 & 0 \\ 0 & 0 & -60 & 120 & -60 & 0 \\ 0 & 0 & 0 & -60 & 120 & -60 \\ 0 & 0 & 0 & 0 & -60 & 60 \end{bmatrix} \begin{Bmatrix} T_1 \\ T_2 \\ T_3 \\ T_4 \\ T_5 \\ T_6 \end{Bmatrix} = \begin{Bmatrix} 24 \\ 48 \\ 48 \\ 48 \\ 48 \\ 24 \end{Bmatrix} \tag{b}$$

The flux boundary condition at node 6 is specified as $q = 1.8(10^6)$ W/m^2. This is characterized by a surface boundary integral that can be written as

$$\int_S [\mathbf{N}]^T q \, dS \tag{c}$$

where the shape function must correspond to the element that defines the surface. In this one-dimensional problem element 5 defines the node that contains the surface. In fact, node 6 is the location of the surface, and node 5 corresponds to a location within the rod. For element 5, nodes 5 and 6, the shape function defining the surface at node 6 is constant, $N_5 = 0$, and $N_6 = 1$. Refer to Fig. 2-5 where element III is a similar element. To define the boundary surface at node 4 let $N_{III3} = 0$ and $N_{III4} = 1$. The surface area defined in Eq. (c) is constant, and after integration Eq. (c) is

$$\int_S [\mathbf{N}]^T q \, dS = qA \begin{Bmatrix} 0 \\ 1 \end{Bmatrix} = -1.8(10^6)(0.4)(10^{-3}) \begin{Bmatrix} 0 \\ 1 \end{Bmatrix} = \begin{Bmatrix} 0 \\ -720 \end{Bmatrix} \tag{d}$$

The flux term is negative because heat is being removed from the system. Equation (d) is added to the right-hand side of Eq. (b), and since the flux is applied at node 6, only that term is modified and appears as

$$[24 \quad 48 \quad 48 \quad 48 \quad 48 \quad -696]^T \tag{e}$$

The matrix on the right-hand side of Eq. (b) is sometimes called the *force matrix*. This terminology is used in matrix analysis of structures and has become standard in all finite element work. The matrix can be thought of as the action matrix, and regardless of the application the matrix contains the terms that cause an action to occur. In this case, the internal heat source and the external flux are the actions.

The nonzero temperature boundary condition at node 1 must be included in the formulation. Recall that in Prob. 2.10 the essential boundary conditions were specified as zero at both ends of the rod problem, and corresponding rows and columns were set equal to zero with a 1 placed on the diagonal of the stiffness matrix. The action matrix was modified by substituting a zero for the first and last nodes. In this problem the temperature at node 1 is specified as 100°C, and the stiffness matrix and force matrix must be modified such that when the set of equations is solved, the result $T_1 = 100$ will be computed.

The finite element statement of the problem given by Eq. (b) can be written in the general form

$$
\begin{bmatrix}
k_{11} & k_{12} & k_{13} & \cdots & k_{1N} \\
k_{21} & k_{22} & k_{23} & \cdots & k_{2N} \\
k_{31} & k_{32} & k_{33} & \cdots & k_{3N} \\
\cdot & \cdot & \cdot & \cdots & \cdot \\
\cdot & \cdot & \cdot & \cdots & \cdot \\
\cdot & \cdot & \cdot & \cdots & \cdot \\
k_{N1} & k_{N2} & k_{N3} & \cdots & k_{NN}
\end{bmatrix}
\begin{Bmatrix}
T_1 \\ T_2 \\ T_3 \\ \cdot \\ \cdot \\ \cdot \\ T_N
\end{Bmatrix}
=
\begin{Bmatrix}
f_1 \\ f_2 \\ f_3 \\ \cdot \\ \cdot \\ \cdot \\ f_N
\end{Bmatrix}
\qquad (f)
$$

If k_{11} is set equal to 1, the remaining first row and first column are set equal to zero, and f_1 is set equal to 100, the result $T_1 = 100$ will be computed. However, the fact that T_1 is nonzero will affect the remaining unknown temperature values. In order to incorporate this effect, visualize the matrices on the left-hand side of Eq. (f) being multiplied together and the T_1 value substituted into the resulting equations. Let k_{11} equal 1 and all other k values in the first row equal 0.

$$
\begin{aligned}
T_1 + \quad 0 \quad + \quad 0 \quad + \cdots + \quad 0 \quad &= \quad 100 \\
k_{22}T_2 + k_{23}T_3 + \cdots + k_{2N}T_N &= f_2 - 100k_{21} \\
k_{32}T_2 + k_{33}T_3 + \cdots + k_{3N}T_N &= f_3 - 100k_{31} \\
\cdot \quad + \quad \cdot \quad + \cdots + \quad \cdot \quad &= \quad \cdot \\
\cdot \quad + \quad \cdot \quad + \cdots + \quad \cdot \quad &= \quad \cdot \\
\cdot \quad + \quad \cdot \quad + \cdots + \quad \cdot \quad &= \quad \cdot \\
k_{N2}T_2 + k_{N3}T_3 + \cdots + k_{NN}T_N &= f_N - 100k_{N1}
\end{aligned}
\qquad (g)
$$

Solving Eqs. (g) will give the correct boundary temperature and will modify the right-hand-side force matrix to include the effect of the boundary temperature. Note that the f_1 term has been set equal to 100, the specified boundary temperature. If a flux boundary condition (natural boundary condition) is specified for node 1 in addition to a temperature boundary condition (essential boundary condition), the boundary value problem will not have proper boundary conditions because both boundary values cannot be included at the same node. The f_1 term corresponding to a flux boundary condition has been replaced by the temperature boundary value.

Equation (f) can be written in the general form

$$
\begin{bmatrix}
1 & 0 & 0 & \cdots & 0 \\
0 & k_{22} & k_{23} & \cdots & k_{2N} \\
0 & k_{32} & k_{33} & \cdots & k_{3N} \\
\cdot & \cdot & \cdot & \cdots & \cdot \\
\cdot & \cdot & \cdot & \cdots & \cdot \\
\cdot & \cdot & \cdot & \cdots & \cdot \\
0 & k_{N2} & k_{N3} & \cdots & k_{NN}
\end{bmatrix}
\begin{Bmatrix}
T_1 \\ T_2 \\ T_3 \\ \cdot \\ \cdot \\ \cdot \\ T_N
\end{Bmatrix}
=
\begin{Bmatrix}
100 \\ f_2 - 100k_{21} \\ f_3 - 100k_{31} \\ \cdot \\ \cdot \\ \cdot \\ f_N - 100k_{N1}
\end{Bmatrix}
\qquad (h)
$$

Rewrite Eq. (b) incorporating the flux boundary condition given by Eq. (e) and the temperature boundary condition given by Eq. (h):

$$
\begin{bmatrix}
1 & 0 & 0 & 0 & 0 & 0 \\
0 & 120 & -60 & 0 & 0 & 0 \\
0 & -60 & 120 & -60 & 0 & 0 \\
0 & 0 & -60 & 120 & -60 & 0 \\
0 & 0 & 0 & -60 & 120 & -60 \\
0 & 0 & 0 & 0 & -60 & 60
\end{bmatrix}
\begin{Bmatrix}
T_1 \\ T_2 \\ T_3 \\ T_4 \\ T_5 \\ T_6
\end{Bmatrix}
=
\begin{Bmatrix}
100 \\ 48 + (60)(100) \\ 48 \\ 48 \\ 48 \\ -696
\end{Bmatrix}
\qquad (i)
$$

The solution of Eq. (i) will give the temperature distribution along the rod. The exact solution is given in Prob. 2.23 as

$$
T = \frac{3.0(10^6)}{6000}(0.2x - 0.5x^2) - \frac{1.8(10^6)x}{6000} + 100
\qquad (j)
$$

The flux distribution is computed using Eq. (2.10):

$$q = -k \frac{dT}{dx}$$

The corresponding finite element formulation for one element, using element 1 as a model, is

$$q = -k[-1/L \quad 1/L] \begin{Bmatrix} T_1 \\ T_2 \end{Bmatrix} \tag{k}$$

The computation must be carried out for each element after Eq. (i) has been solved. Note that the flux will be constant for each element and that the element may be termed a constant-flux element. In other words, the two-term interpolation formulas used thus far will always give constant results for the first derivative of the unknown functions. The exact solution for the flux is (see Prob. 2.23)

$$q = -k \frac{dT}{dx} = -6000 \left[\frac{3(10^6)}{6000} (0.2 - x) - \frac{1.8(10^6)}{6000} \right] \tag{l}$$

The finite element and exact solutions are compared in Table 2.3. The approximate and exact solutions for temperature are the same at each node point in this example. The exact solution for the flux has been computed at each node point and at the center of each element. The finite element solution for flux is constant for each element and agrees with the exact solution at the midpoint of each element. The finite element solution for flux can be improved by using more elements.

Table 2.3   Comparison of Finite Element and Exact Solutions

Node	Temperature (°C) Finite element	Exact	Flux (10^6 W/m^2) Finite element	Exact
1	100	100	(Element 1) −1.26	−1.20 at $x = 0.00$ −1.26 at $x = 0.02$
2	91.6	91.6	(Element 2) −1.38	−1.34 at $x = 0.04$ −1.38 at $x = 0.06$
3	82.4	82.4	(Element 3) −1.50	−1.44 at $x = 0.08$ −1.50 at $x = 0.10$
4	72.4	72.4	(Element 4) −1.62	−1.56 at $x = 0.12$ −1.62 at $x = 0.14$
5	61.6	61.6	(Element 5) −1.74	−1.68 at $x = 0.16$ −1.74 at $x = 0.18$
6	50.0	50.0		−1.80 at $x = 0.20$

2.14. The solid material shown in Fig. 2-6 is subjected to a temperature differential that can be described using mixed boundary conditions of the type described by Eq. (2.28). Use five linear finite elements of equal length and compute the steady-state temperature distribution through the thickness of the material. Assume, $k = 250$ W/(m·K), $W = 0.1$ m, $h_1 = 2000$ W/(m^2·K), $h_2 = 5000$ W/(m^2·K), $T_1^\infty = 100°C$, and $T_2^\infty = 50°C$.

Heat is flowing into the material at $x = 0$ and can be described as

$$-k \frac{dT(0)}{dx} + h_1[T(0) - T_1^\infty] = 0 \tag{a}$$

Similarly, heat is flowing out of the material at $x = 0.1$ and can be described as

$$k \frac{dT(0.1)}{dx} + h_2[T(0.1) - T_2^\infty] = 0 \tag{b}$$

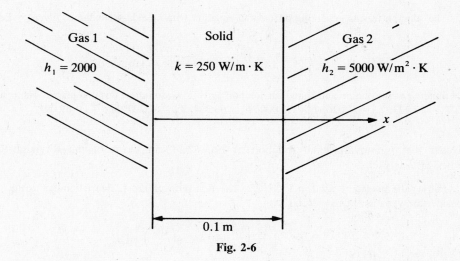

Fig. 2-6

The local finite element matrix and final global matrix are constructed using the same procedure as in Prob. 2.13. The area is constant and can be assumed as unity. Then, $Ak/L = 250/0.02 = 12{,}500$, and the global matrix before substituting boundary conditions is

$$12{,}500 \begin{bmatrix} 1 & -1 & 0 & 0 & 0 & 0 \\ -1 & 2 & -1 & 0 & 0 & 0 \\ 0 & -1 & 2 & -1 & 0 & 0 \\ 0 & 0 & -1 & 2 & -1 & 0 \\ 0 & 0 & 0 & -1 & 2 & -1 \\ 0 & 0 & 0 & 0 & -1 & 1 \end{bmatrix} \begin{Bmatrix} T_1 \\ T_2 \\ T_3 \\ T_4 \\ T_5 \\ T_6 \end{Bmatrix} = \begin{Bmatrix} q_1 \\ q_2 \\ q_3 \\ q_4 \\ q_5 \\ q_6 \end{Bmatrix} \qquad (c)$$

Use Eq. (a) and write the boundary condition at $x = 0$ (node 1) as

$$q_1 = h_1(T_1^\infty - T_1)$$

or, for element 1 and a constant area assumed to be unity,

$$\int_s [\mathbf{N}]^T q \, dS = q_1 \begin{Bmatrix} 1 \\ 0 \end{Bmatrix} = \begin{Bmatrix} h_1 T_1^\infty - h_1 T_1 \\ 0 \end{Bmatrix} = \begin{Bmatrix} 20{,}000 - 2000 T_1 \\ 0 \end{Bmatrix} \qquad (d)$$

Similarly, Eq. (b) is used to include the boundary condition at node 6 as

$$q_6 = h_2(T_2^\infty - T_6)$$

and for element 5 becomes

$$q_6 \begin{Bmatrix} 0 \\ 1 \end{Bmatrix} = \begin{Bmatrix} 0 \\ h_2 T_2^\infty - h_2 T_6 \end{Bmatrix} = \begin{Bmatrix} 0 \\ 25{,}000 - 5000 T_6 \end{Bmatrix} \qquad (e)$$

Combine Eqs. (c)–(e) with 10^4 factored out of the matrix to obtain

$$10^4 \begin{bmatrix} 1.45 & -1.25 & 0 & 0 & 0 & 0 \\ -1.25 & 2.50 & 0 & 0 & 0 & 0 \\ 0 & -1.25 & 2.50 & -1.25 & 0 & 0 \\ 0 & 0 & -1.25 & 2.50 & -1.25 & 0 \\ 0 & 0 & 0 & -1.25 & 2.50 & -1.25 \\ 0 & 0 & 0 & 0 & -1.25 & 1.75 \end{bmatrix} \begin{Bmatrix} T_1 \\ T_2 \\ T_3 \\ T_4 \\ T_5 \\ T_6 \end{Bmatrix} = \begin{Bmatrix} 2(10^5) \\ 0 \\ 0 \\ 0 \\ 0 \\ 2.5(10^5) \end{Bmatrix}$$

The solution for the governing differential equation with boundary conditions given by Eqs. (a) and (b) is

$$T(x) = T_1^\infty - (T_1^\infty - T_2^\infty) \frac{(x/k) + (1/h_1)}{(1/h_1) + (1/h_2) + (L/k)} \qquad (f)$$

The finite element solution and analytical solution agree to within three decimal places. The results at each node are $T_1 = 77.273$, $T_2 = 73.636$, $T_3 = 70.000$, $T_4 = 66.364$, $T_5 = 62.727$, and $T_6 = 59.091$.

2.15. Discuss the pseudovariational function of Eq. (2.24) in terms of the Rayleigh-Ritz method of numerical analysis.

Follow the procedure used in Prob. 2.2. The first term of Eq. (2.24) is similar to Eq. (a) of Prob. 2.2. Consider the variation of the second term

$$\int_0^L \frac{u}{2} \left[\frac{dC}{dx} \delta C + C \, \delta\left(\frac{dC}{dx}\right) \right] A \, dx$$

Integrate the second term by parts:

$$\int_0^L \frac{u}{2} \left(\frac{dC}{dx} \delta C - \frac{dC}{dx} \delta C \right) A \, dx + \frac{Au}{2} C \, \delta C \bigg|_0^L$$

and this term is zero. Similarly, assume homogeneous boundary conditions $C(0) = C(L) = 0$ and an approximate solution similar to that of Prob. 2.3,

$$C = c_3(x^2 - Lx) \qquad \text{and} \qquad \frac{dC}{dx} = c_3(2x - L)$$

Substituting into $\int_0^L \frac{1}{2} uC(dC/dx)A \, dx$ will also give zero. It follows that the function given by Eq. (2.24) is not valid for a Rayleigh-Ritz analysis.

2.16. Use the pseudovariational function of Eq. (2.24) to formulate the local stiffness matrix corresponding to mass transport as defined by Eq. (2.18). Assume linear shape functions and an element length of L.

The matrix method used in Probs. 2.9 and 2.11 must be used to formulate the local stiffness matrix because the Rayleigh-Ritz method of solution will give erroneous results. Understanding the reasons why this method fails to give proper Rayleigh-Ritz results requires some background in the mathematics of finite elements and variational principles and the topic will be discussed in Chap. 5.

Equation (2.24) is rewritten as follows for ready reference:

$$J_2(C) = \int_V \frac{1}{2} \left[D\left(\frac{dC}{dx}\right)^2 + Cu\frac{dC}{dx} + K_r C^2 - 2mC \right] dV \qquad (a)$$

The volume integral is rewritten in terms of an area assumed to be constant and the element length L. Equation (a) is written as a matrix equation in terms of shape functions following the method used in Prob. 2.11:

$$J(C) = \frac{A}{2} \int_0^L \{C\}^T \left[\frac{dN}{dx}\right]^T [D] \left[\frac{dN}{dx}\right] \{C\} \, dx$$

$$+ \frac{A}{2} \int_0^L \{C\}^T [N]^T [u] \left[\frac{dN}{dx}\right] \{C\} \, dx + \frac{A}{2} \int_0^L \{C\}^T [N]^T [K_r][N]\{C\} \, dx$$

$$- A \int_0^L \{C\}[N]^T m \, dx$$

Minimize $J(C)$ with respect to $\{C\}$:

$$\frac{\partial J(C)}{\partial \{C\}} = \int_0^L \left[\frac{dN}{dx}\right]^T [D]\left[\frac{dN}{dx}\right]\{C\}\, dx$$

$$+ \int_0^L [N]^T [u]\left[\frac{dN}{dx}\right]\{C\}\, dx + \int_0^L [N]^T [K_r][N]\{C\}\, dx - \int_0^L [N]^T m\, dx = 0$$

The first, third, and fourth terms have been dealt with in Probs. 2.9 and 2.11. Note that the sign of the fourth term is opposite that of the corresponding term in Eq. (c) of Prob. 2.11 because the signs are opposite in the governing differential equations. There remains to be developed the local element for the second term as the transpose of the shape function times the velocity matrix times the shape function derivative matrix:

$$A\int_0^L \left\{\begin{matrix}(L-x)/L \\ x/L\end{matrix}\right\}[u][-1/L \quad 1/L]\left\{\begin{matrix}C_1 \\ C_2\end{matrix}\right\} dx = \frac{Au}{2}\begin{bmatrix}-1 & 1 \\ -1 & 1\end{bmatrix}\left\{\begin{matrix}C_1 \\ C_2\end{matrix}\right\} \qquad (b)$$

Note that Eq. (b) is not symmetric, as were the previous local stiffness matrices that were derived. Combine Eq. (b) with the stiffness matrices of the form of Eq. (d) of Prob. 2.11.

$$\left[\begin{bmatrix}D/L & -D/L \\ -D/L & D/L\end{bmatrix} + \begin{bmatrix}-u/2 & u/2 \\ -u/2 & u/2\end{bmatrix} + \begin{bmatrix}K_rL/3 & K_rL/6 \\ K_rL/6 & K_rL/3\end{bmatrix}\right]A\left\{\begin{matrix}C_1 \\ C_2\end{matrix}\right\} = A\left\{\begin{matrix}mL/2 \\ mL/2\end{matrix}\right\} \qquad (c)$$

The term containing the velocity u is usually referred to a convection term. When u is large compared to D, the analysis will become unstable. This is not a fault of the finite element formulation but occurs in numerical analysis regardless of the method of analysis. The examples in this text are limited to problems such as groundwater movement where the velocity term is small.

2.17. Use the results of Prob. 2.16 to obtain a three-element solution for the differential equation

$$u\frac{dC}{dx} - D\frac{d^2C}{dx^2} = m$$

Assume boundary conditions of $C(x = 0) = C(x = 3L) = 0$ and obtain an exact solution. Assume $u = D = L = A = 1$ and compare the two solutions.

The exact solution is

$$C = -3Lm\frac{1 - e^{ux/D}}{u(1 - e^{3uL/D})} + \frac{mx}{u} \qquad (a)$$

Substitute into Eq. (c) of Prob. 2.16 with $K_r = 0$, and the local element model becomes

$$\left[\begin{bmatrix}D/L & -D/L \\ -D/L & D/L\end{bmatrix} + \begin{bmatrix}-u/2 & u/2 \\ -u/2 & u/2\end{bmatrix}\right]A\left\{\begin{matrix}C_1 \\ C_2\end{matrix}\right\} = \left\{\begin{matrix}1 \\ 1\end{matrix}\right\}\frac{mLA}{2} \qquad (b)$$

The global matrix is constructed using Eq. (*b*) and after substituting physical parameters becomes

$$
\begin{bmatrix}
0.5 & -0.5 & 0 & 0 \\
-1.5 & 2.0 & -0.5 & 0 \\
0 & -1.5 & 2.0 & -0.5 \\
0 & 0 & -1.5 & 1.5
\end{bmatrix}
\begin{Bmatrix}
C_0 \\
C_L \\
C_{2L} \\
C_{3L}
\end{Bmatrix}
=
\begin{Bmatrix}
m/2 \\
m \\
m \\
m/2
\end{Bmatrix}
$$

Substituting boundary conditions and solving for the unknowns give $C_{2L} = 0.769m$ and $C_{3L} = 1.077m$. The exact solution is obtained using Eq. (*a*) as $C_{2L} = 0.730m$ and $C_{3L} = 0.996m$.

2.18. A one-dimensional aquifer is modeled using a controlled laboratory experiment. The aquifer has constant area and length 1 m and contaminated fluid moving through it. The elevation head at $x = 0$ is 0.3 m, and at $x = 1$ m is 0.15 m. A chemical contaminant is measured to have a concentration of zero at $x = 0$ and 10 mg/m^3 at $x = 1$ m and also reacts with its surroundings with rate constant $K_r = 3.0(10^{-8})$ as it moves through the aquifer. It is known that the chemical substance will diffuse in the medium with diffusion constant $D = 1.0(10^{-8})$. The hydraulic conductivity of the material is $K = 1.0(10^{-7})$. Compute the steady-state distribution of chemical contaminant in the aquifer using five linear finite elements.

The solution of this problem is developed in two steps. The velocity of the fluid u is computed using the theory of potential flow as outlined in Sec. 2.2. The velocity of the fluid is used in the computation for mass transport. The analytical solution of Eq. (*2.15*) with the boundary conditions given above is elementary and would result in a linear variation for the potential and a constant velocity. However, the finite element solution will be formulated to illustrate the entire solution process. The local stiffness matrix of Prob. 2.9 will be used as a model. The area is constant and can be factored out of the matrix equation. The constitutive parameter E in Eq. (*e*) of Prob. 2.9 will be unity, or the hydraulic conductivity K can be used in place of E. The result will be the same since the equation to be solved is homogeneous:

$$
\frac{d^2\phi}{dx^2} = -K\frac{d^2h}{dx^2} = 0
$$

The length of an element is $L = 1 \text{ m}/5 = 0.2$ m. The right-hand side of the local stiffness matrix is zero. The final global stiffness matrix after substituting the nonzero essential boundary conditions following the method used in Prob. 2.13 becomes

$$
\begin{bmatrix}
1.0 & 0 & 0 & 0 & 0 & 0 \\
0 & 10.0 & -5.0 & 0 & 0 & 0 \\
0 & -5.0 & 10.0 & -5.0 & 0 & 0 \\
0 & 0 & -5.0 & 10.0 & -5.0 & 0 \\
0 & 0 & 0 & -5.0 & 10.0 & 0 \\
0 & 0 & 0 & 0 & 0 & 1.0
\end{bmatrix}
\begin{bmatrix}
h_1 \\
h_2 \\
h_3 \\
h_4 \\
h_5 \\
h_6
\end{bmatrix}
=
\begin{bmatrix}
0.3 \\
1.5 \\
0.0 \\
0.0 \\
0.75 \\
0.15
\end{bmatrix}
$$

The solution gives the exact answer as $h_1 = 0.3$, $h_2 = 0.27$, $h_3 = 0.24$, $h_4 = 0.21$, $h_5 = 0.18$, and $h_6 = 0.15$. The velocity is constant and is computed using Eq. (*2.14*) following the method used in Prob. 2.13. Computations for element 1 are

$$
u = -\frac{(10^{-7})(0.27 - 0.30)}{0.2} = 1.5(10^{-8}) \tag{a}
$$

Use Eq. (c) of Prob. 2.15 to formulate the local stiffness matrix for mass transport for element 1 as

$$\left[(10^{-8}) \begin{bmatrix} 5.0 & -5.0 \\ -5.0 & 5.0 \end{bmatrix} + (10^{-8}) \begin{bmatrix} -0.75 & 0.75 \\ -0.75 & 0.75 \end{bmatrix} + (10^{-8}) \begin{bmatrix} 0.2 & 0.1 \\ 0.1 & 0.2 \end{bmatrix} \right] \begin{Bmatrix} C_1 \\ C_2 \end{Bmatrix}$$

or

$$\begin{bmatrix} 4.45(10^{-8}) & -4.15(10^{-8}) \\ -5.65(10^{-8}) & 5.95(10^{-8}) \end{bmatrix} \begin{Bmatrix} C_1 \\ C_2 \end{Bmatrix} \tag{b}$$

The global matrix is assembled using Eq. (b) and the boundary conditions $C(x=0) = 0$ and $C(x=1.) = 10$. as

$$(10^{-7}) \begin{bmatrix} 1.(10^7) & 0 & 0 & 0 & 0 & 0 \\ 0 & 1.040 & -0.415 & 0 & 0 & 0 \\ 0 & -0.565 & 1.040 & -0.415 & 0 & 0 \\ 0 & 0 & -0.565 & 1.040 & -0.415 & 0 \\ 0 & 0 & 0 & -0.565 & 1.040 & 0 \\ 0 & 0 & 0 & 0 & 0 & 1.(10^7) \end{bmatrix} \begin{Bmatrix} C_1 \\ C_2 \\ C_3 \\ C_4 \\ C_5 \\ C_6 \end{Bmatrix} = \begin{Bmatrix} 0.0 \\ 0.0 \\ 0.0 \\ 0.0 \\ 0.415(10^{-6}) \\ 10.0 \end{Bmatrix}$$

The analytical solution for a length L and boundary conditions $C(x=0) = 0$ and $C(x=L) = C_L$ is

$$C = \frac{C_L e^{\alpha(x-L)} \sinh(\beta x)}{\sinh(\beta L)}$$

$$\alpha = \frac{u}{2D} \quad \text{and} \quad \beta = \left(\alpha^2 + \frac{K_r}{D} \right)^{1/2}$$

The finite element and analytical solutions are tabulated in Table 2.4.

**Table 2.4　Comparison of Finite Element and Ana-
lytical Solutions for C**

Node	x	Finite element	Analytical
1	0.0	0.0	0.0
2	0.2	0.6392	0.6576
3	0.4	1.6019	1.6383
4	0.6	3.1442	3.1937
5	0.8	5.6985	5.7446
6	1.0	10.0	10.0

2.19. Obtain a finite element solution for the axisymmetric coaxial cable illustrated in Fig. 2-7. Compare results using four linear elements (two equal-length elements in each segment of the cable) with six linear elements, three equal-length elements in each segment of the cable. Assume the inside radius $r_a = 5$ mm, the outside radius $r_c = 25$ mm, and the interface radius $r_b = 10$ mm. Assume for illustration that the permittivity of the core is $\epsilon_1 = 0.5$, and for the outer layer $\epsilon_2 = 2.0$. Assume a charge density of $\rho_1 = 100$ for the core and $\rho_2 = 0$ for the outer layer. Assume boundary conditions of $\phi_{r=a} = 500$ and $\phi_{r=c} = 0$. Refer to Prob. 2.28 and compare your results with the exact solution.

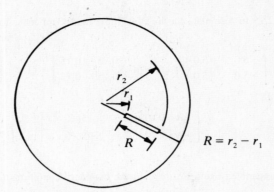

Axisymmetric radial element

$$R = r_2 - r_1$$

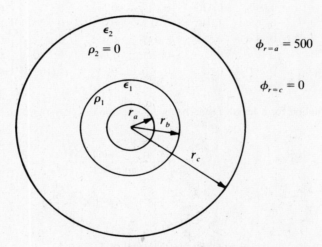

Fig. 2-7   Coaxial cable.

The variational function given by Eq. (2.31) is written in terms of shape functions following the method used in Prob. 2.9 and after minimizing with respect to ϕ appears as

$$\int_{r_1}^{r_2} \left[2\pi r \left\{ \begin{matrix} -1/R \\ 1/R \end{matrix} \right\} [\epsilon][-1/R \quad 1/R] \left\{ \begin{matrix} \phi_1 \\ \phi_2 \end{matrix} \right\} - 2\pi r \rho \left\{ \begin{matrix} (r_2 - r)/R \\ (r - r_1)/R \end{matrix} \right\} \right] dr = 0 \qquad (a)$$

where the shape functions are

$$N_1 = \frac{r_2 - r}{R} \qquad \text{and} \qquad N_2 = \frac{r - r_1}{R}$$

and $R = r_2 - r_1$, the length of a radial element. In this formulation the limits of integration should correspond to the element being constructed. This is in contrast to Prob. 2.9 where the limits used were 0 to L, the length of an element. As the radius increases, the volume of material defined by the integration increases. The matrix multiplication indicated by Eq. (a) is completed, and the integration is in terms of r and r^2 and gives the local stiffness matrix as

$$\pi\epsilon \left\{ \frac{r_1 + r_2}{R} \right\} \begin{bmatrix} 1 & -1 \\ -1 & 1 \end{bmatrix} \left\{ \begin{matrix} \phi_1 \\ \phi_2 \end{matrix} \right\} = \frac{2\pi\rho}{R} \left\{ \begin{matrix} r_2^3/6 - r_2 r_1^2/2 + r_1^3/3 \\ r_2^3/3 - r_1 r_2^2/2 + r_1^3/6 \end{matrix} \right\} \qquad (b)$$

Note that according to the right-hand side of Eq. (b) the charge density is not distributed equally between the two nodes. The total stiffness matrix and force matrix for the four-element solution, using Eq. (b) as a model, is computed as

$$\begin{bmatrix} 1.0 & 0.0 & 0.0 & 0.0 & 0.0 \\ 0.0 & 18.850 & -10.996 & 0.0 & 0.0 \\ 0.0 & -10.996 & 34.034 & -23.038 & 0.0 \\ 0.0 & 0.0 & -23.038 & 58.643 & 0.0 \\ 0.0 & 0.0 & 0.0 & 0.0 & 1.0 \end{bmatrix} \begin{bmatrix} \phi_1 \\ \phi_2 \\ \phi_3 \\ \phi_4 \\ \phi_5 \end{bmatrix} = \begin{bmatrix} 500.000 \\ 15707.962 \\ 7199.484 \\ 0.0 \\ 0.0 \end{bmatrix} \qquad (c)$$

The exact solution is computed using part (c) of Prob. 2.28. Substituting parameters into the matrix equation of part (c) gives an equation that is solved for the constants of integration:

$$\begin{bmatrix} 1.609 & 1.0 & 0.0 & 0.0 \\ 2.302 & 1.0 & -2.302 & -1.0 \\ 0.05 & 0.0 & -0.20 & 0.0 \\ 0.0 & 0.0 & 3.219 & 1.0 \end{bmatrix} \begin{bmatrix} C_1 \\ C_2 \\ C_3 \\ C_4 \end{bmatrix} = \begin{bmatrix} 1750.0 \\ 5000.0 \\ 500.0 \\ 0.0 \end{bmatrix} \qquad (d)$$

The solution of Eq. (d) gives $C_1 = 6009.762$, $C_2 = -7919.703$, $C_3 = -997.560$, $C_4 = 3211.147$. Substituting the constants of integration into the solutions given in part (a) of Prob. 2.28 gives the analytical results of Table 2.5. The six-element (seven-node) finite element solution is obtained in a manner similar to the four-element solution; details will not be given. The final results are shown in Table 2.5.

Table 2.5 Comparison of Solutions for ϕ in a Coaxial Cable

r	Four elements	Six elements	Analytical
5.0	500.0	500.0	500.0
6.67	—	1246.54	1259.39
7.50	1347.35	—	1376.88
8.33	—	1333.28	1350.39
10.00	881.17	899.18	918.29
15.00	—	503.04	509.71
17.50	346.17	—	355.93
20.00	—	220.08	222.73
25.00	0.0	0.0	0.0

The results are not as accurate as those in some of the previous finite element examples. A solution for a coaxial cable made of one material using six elements would be more accurate. The change in material properties causes a loss of accuracy. However, with the addition of a few elements the accuracy will increase quite rapidly. This example, with the analytical solution, can be utilized to test the accuracy of finite element solutions and establish the optimum number of elements to be used.

2.20. Compare the formulation of one-dimensional finite element problems using two-node linear elements versus three-node quadratic elements.

The two-node linear element was discussed in Prob. 2.8 and illustrated in Fig. 2-5. Each element is made up of two shape functions with the property that the function defined at node 1, element I, has a magnitude of 1 at that node and is 0 at the remaining node. See, for instance, Fig. 2-5(b); the two shape functions combine to describe the element.

A three-node quadratic element has similar properties. The quadratic function must span two spaces and connect three nodes. The two spaces define the element. The two-node elements that have been discussed were assigned a length of L because it was convenient for computation. A three-node element can have a length of L or $2L$ or any convenient length. However, the limits of integration must correspond to the element length when the local element is constructed.

Assume a one-dimensional space of arbitrary length a. Also, assume that two three-node elements will be used to model the space. Four divisions are required as shown in Fig. 2-8, two for each element. The function is defined over element I using the general statement

$$f_{\text{I}} = N_{\text{I}1}f_1 + N_{\text{I}2}f_2 + N_{\text{I}3}f_3 \tag{a}$$

Similarly, for element II,

$$f_{\text{II}} = N_{\text{II}3}f_3 + N_{\text{II}4}f_4 + N_{\text{II}5}f_5 \tag{b}$$

The shape functions are illustrated in Fig. 2-8 and can be compared with Fig. 2-5 for linear elements. It should be obvious that for cubic interpolation formulas there are four nodes and the element will span three spaces.

Equations (a) and (b) can be written in matrix format as

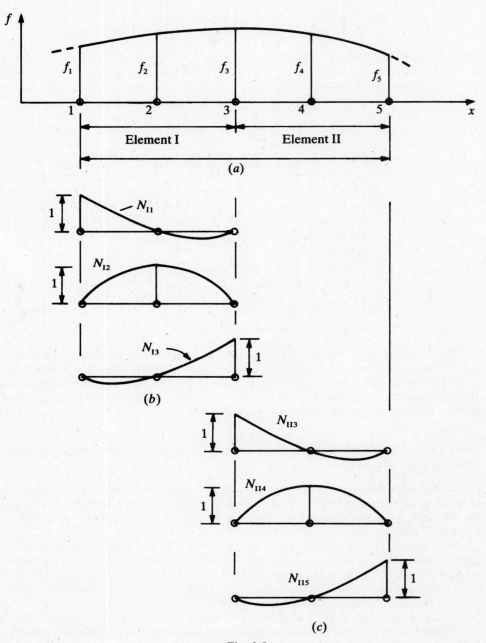

Fig. 2-8

$$f_I = [N_{I1} \quad N_{I2} \quad N_{I3}] \begin{Bmatrix} f_1 \\ f_2 \\ f_3 \end{Bmatrix} \quad \text{and} \quad f_{II} = [N_{II3} \quad N_{II4} \quad N_{II5}] \begin{Bmatrix} f_3 \\ f_4 \\ f_5 \end{Bmatrix}$$

or

$$f = [N]\{f\} \tag{c}$$

The superposition of shape functions follows the same process for each element. Node connectivity occurs at node 3 for this illustration. Once the shape function has been written in the form of Eq. (e), the procedure for formulating the local stiffness matrix is identical with that in all of the preceding examples. The shape functions are approximated using interpolation formulas such as

$$f_1 = A + Bx + Cx^2 \qquad f_2 = A + Bx + Cx^2 \qquad f_3 = A + Bx + Cx^2$$

The method used in Prob. 2.4 can be used to derive a general shape function, however, the algebra becomes much more cumbersome. Refer to Prob. 2.29.

2.21. Derive the local stiffness matrix for a rod subject to an axial force using the direct method.

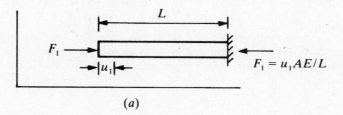

(a)

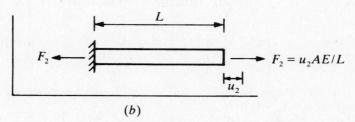

(b)

Fig. 2-9 Rod finite element.

Assume a rod of length L as shown in Fig. 2-9(a). Let the right end be constrained and a positive displacement u_1 applied at the left end. The corresponding force, using Eq. (2.32), is

$$F_1 = u_1 \frac{AE}{L} \tag{a}$$

For equilibrium the reaction at the right end is equal and opposite, or

$$F_1 = -u_1 \frac{AE}{L}$$

Let the rod be constrained as shown in Fig. 2-9(b) and a positive displacement u_2 applied at the right end. The corresponding force on the right is positive and is given by

$$F_2 = u_2 \frac{AE}{L} \tag{b}$$

Similarly, the reaction on the left is $F_2 = -u_2 AE/L$.

The rod is in force equilibrium, which means that the forces at either end of the rod element are equal and opposite in direction. Imagine that the results in Figs. 2-9(a) and (b) can be added to give a general case representing a rod that displaces u_1 at the left end and u_2 at the right end as the result of an external nodal force of P. The result is

$$u_1 \frac{AE}{L} - u_2 \frac{AE}{L} = P$$

$$-u_1 \frac{AE}{L} + u_2 \frac{AE}{L} = P \tag{c}$$

In matrix format the result is

$$\frac{AE}{L} \begin{bmatrix} 1 & -1 \\ -1 & 1 \end{bmatrix} \begin{Bmatrix} u_1 \\ u_2 \end{Bmatrix} = \begin{Bmatrix} P \\ P \end{Bmatrix} \tag{d}$$

Equation (d) above can be compared with Eq. (e) of Prob. 2.9. The force matrix above represents boundary loadings, the axial force applied at either end of a bar in static equilibrium. The force matrix of Prob. 2.9 represents an internal distributed body force type of loading. The local stiffness matrix is the same using either method of derivation.

Supplementary Problems

2.22. Solve Eq. (2.11) assuming a rod of length L, a constant heat source Q, and boundary conditions $T(x = 0) = T_0$ and $q(x = L) = -q_L$ (removing heat).

2.23. Derive a linear interpolation polynomial for a function $T(\theta)$ in polar coordinates for a constant radial coordinate and an arc length of $\alpha = \theta_2 - \theta_1$.

2.24. Assume $T = C \sin(\pi x/L)$, where C is a constant, and use the Rayleigh-Ritz method to solve Prob. 2.3.

2.25. Obtain a 10-element, 11-node solution for the cable deflection described in Prob. 2.12 and compare it with the exact solution given in Prob. 2.1.

2.26. Obtain a 10-element, 11-node solution for the mass transport described in Prob. 2.18.

2.27. Show that the variational function given by Eq. (2.31) gives the governing differential equation, Eq. (2.29).

2.28. Refer to Fig. 2-7 and Eq. (2.30) and obtain a general solution for the coaxial cable.

(a) Establish and solve the governing differential equation for each section of the cable.

(b) Establish the boundary conditions and continuity conditions.

(c) Formulate the equations of part (b) as a matrix equation that can be solved for the constants of integration.

2.29. Derive the shape functions for a general three-node quadratic element in terms of x_1, x_2, and x_3 as shown in Fig. 2-10.

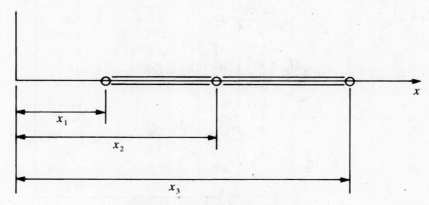

Fig. 2-10 Three-node element.

2.30. Refer to Prob. 2.29 and Fig. 2-10 and let $x_1 = -L$, $x_2 = 0$, and $x_3 = L$. Use the results of Prob. 2.29 and derive the corresponding shape functions.

2.31. Refer to Prob. 2.30 and derive the local stiffness matrix for an element of length $2L$ with coordinates $(-L, 0, L)$. See Prob. 2.9.

2.32. Use the three-node element of length $2L$ defined in Prob. 2.30 and derive the corresponding force matrix that defines the distribution of body force for each node. See Probs. 2.9 and 2.19.

Answers to Supplementary Problems

2.22. $T(x) = Q(xL - x^2/2)/k - q_L x/k + T_0$.

2.23. $T = (\theta_2 - \theta)T_1/\alpha + (\theta - \theta_1)T_2/\alpha$.

2.24. $T = (4QL^2/\pi^3 k) \sin(\pi x/L)$; at $x = L/2$, $T = 0.129 QL^2/k$ compares with the exact answer of $T = 0.125 QL^2/k$.

2.25.

Cable Deflection			
Node	x	**Finite element**	**Exact**
1	0	0.0	0.0
2	12	1.0913	1.0853
3	24	1.8264	1.8173
4	36	2.2953	2.2848
5	48	2.5555	2.5444
6	60	2.6389	2.6276

2.26.

Mass Transport			
Node	x	Finite element	Exact
1	0.0	0.0	0.0
2	0.1	0.2975	0.2997
3	0.2	0.6531	0.6576
4	0.3	1.0881	1.0948
5	0.4	1.6294	1.6383
6	0.5	2.3121	2.3229
7	0.6	3.1816	3.1937
8	0.7	4.2965	4.3090
9	0.8	5.7333	5.7446
10	0.9	7.5914	7.5990
11	1.0	10.0	10.0

2.28. (a)
$$\frac{\epsilon_1}{r}\left[\frac{d}{dr}\left(\frac{r\,d\phi_1}{dr}\right)\right] = -\rho_1 \qquad a \le r \le b$$

$$\frac{d\phi_1}{dr} = -\frac{\rho_1 r^3}{2\epsilon_1} + \frac{C_1}{r}$$

$$\phi_1 = -\frac{\rho_1 r^2}{4\epsilon_1} + C_1 \ln r + C_2$$

and
$$\left(\frac{\epsilon_2}{r}\right)\left[\frac{d}{dr}\left(\frac{r\,d\phi_2}{dr}\right)\right] = 0 \qquad b \le r \le c$$

$$\frac{d\phi_2}{dr} = \frac{C_3}{r}$$

$$\phi_2 = C_3 \ln r + C_4$$

(b)
$$\phi_1(r = a) = \phi_a \qquad \text{or} \qquad \phi_a = \frac{-\rho_1 a^2}{4\epsilon_1} + C_1 \ln a + C_2$$

$$\phi_2(r = c) = 0 \qquad \text{or} \qquad 0 = C_3 \ln c + C_4$$

The electric potential is continuous at $r = b$, or

$$\frac{-\rho_1 b^2}{4\epsilon_1} + C_1 \ln b + C_2 = C_3 \ln b + C_4$$

The electric displacement is continuous at $r = b$, or

$$\epsilon_1\left[\frac{-\rho_1 b}{2\epsilon_1} + \frac{C_1}{b}\right] = \frac{\epsilon_2 C_3}{b}$$

(c)
$$\begin{bmatrix} \ln a & 1 & 0 & 0 \\ \ln b & 1 & -\ln b & -1 \\ \epsilon_1/b & 0 & -\epsilon_2/b & 0 \\ 0 & 0 & \ln c & 1 \end{bmatrix}\begin{Bmatrix} C_1 \\ C_2 \\ C_3 \\ C_4 \end{Bmatrix} = \begin{Bmatrix} \phi_a + \rho_1 a^2/4\epsilon_1 \\ \rho_1 b^2/4\epsilon_1 \\ \rho_1 b/2 \\ 0 \end{Bmatrix}$$

2.29.
$$N_1 = \frac{(x - x_2)(x - x_3)}{(x_1 - x_2)(x_1 - x_3)}$$

$$N_2 = \frac{(x - x_1)(x - x_3)}{(x_2 - x_1)(x_2 - x_3)}$$

$$N_3 = \frac{(x - x_1)(x - x_2)}{(x_3 - x_1)(x_3 - x_2)}$$

2.30. $N_1 = (x^2 - xL)/2L^2, N_2 = (L^2 - x^2)/L^2, N_3 = (x^2 + xL)/2L^2.$

2.31.
$$\frac{AE}{L} \begin{bmatrix} \frac{7}{6} & -\frac{4}{3} & \frac{1}{6} \\ -\frac{4}{3} & \frac{8}{3} & -\frac{4}{3} \\ \frac{1}{6} & -\frac{4}{3} & \frac{7}{6} \end{bmatrix} \begin{Bmatrix} u_1 \\ u_2 \\ u_3 \end{Bmatrix}$$

2.32. Distribution to node 1 = $AfL/3$, distribution to node 2 = $4AfL/3$, distribution to node 3 = $AfL/3$. (Note that the node distributions add to give $2AfL$, the total body force on an element.)

Chapter 3

Two-Dimensional Finite Elements

3.1. INTRODUCTION

The one-dimensional concepts of Chap. 2 will be extended to two dimensions in this chapter. The differential equations and boundary conditions that govern the various engineering theories will be written in a general format in two dimensions. In addition, the equations governing planar problems in the theory of elasticity will be introduced.

Two fundamental finite elements will be discussed. The four-node quadrilateral element and the three-node triangular element are the most elementary two-dimensional elements used for computation. However, with these two elements it is possible to model any two-dimensional physical problem. The quadrilateral element of this chapter is rectangular and must conform to a cartesian global coordinate system. The triangular element can be used to model a two-dimensional problem with a curved boundary by approximating the curve with a series of straight lines. Both these elements can be extended to special situations such as axisymmetric problems. The four-node element can be formulated in cylindrical coordinates for nonaxisymmetric situations, but each element must conform to the cylindrical coordinate system.

The variational functions of Chap. 2 have two-dimensional counterparts. Again, as in Chap. 2, the variational statement of the problem will serve as the basis for developing two-dimensional finite elements.

A brief discussion of transformation matrices is included near the end of the chapter. Two types of transformations will be discussed: One deals with two-dimensional boundary conditions for applications in elasticity, and the other is a general transformation for renumbering or resequencing the nodes or degrees of freedom for an element.

3.2. TWO-DIMENSIONAL BOUNDARY-VALUE PROBLEMS

The various physical problems described in Chap. 2 can be extended to two dimensions and, with the exception of Eq. (2.18), can be written in the general form

$$\frac{\partial}{\partial x}\left[\alpha_x\frac{\partial\phi(x,y)}{\partial x}\right] + \frac{\partial}{\partial y}\left[\alpha_y\frac{\partial\phi(x,y)}{\partial y}\right] + \beta\phi(x,y) = f(x,y) \tag{3.1}$$

where α_x, a_y, and β are known parameters, such as material constants, and in general can be functions of x and y. However, these parameters are assumed not to vary within an element, and any global variation can be modeled as a change from element to element. The two-dimensional counterpart for cable deflection is interpreted as the deflection of a flexible membrane, with $\alpha_x = \alpha_y = T$ the membrane tension, $\beta = -k$ the elastic foundation modulus, and f of Eq. (3.1) taken as $-f$, the pressure normal to the membrane in the direction of the deflection w.

The two-dimensional equation for mass transport is similar to Eq. (2.18) and is

$$u_x\frac{\partial C(x,y)}{\partial x} + u_y\frac{\partial C(x,y)}{\partial y} - \frac{\partial}{\partial x}\left[D_x\frac{\partial C(x,y)}{\partial x}\right] - \frac{\partial}{\partial y}\left[D_y\frac{\partial C(x,y)}{\partial y}\right] + K_rC(x,y) = m \tag{3.2}$$

where u_x, u_y, D_x, and D_y are velocities and diffusivities in the x and y directions, respectively.

3.3. CONNECTIVITY AND NODAL COORDINATES

Node numbering for three- and four-node elements is shown in Fig. 3-1. Node numbering for the purpose of deriving a local stiffness matrix always employs consecutive numbers beginning with 1. It is accepted practice in all finite element analyses to number the element nodes in a *counterclockwise* direction. The rectangular local elements derived in this chapter should be numbered beginning with the smallest coordinate value of x and y. The sequence of numbering for higher-order elements with midside and interior nodes is not standardized; the corner nodes may be numbered first and the midside nodes given higher numbers, or all nodes may be numbered in order. No matter what the sequence, the numbering proceeds counterclockwise. A finite element model is a geometric picture of a prototype, and each node must have an identifier (node number) and a coordinate location corresponding to that identifier. Each element of the model must have an identifier (element number) and a connectivity array. In a computer code there are usually two arrays for describing the geometry of the model, nodal coordinates and connectivity. In two-dimensional problems the nodal coordinate array has dimensions of $N_{node} \times 2$, where N_{node} is the number of nodes. The connectivity array is the same as discussed in the previous chapter, $N_{el} \times N_{node}$. In application the connectivity array may contain additional information, such as a material identifier. A prototype to be modeled may be made of one or more materials, such as in Prob. 2.19, and a material identifier (material number) is required for each element.

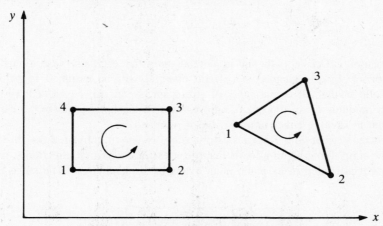

Fig. 3-1 Counterclockwise node numbering.

3.4. THEORY OF ELASTICITY

The finite element method has been used extensively for solution of the equations for two-dimensional elasticity. The elasticity problem is formulated in terms of displacements, and all finite element nodes must have two degrees of freedom to represent a displacement in each coordinate direction. The other problems of Chap. 2 (temperature distribution, mass transport, and so on) are classified as scalar field problems and have one unknown per finite element node in either one or two dimensions. The deflection of a cable in one dimension and the deflection of a membrane in two dimensions are formulated in terms of displacement. Displacement, in this context, is a vector quantity because it has magnitude and direction. However, the vector of displacement has only one component, and its direction is known before the problem is solved. It follows that formulation of the finite element is the same for a scalar and for a single-component vector. The first derivative of the primary variable in a scalar field problem is directional, and it follows that in two or three dimensions is a vector quantity. The first derivative is directional in one dimension but requires no special consideration since the direction is known.

The first derivatives of the two-component vector displacement in two-dimensional elasticity correspond to the various strains, and the strains can be related linearly to the stresses. Stress and strain are second-order tensor quantities and as such have certain mathematical and geometric properties. This means that the finite element formulation for an elasticity problem will be more involved than the formulation of scalar field problems. However, the formulation still follows the fundamental concepts established in Chap. 2.

The strain-displacement equations of the theory of elasticity in two dimensions relate the displacements u and v in the x- and y-coordinate directions to the corresponding strains:

$$\epsilon_{xx} = \frac{\partial u}{\partial x} \qquad \epsilon_{yy} = \frac{\partial v}{\partial y} \qquad \epsilon_{xy} = \frac{\partial u}{\partial y} + \frac{\partial v}{\partial x} \tag{3.3}$$

The normal strains ϵ_{xx} and ϵ_{yy} define deformation per unit length in the x and y directions, respectively. Shear strain ϵ_{xy} defines the relative angular deformation of an element of material. The definition given by Eqs. (3.3) is referred to as the engineering definition of strain. The equations of equilibrium define the balance of forces on a differential element. While they will not be used directly in this chapter to formulate the finite element, they must be satisfied when analytical solutions are attempted. These equations define a relationship among normal stress, shear stress, and body force:

$$\frac{\partial \sigma_{xx}}{\partial x} + \frac{\partial \sigma_{xy}}{\partial y} + f_x = 0 \tag{3.4}$$

$$\frac{\partial \sigma_{xy}}{\partial x} + \frac{\partial \sigma_{yy}}{\partial y} + f_y = 0 \tag{3.5}$$

The constitutive relation is Hooke's law and in general states that each stress component can be related to each strain component. In this chapter the elastic material will be assumed to be homogeneous and isotropic, and it follows that all components of stress and strain can be related using the engineering constants, Young's modulus E and Poisson's ratio ν. Two-dimensional problems in cartesian coordinates are usually assumed to involve either *plane stress* or *plane strain*.

Plane stress occurs when the thickness dimension is small compared to the length and width dimensions, and the simplifying assumption is that the stress in the normal direction (z axis) is zero. A thin plate or disk of material loaded in its plane is an example of plane stress. The stress-strain relations are

$$\sigma_{xx} = \frac{E}{1 - \nu^2} (\epsilon_{xx} + \nu\epsilon_{yy}) \tag{3.6}$$

$$\sigma_{yy} = \frac{E}{1 - \nu^2} (\epsilon_{yy} + \nu\epsilon_{xx}) \tag{3.7}$$

$$\sigma_{xy} = G\epsilon_{xy} = \frac{E}{2(1 + \nu)} \epsilon_{xy} \tag{3.8}$$

where G is the shear modulus and can be defined in terms of E and ν:

$$G = \frac{E}{2(1 + \nu)} \tag{3.9}$$

Plane strain can be assumed when the length dimension is large compared with the cross section of a body such as a gun barrel. A proper assumption is that the displacement and $\partial/\partial z$ are zero in the z direction. It follows that the stress-strain relations for plane strain are

$$\sigma_{xx} = \frac{E}{(1 + \nu)(1 - 2\nu)} [(1 - \nu)\epsilon_{xx} + \nu\epsilon_{yy}] \tag{3.10}$$

$$\sigma_{yy} = \frac{E}{(1 + \nu)(1 - 2\nu)} [(1 - \nu)\epsilon_{yy} + \nu\epsilon_{xx}] \tag{3.11}$$

$$\sigma_{zz} = \nu(\sigma_{xx} + \sigma_{yy}) \tag{3.12}$$

and the relation between shear stress and shear strain is the same as in Eq. (3.8).

3.5. VARIATIONAL FUNCTIONS

Finite elements can be derived for two-dimensional problems using variational functions following the methods of the previous chapter. However, a variational function must exist for the governing differential equation. Variational functions for two-dimensional problems can be obtained as an extension of the one-dimensional function in the same way that the governing differential equation is an extension of its one-dimensional counterpart. In other words, if the variational function exists in one dimension, it also exists in two or three dimensions.

The variational function corresponding to plane elasticity problems can be written in a general form using cartesian tensor notation in terms of the stress tensor and strain tensor:

$$J(u) = \int_V \left(\frac{1}{2}\, \sigma_{kj}\epsilon_{kj} - f_k u_k \right) dV - \int_S T_k u_k\, dS \tag{3.13}$$

Equation (3.13) is analogous to the *principle of minimum potential energy* sometimes discussed in mechanics of materials and theory of elasticity.

3.6. TRIANGULAR ELEMENTS AND AREA COORDINATES

Finite element concepts for two-dimensional problems are developed in detail in this chapter using four-node rectangular elements. In analysis applications the three-node triangular element is probably the more popular element. The obvious limitation of the four-node rectangular element is that it must conform to an application that can be approximated using rectangles, whereas the three-node element can be used to approximate any shape. Both elements can be inferior when compared with elements of higher order, however, they are the obvious starting point from an instructional point of view. Area integration for rectangular elements is elementary, and that is the fundamental reason for emphasizing these elements for the derivation of local finite element stiffness matrices.

The reader, after studying Chap. 2, should be aware of the fact that derivatives of shape functions are a significant feature of finite elements. Three-node triangular elements are linear in both x and y, while four-node rectangular elements contain products of x and y. First derivatives for the three-node element are constant, and the elements are often called constant-strain triangles (CSTs). A more appropriate name would be constant-gradient triangles, but the terminology comes from early applications in the theory of elasticity and has become standard. First derivatives of the shape functions for four-node elements are functions of x or y.

Area integration for three-node triangular elements is deceptively simple because the first derivatives are constant and area integration reduces to evaluating the area of the associated triangle. The stiffness matrix for a triangular element is often derived using area coordinates, and area integration is carried out using a specialized type of numerical integration; as such, derivation of the stiffness matrix can be somewhat obscure. Triangular elements, area coordinates, and their numerical integration are discussed after the fundamental variational functions are developed in matrix format and their use illustrated using rectangular elements. Once the variational function is written in matrix format and minimized, it is applicable for deriving any finite element since the shape functions actually define the process for deriving the stiffness matrix.

3.7. TRANSFORMATIONS

Matrix transformations are computer manipulations that modify a matrix based upon some type of constraint or condition. The matrix that is modified is usually the stiffness matrix. However, under some circumstances it may be efficient to modify the force matrix with machine computations. In this discussion two types of matrix transformations will be addressed. The first is based upon the standard vector transformation ilustrated in Fig. 3-2, and the second modifies the local stiffness matrix when the node numbering sequence must be changed after the local element is formulated. This situation will occur frequently in later chapters.

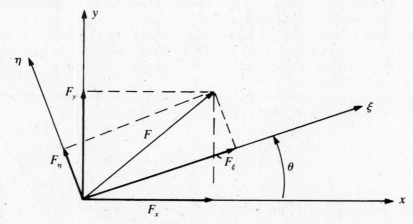

Fig. 3-2   Vector transformation.

The vector transformation defined by Fig. 3-2 can be used to express the vector components of F with ξ, η coordinates in terms of components in the x, y coordinate system:

$$\begin{Bmatrix} F_\xi \\ F_\eta \end{Bmatrix} = \begin{bmatrix} \cos\theta & \sin\theta \\ -\sin\theta & \cos\theta \end{bmatrix} \begin{Bmatrix} F_x \\ F_y \end{Bmatrix} = [\mathrm{T}] \begin{Bmatrix} F_x \\ F_y \end{Bmatrix} \qquad (3.14)$$

where $[\mathrm{T}]$ is the transformation matrix. The vector can define force, displacement, or any other vector quantity. For many applications in finite element computations, the transformation is from the ξ, η system to the local x, y system:

$$\begin{Bmatrix} F_x \\ F_y \end{Bmatrix} = \begin{bmatrix} \cos\theta & -\sin\theta \\ \sin\theta & \cos\theta \end{bmatrix} \begin{Bmatrix} F_\xi \\ F_\eta \end{Bmatrix} = [\mathrm{T}]^T \begin{Bmatrix} F_\xi \\ F_\eta \end{Bmatrix} \qquad (3.15)$$

where $[\mathrm{T}]^T$ is the transpose of $[\mathrm{T}]$.

3.8. CYLINDRICAL COORDINATES

The three-dimensional counterpart of Eq. (3.1) in cylindrical r, θ, z coordinates can be written in a general form. The three-dimensional equations are given but will be used in various two-dimensional forms in subsequent applied problems. The material constants will be assumed to be independent of the coordinates, and again, any spatial dependence can be included in the finite element formulation:

$$\alpha_r \left[\frac{\partial^2 \phi(r, \theta, z)}{\partial r^2} + \frac{1}{r} \frac{\partial \phi(r, \theta, z)}{\partial r} \right] + \alpha_\theta \frac{1}{r^2} \frac{\partial^2 \phi(r, \theta, z)}{\partial \theta^2} + \alpha_z \frac{\partial^2 \phi(r, \theta, z)}{\partial z^2} + \beta \phi(r, \theta, z) = f(r, \theta, z) \quad (3.16)$$

The three-dimensional strain-displacement equations of elasticity in cylindrical coordinates, where u, v, and w are the displacements in the r, θ, and z directions, respectively, are

$$\epsilon_r = \frac{\partial u}{\partial r} \qquad \epsilon_\theta = \frac{1}{r}\frac{\partial v}{\partial \theta} + \frac{u}{r} \qquad \epsilon_z = \frac{\partial w}{\partial z}$$

$$\epsilon_{r\theta} = \frac{1}{r}\frac{\partial u}{\partial \theta} + \frac{\partial v}{\partial r} - \frac{v}{r} \qquad \epsilon_{rz} = \frac{\partial w}{\partial r} + \frac{\partial u}{\partial z} \qquad \epsilon_{\theta z} = \frac{\partial v}{\partial z} + \frac{1}{r}\frac{\partial w}{\partial \theta} \qquad (3.17)$$

The shear strains of Eq. (3.17) are specifically the engineering definition of strain. The three-dimensional definition can be reduced to two dimensions using r, z coordinates or r, θ coordinates. The three-dimensional stress-strain equations for isotropic elasticity are similar to the plane strain equations:

$$\sigma_{rr} = \frac{E}{(1+\nu)(1-2\nu)}\left[(1-\nu)\epsilon_{rr} + \nu(\epsilon_{\theta\theta} + \epsilon_{zz})\right]$$

$$\sigma_{\theta\theta} = \frac{E}{(1+\nu)(1-2\nu)}\left[(1-\nu)\epsilon_{\theta\theta} + \nu(\epsilon_{rr} + \epsilon_{zz})\right]$$

$$\sigma_{zz} = \frac{E}{(1+\nu)(1-2\nu)}\left[(1-\nu)\epsilon_{zz} + \nu(\epsilon_{\theta\theta} + \epsilon_{rr})\right]$$

$$\sigma_{r\theta} = G\epsilon_{r\theta} \qquad \sigma_{rz} = G\epsilon_{rz} \qquad \sigma_{\theta z} = G\epsilon_{\theta z} \qquad (3.18)$$

Solved Problems

3.1. A rectangular finite element with dimensions $a \times b$ is defined in an x, y coordinate system in Fig. 3-3. Assume a function of the form

$$\phi = A + Bx + Cy + Dxy \qquad (a)$$

and derive the shape functions for the rectangular element. Write the final result in the form

$$\phi = N_1\phi_1 + N_2\phi_2 + N_3\phi_3 + N_4\phi_4 \qquad (b)$$

The boundary conditions for the element shown in Fig. 3-3 are

$$\phi(0,0) = \phi_1 \qquad \phi(a,0) = \phi_2 \qquad \phi(a,b) = \phi_3 \qquad \phi(0,b) = \phi_4 \qquad (c)$$

Substituting Eqs. (c) into Eq. (a) gives four equations in terms of the unknown constants:

$$\phi_1 = A \qquad\qquad \phi_2 = A + Ba$$

$$\phi_3 = A + Ba + Cb + Dab \qquad \phi_4 = A + Cb$$

It follows that

$$A = \phi_1 \qquad B = \frac{\phi_2 - \phi_1}{a} \qquad C = \frac{\phi_4 - \phi_1}{b} \qquad D = \frac{\phi_1 - \phi_2 + \phi_3 - \phi_4}{ab} \qquad (d)$$

Substitute Eqs. (d) into Eq. (a) and rearrange to obtain the form of Eq. (b). The shape functions can be written as

$$N_1 = \frac{(a-x)(b-y)}{ab} \qquad N_2 = \frac{x(b-y)}{ab}$$

$$N_3 = \frac{xy}{ab} \qquad\qquad N_4 = \frac{y(a-x)}{ab} \qquad (e)$$

The shape function for node 1 has a magnitude of 1 at node 1 and is 0 at the remaining nodes. All shape functions have this property, as illustrated in Fig. 3-3.

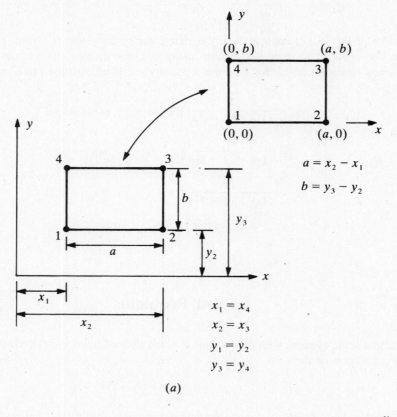

(a)

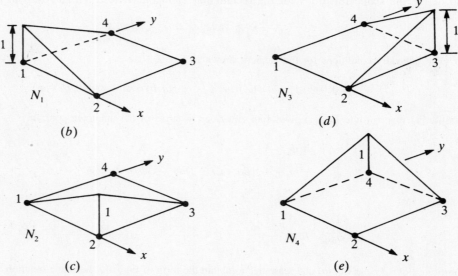

(b)   (d)

(c)   (e)

Fig. 3-3   Rectangular four-node finite element.

3.2.   A three-node triangular element is defined in Fig. 3-4 in an x, y coordinate system with nodes 1, 2, and 3 located at (x_1, y_1), (x_2, y_2), and (x_3, y_3), respectively, in the global system. Derive shape functions in terms of global coordinates.

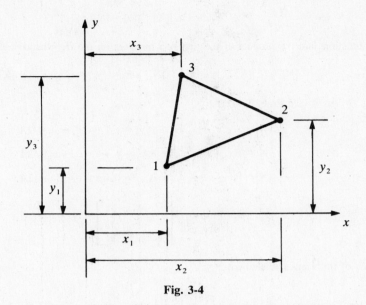

Fig. 3-4

Assume an interpolation function to represent the variation of the unknown quantity

$$\phi = C_1 + C_2 x + C_3 y \tag{a}$$

Write the interpolation function as a matrix equation

$$\phi = \begin{bmatrix} 1 & x & y \end{bmatrix} \begin{Bmatrix} C_1 \\ C_2 \\ C_3 \end{Bmatrix} \quad \text{or in matrix format} \quad \phi = [\alpha]\{C\} \tag{b}$$

The boundary conditions are in terms of nodal point values of ϕ:

$$\phi(x_1, y_1) = \phi_1 \qquad \phi(x_2, y_2) = \phi_2 \qquad \phi(x_3, y_3) = \phi_3 \tag{c}$$

Substitute Eqs. (c) into Eq. (a) to obtain three equations that can be solved for C_1, C_2, and C_3:

$$\begin{aligned} \phi_1 &= C_1 + C_2 x_1 + C_3 y_1 \\ \phi_2 &= C_1 + C_2 x_2 + C_3 y_2 \\ \phi_3 &= C_1 + C_2 x_3 + C_3 y_3 \end{aligned} \quad \text{or} \quad \begin{Bmatrix} \phi_1 \\ \phi_2 \\ \phi_3 \end{Bmatrix} = \begin{bmatrix} 1 & x_1 & y_1 \\ 1 & x_2 & y_2 \\ 1 & x_3 & y_3 \end{bmatrix} \begin{Bmatrix} C_1 \\ C_2 \\ C_3 \end{Bmatrix} \tag{d}$$

Write Eq. (d) in matrix form as

$$\{\phi\} = [X]\{C\} \tag{e}$$

and solve for $\{C\}$ as

$$\{C\} = [X]^{-1}\{\phi\} \tag{f}$$

Substitute Eq. (f) into Eq. (b):

$$\phi = [\alpha][X]^{-1}\{\phi\} = [N]\{\phi\} \qquad (g)$$

The shape functions are the product of the first two matrices on the right-hand side of Eq. (g).

$$[N] = [\alpha][X]^{-1} \quad \text{or} \quad [N_1 \quad N_2 \quad N_3] = [1 \quad x \quad y][X]^{-1} \qquad (h)$$

Solving Eq. (h) gives (see Prob. 3.30)

$$N_1 = [(x_2 y_3 - x_3 y_2) + x(y_2 - y_3) + y(x_3 - x_2)]/2A$$

$$N_2 = [(x_3 y_1 - x_1 y_3) + x(y_3 - y_1) + y(x_1 - x_3)]/2A \qquad (i)$$

$$N_3 = [(x_1 y_2 - x_2 y_1) + x(y_1 - y_2) + y(x_2 - x_1)]/2A$$

where

$$A = \frac{1}{2} \det \begin{bmatrix} 1 & x_1 & y_1 \\ 1 & x_2 & y_2 \\ 1 & x_3 & y_3 \end{bmatrix} \qquad (j)$$

A is the area of the triangular element.

3.3. A quadrilateral element is shown in an x, y coordinate system in Fig. 3-5. The nodes are located at the coordinate points as illustrated, and a temperature distribution has been computed at each node as $T_1 = 100°$, $T_2 = 60°$, $T_3 = 50°$, and $T_4 = 90°$. Use the shape function derived in Prob. 3.1 and compute the temperature at $x = 2.5$ and $y = 2.5$.

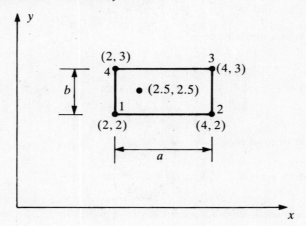

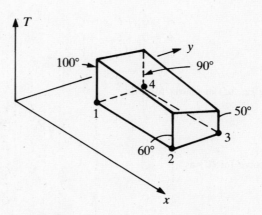

Fig. 3-5

Use Eqs. (b) and (e) of Prob. 3.1. Note that $a = 2.0$ and $b = 1.0$ and that the location of point $(2.5, 2.5)$ is within the element at $(a/4, b/2)$. The shape functions are computed as

$$N_1 = \frac{(a - a/4)(b - b/2)}{ab} = \frac{3}{8} \qquad N_2 = \frac{(a/4)(b - b/2)}{ab} = \frac{1}{8} \qquad N_3 = \frac{(a/4)(b/2)}{ab} = \frac{1}{8} \qquad N_4 = \frac{(a - a/4)(b/2)}{ab} = \frac{3}{8}$$

The temperature is computed as

$$T(2.5, 2.5) = \tfrac{3}{8}(100) + \tfrac{1}{8}(60) + \tfrac{1}{8}(50) + \tfrac{3}{8}(90) = 85°$$

3.4. Deduce the variational function that corresponds to two-dimensional heat transfer. Write the variational function in terms of linear four-node shape functions and arrive at a general matrix statement corresponding to the variational function.

Use Eq. (2.23) and extend the one-dimensional function to two dimensions:

$$J(T) = \int_A \left[\frac{k_x}{2} \left(\frac{\partial T}{\partial x} \right)^2 + \frac{k_y}{2} \left(\frac{\partial T}{\partial y} \right)^2 - QT \right] t \, dx \, dy \qquad (a)$$

A constant thickness t has been assumed, and the thermal conductivities k_x and k_y can have different values in the x and y directions. The shape functions representing the temperature distribution within an element have the form of Eq. (b) of Prob. 3.1, and in matrix form the temperature is expressed as

$$T = [N_1 \quad N_2 \quad N_3 \quad N_4] \begin{Bmatrix} T_1 \\ T_2 \\ T_3 \\ T_4 \end{Bmatrix} = [N]\{T\} \qquad (b)$$

where only the shape functions are functions of x and y. Define operator matrices that represent the partial derivatives occurring in Eq. (a):

$$[L_x] = \left[\frac{\partial}{\partial x} \right] \qquad \text{and} \qquad [L_y] = \left[\frac{\partial}{\partial y} \right] \qquad (c)$$

Let $[L_x]$ operate on $[N]$ or $[L_x][N]$, and the result will be a 1×1 matrix times a 1×4 matrix, giving a 1×4 matrix that defines the partial derivatives of the shape functions with respect to x. Similarly, $[L_y][N]$ is a 1×4 matrix that defines the partial derivatives with respect to y. For the present let k_x and k_y each be represented as 1×1 matrices. Equation (a) can now be written as a matrix equation

$$J(T) = \int_A \left(\frac{1}{2} \{T\}^T [N]^T [L_x]^T [k_x][L_x][N]\{T\} + \frac{1}{2} \{T\}^T [N]^T [L_y]^T [k_y][L_y][N]\{T\} - \{T\}^T [N]^T Q \right) t \, dx \, dy \qquad (d)$$

The final result after multiplying all the matrix terms is a 1×1 matrix and indicates a scalar quantity. The function is in a form that can be minimized with respect to $\{T\}$. The first two terms would result in two 4×4 matrices that would be added. The last term would be a 4×1 matrix showing the distribution of Q to each of the nodes. The form of Eq. (d) illustrates the process of formulating the finite element. However, a more compact form is desirable for computation. Rather than define $[L_x]$ and $[L_y]$ separately, one operator matrix can be defined as

$$[L] = \begin{bmatrix} \partial/\partial x \\ \partial/\partial y \end{bmatrix}$$

And one matrix can be used to define the partial derivatives as [L] operating on [N]:

$$[L][N] = \begin{bmatrix} \partial/\partial x \\ \partial/\partial y \end{bmatrix} [N_1 \quad N_2 \quad N_3 \quad N_4] = \begin{bmatrix} \partial N_1/\partial x & \partial N_2/\partial x & \partial N_3/\partial x & \partial N_4/\partial x \\ \partial N_1/\partial y & \partial N_2/\partial y & \partial N_3/\partial y & \partial N_4/\partial y \end{bmatrix} \qquad (e)$$

The last term in Eq. (e) is commonly referred to as the B matrix. The variational function can be written in a more compact form by defining a thermal conductivity matrix

$$[k] = \begin{bmatrix} k_x & 0 \\ 0 & k_y \end{bmatrix}$$

and

$$J(T) = \int_A \left(\frac{1}{2} \{T\}^T [N]^T [L]^T [k][L][N]\{T\} - \{T\}^T [N]^T Q \right) t \, dx \, dy$$

or

$$J(T) = \int_A \left(\frac{1}{2} \{T\}^T [B]^T [k][B]\{T\} - \{T\}^T [N]^T Q \right) t \, dx \, dy \qquad (f)$$

Equation (f) is an equivalent matrix statement of the variational function.

3.5.   Derive a local stiffness matrix for heat transfer using shape functions for a four-node quadrilateral element.

The variational function is written in matrix format, as in Eq. (f) of Prob. 3.4, and will be minimized with respect to the unknown variable that in this case is {T}. It follows that

$$\int_A [B]^T [k][B]\{T\} t \, dx \, dy = \int_A [N]^T Q t \, dx \, dy \qquad (a)$$

The first three matrices on the left-hand side of Eq. (a) are multiplied together to give a 4×4 local stiffness matrix. The local stiffness matrix is defined as

$$[K] = \int_A [B]^T [k][B] t \, dx \, dy \qquad (b)$$

For illustration, the first term in the stiffness matrix appears as

$$K_{11} = \int_0^a \int_0^b \left(\frac{\partial N_1}{\partial x} k_x \frac{\partial N_1}{\partial x} + \frac{\partial N_1}{\partial y} k_y \frac{\partial N_1}{\partial y} \right) t \, dx \, dy \qquad (c)$$

The derivatives of the shape functions are required before an attempt is made to evaluate the integral and are listed below for ready reference. Refer to Eq. (e) of Prob. 3.1.

$$\frac{\partial N_1}{\partial x} = \frac{-(b-y)}{ab} \qquad \frac{\partial N_1}{\partial y} = \frac{-(a-x)}{ab}$$

$$\frac{\partial N_2}{\partial x} = \frac{b-y}{ab} \qquad \frac{\partial N_2}{\partial y} = \frac{-x}{ab}$$

$$\frac{\partial N_3}{\partial x} = \frac{y}{ab} \qquad \frac{\partial N_3}{\partial y} = \frac{x}{ab}$$ (d)

$$\frac{\partial N_4}{\partial x} = \frac{-y}{ab} \qquad \frac{\partial N_4}{\partial y} = \frac{a-x}{ab}$$

The proper derivatives are substituted into Eq. (c) to give, for example, the first term in the stiffness matrix:

$$K_{11} = \int_0^a \int_0^b \frac{t}{a^2 b^2} [(b-y)^2 k_x + (a-x)^2 k_y] \, dx \, dy$$

Integrating and substituting limits gives

$$K_{11} = \frac{(b^2 k_x + a^2 k_y)t}{3ab}$$

The remaining terms in the stiffness matrix are evaluated in a similar manner. The local stiffness matrix appears as follows and the reader may verify the results as an exercise:

$$[K] = \frac{t}{6ab} \begin{bmatrix} 2b^2 k_x + 2a^2 k_y & -2b^2 k_x + a^2 k_y & -b^2 k_x - a^2 k_y & b^2 k_x - 2a^2 k_y \\ & 2b^2 k_x + 2a^2 k_y & b^2 k_x - 2a^2 k_y & -b^2 k_x - a^2 k_y \\ \text{Symmetric} & & 2b^2 k_x + 2a^2 k_y & -2b^2 k_x + a^2 k_y \\ & & & 2b^2 k_x + 2a^2 k_y \end{bmatrix} \quad (e)$$

The local stiffness matrix is symmetric, and some terms are repeated in the matrix. For instance, all diagonal terms are the same. The term on the right-hand side of Eq. (a) represents the distribution of the heat-source term.

$$\int_0^a \int_0^b \frac{t}{ab} \begin{Bmatrix} (a-x)(b-y) \\ x(b-y) \\ xy \\ y(a-x) \end{Bmatrix} Q \, dx \, dy = \begin{Bmatrix} ab/4 \\ ab/4 \\ ab/4 \\ ab/4 \end{Bmatrix} tQ \qquad (f)$$

The heat source is distributed equally to the four nodes.

3.6. Compute the steady-state temperature distribution for the plate illustrated in Fig. 3-6. A constant temperature of $T_0 = 100$ is maintained along the edge $y = W$, and all other edges have zero temperature. For computation, assume the thermal conductivities are $k_x = k_y = 1$. Compare the numerical result with the exact solution. Assume $W = L = 1$ and a thickness of $t = 1$ since the solution is independent of t.

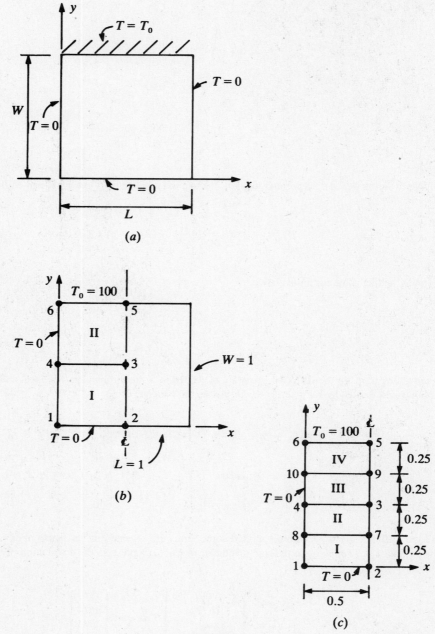

Fig. 3-6

The governing differential equation, with $k_x = k_y$, is Laplace's equation, and an analytical solution can be obtained using separation of variables. In this case Laplace's equation is

$$\frac{\partial^2 T}{\partial x^2} + \frac{\partial^2 T}{\partial y^2} = 0$$

subject to $T(0, y) = T(x, 0) = T(L, y) = 0$ and $T(x, W) = T_0 = 100$. The solution is

$$T = T_0 \frac{2}{\pi} \sum_{n=1,3,5}^{\infty} \frac{(-1)^{n+1}}{n} \frac{\sinh(n\pi y/L)}{\sinh(n\pi W/L)} \sin\left(\frac{n\pi x}{L}\right)$$ (a)

Three terms $(n = 1, 3, 5)$ in the series solution given by Eq. (a) give the following results: $T(0.5, 0.5) = 25.0$, $T(0.5, 0.25) = 9.54$, and $T(0.5, 0.75) = 54.12$. These results are compared with a two-element solution and a four-element solution. The plate, as shown in Fig. 3.6(b), illustrates the use of symmetry in a two-dimensional problem. The center of the plate along nodes 2, 3, and 5 is an axis of symmetry; the two elements are equivalent to a four-element formulation of the entire plate. All nodes except node 3 are boundary nodes, and the two-element formulation has only one unknown. The solution at node 3 is $T_3 = 37.50$. This compares with the analytical solution of $T(0.5, 0.5)$.

The division of the half-plate into four elements is shown in Fig. 3-6(c), and again symmetry is used to advantage along the center of the plate. There are three unknowns in the formulation, and the solution process follows the two-element solution. The final results are computed as $T_3 = 27.81$, $T_7 = 9.73$, and $T_9 = 69.74$ and can be compared with the analytical solution. The temperature at any coordinate location within the plate can be computed using the method illustrated in Prob. 3.3. The solution for this steady-state problem, since there is no heat-source term, is independent of the thermal conductivity. The thermal conductivity was assumed as 1.0 since it must have a nonzero value in the matrix formulation.

3.7. Construct the nodal coordinate and element connectivity arrays for the finite element models used in Prob. 3.6. Construct the corresponding stiffness matrices.

The two-element model will be analyzed first. The nodal coordinate array and connectivity array (Fig. 3-6) appear as shown in Tables 3.1 and 3.2, respectively.

Table 3.1

Node	X	Y
1	0.0	0.0
2	0.5	0.0
3	0.5	0.5
4	0.0	0.5
5	0.5	1.0
6	0.0	1.0

Table 3.2

Global element	Local element			
	Node 1	Node 2	Node 3	Node 4
I	1	2	3	4
II	4	3	5	6

The local elements for the two-element model are identical, with $a = b = 0.50$. Substituting into Eq. (e) of Prob. 3.5 gives the local stiffness matrix:

$$[K] = \begin{bmatrix} 0.667 & -0.167 & -0.333 & -0.167 \\ -0.167 & 0.667 & -0.167 & -0.333 \\ -0.333 & -0.167 & 0.667 & -0.167 \\ -0.167 & -0.333 & -0.167 & 0.667 \end{bmatrix}$$

The two local stiffness matrices combine according to the connectivity array. The boundary conditions are all of the essential type (specified values of T), and initially the force matrix is zero. The global stiffness matrix is 6×6, and the reader should verify that the two local stiffness matrices combine to give

$$[K] = \begin{bmatrix} & 1 & 2 & 3 & 4 & 5 & 6 \\ & 0.667 & -0.167 & -0.333 & -0.167 & 0.000 & 0.000 \\ & -0.167 & 0.667 & -0.167 & -0.333 & 0.000 & 0.000 \\ & -0.333 & -0.167 & 1.333 & -0.333 & -0.167 & -0.333 \\ & -0.167 & -0.333 & -0.333 & 1.333 & -0.333 & -0.167 \\ & 0.000 & 0.000 & -0.167 & -0.333 & 0.667 & -0.167 \\ & 0.000 & 0.000 & -0.333 & -0.167 & -0.167 & 0.667 \end{bmatrix}$$

The numbers across the top of the matrix correspond to the nodes of Fig. 3-6(*b*). Nodes 3 and 4 have contributions from both elements, whereas the remaining nodes have contributions from only one element.

Substitution of nonzero boundary conditions into the matrix equation was illustrated in Prob. 2.13. The same method is used to reduce the global stiffness matrix to the following matrix equation. Again, the reader should verify the result.

$$\begin{bmatrix} 1.000 & 0.000 & 0.000 & 0.000 & 0.000 & 0.000 \\ 0.000 & 1.000 & 0.000 & 0.000 & 0.000 & 0.000 \\ 0.000 & 0.000 & 1.333 & 0.000 & 0.000 & 0.000 \\ 0.000 & 0.000 & 0.000 & 1.000 & 0.000 & 0.000 \\ 0.000 & 0.000 & 0.000 & 0.000 & 1.000 & 0.000 \\ 0.000 & 0.000 & 0.000 & 0.000 & 0.000 & 1.000 \end{bmatrix} \begin{bmatrix} T_1 \\ T_2 \\ T_3 \\ T_4 \\ T_5 \\ T_6 \end{bmatrix} = \begin{bmatrix} 0.0 \\ 0.0 \\ 50 \\ 0.0 \\ 100 \\ 100 \end{bmatrix}$$

The solution for these equations agrees with the result given in Prob. 3.6.

The division of the plate into four elements increases the number of nodes to 10. The node numbering shown in Fig. 3-6 is not an optimum node numbering scheme, but it uses the same numbering as the two-element model, and a more random numbering scheme would better illustrate connectivity. The nodal coordinate and connectivity arrays are given in Tables 3.3 and 3.4, respectively. Again, all local stiffness matrices are the same. With $a = 0.50$ and $b = 0.25$, they are computed as

Table 3.3

Node	X	Y
1	0.00	0.00
2	0.50	0.00
3	0.50	0.50
4	0.00	0.50
5	0.50	1.00
6	0.00	1.00
7	0.50	0.25
8	0.00	0.25
9	0.50	0.75
10	0.00	0.75

Table 3.4

Global element	Local element			
	Node 1	Node 2	Node 3	Node 4
I	1	2	7	8
II	8	7	3	4
III	4	3	9	10
IV	10	9	5	6

$$[K] = \begin{bmatrix} 0.8333 & 0.1667 & -0.4167 & -0.5833 \\ 0.1667 & 0.8333 & -0.5833 & -0.4167 \\ -0.4167 & -0.5833 & 0.8333 & 0.1667 \\ -0.5833 & -0.4167 & 0.1667 & 0.8333 \end{bmatrix}$$

The global stiffness matrix will be 10×10. The connectivity array and Fig. 3-6(c) show that nodes 3, 4, and 7–10 will have contributions from two elements. The global stiffness matrix, before substituting boundary conditions, is the symmetric matrix

	1	2	3	4	5	6	7	8	9	10
	0.8333	0.1667	0.0000	0.0000	0.0000	0.0000	−0.4167	−0.5833	0.0000	0.0000
		0.8333	0.0000	0.0000	0.0000	0.0000	−0.5833	−0.4167	0.0000	0.0000
			1.6667	0.3333	0.0000	0.0000	−0.5833	−0.4167	−0.5833	−0.4167
				1.6667	0.0000	0.0000	−0.4167	−0.5833	−0.4167	−0.5833
					0.8333	0.1667	0.0000	0.0000	−0.5833	−0.4167
						0.8333	0.0000	0.0000	−0.4167	−0.5833
							1.6667	0.3333	0.0000	0.0000
								1.6667	0.0000	0.0000
									1.6667	0.3333
										1.6667

The numbers across the top of the matrix correspond to the node numbers. The substitution of boundary conditions follows the method of previous solutions and will not be illustrated. The reader may verify that the boundary conditions reduce the matrix to three unknowns and confirm the results for temperature given in Prob. 3.6.

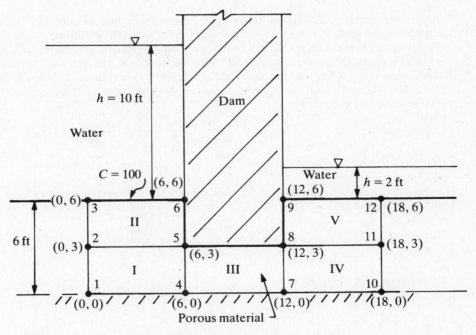

Fig. 3-7 Confined flow.

3.8. Formulate and obtain a finite element solution for the confined-flow situation illustrated in Fig. 3-7. Water, 10 ft deep upstream and 2 ft deep downstream, is confined using a dam that extends 3 ft into a porous material. Assume the material is homogeneous with hydraulic conductivity $K = 10(10^{-5})$. Also, a chemical species exists upstream in a concentration of 100 ppm. The species will diffuse within the porous medium with diffusion constant $D_x = D_y = 10(10^{-4})$ and reacts chemically with rate constant $K_r = 20(10^{-6})$. Compute the steady-state distribution of the chemical species considering (a) steady-state diffusion, (b) steady-state diffusion with convection. Then, assume the chemical species is measured downstream and has a concentration of 50 ppm. Compute the steady-state distribution of the chemical species considering (c) steady-state diffusion with convection and (d) steady-state diffusion with convection and chemical reaction. The mass transport analysis illustrated in this problem is satisfactory for the confined-flow type of formulation. The convection terms do not dominate the analysis. The reader should be cautious when applying this analysis to a situation when the convection terms dominate the analysis; the solution can become unstable. (Note that the material parameters in this problem, such as the reaction rate term, are chosen to illustrate the analysis rather than to represent a physical problem.)

The results derived in Probs. 3.5 and 3.33 will be used to construct the stiffness matrices. The coordinate number and connectivity arrays are given later in the solution to Prob. 3.35.

(a)   The steady-state solution for diffusion alone can be computed numerically using boundary conditions of $C = 100$ at nodes 3 and 6. The result will be 100 at all nodes and is independent of the diffusion constant. The reader can verify this result as a check that the global stiffness matrix is formulated correctly.

(b)   The velocity of the water as it moves through the soil medium is computed using the theory of potential flow (see Chap. 2). A homogeneous material means that $K_x = K_y = K$, and the solution for the potential is obtained from

$$\frac{\partial^2 \phi}{\partial x^2} + \frac{\partial^2 \phi}{\partial y^2} = 0 \qquad \text{or} \qquad \frac{\partial^2 h}{\partial x^2} + \frac{\partial^2 h}{\partial y^2} = 0 \qquad (a)$$

and is independent of K. Equation (e) of Prob. 3.5 is used to formulate the local stiffness matrix. The elements in this application are arranged in such a manner that all local stiffness matrices will be the same. Obviously, all elements do not have to be the same. The piezometric head in this case is computed using $y = 0$ as a datum and on the upstream side is the elevation head plus the pressure head or 16 ft. On the downstream side $h = 8$ ft. The boundary conditions are $h = 16$ at nodes 1, 2, 3, and 6, and $h = 8$ at nodes 9–12. The solution for ϕ at the remaining nodes is $h_4 = 13.803$, $h_5 = 14.379$, $h_7 = 10.197$, and $h_8 = 9.621$. The velocities are computed as

$$u_x = -K\frac{dh}{dx} \qquad \text{and} \qquad u_y = -K\frac{dh}{dy} \qquad (b)$$

Using Eq. (e) of Prob. 3.4, the finite element equivalent for computing the velocity at the center of an element is written

$$\{u\} = -[K][L][N]\{h\}$$

or, following Eqs. (d) of Prob. 3.5,

$$\begin{Bmatrix} u_x \\ u_y \end{Bmatrix} = \frac{-1}{ab} \begin{bmatrix} K_{11} & 0 \\ 0 & K_{22} \end{bmatrix} \begin{bmatrix} -b+y & b-y & y & -y \\ -a+x & -x & x & a-x \end{bmatrix} \begin{Bmatrix} h_1 \\ h_2 \\ h_3 \\ h_4 \end{Bmatrix} \qquad (c)$$

In this example the velocities are computed at the center of the element, and it follows that $x = a/2$ and $y = b/2$ in Eq. (c). The matrix of h values corresponds to the connectivity table of Prob. 3.34. The results for velocity are tabulated in Table 3.5. The concentration is computed using the local stiffness matrix for the second-derivative terms, Eq. (e) of Prob. 3.5, combined with the results given in Prob. 3.33 for the transport terms. The boundary conditions remain the same as in part (a), $C_3 = C_6 = 100$. The final results can be verified as 100 at each node.

Table 3.5	Velocities at the Center of
Each Element (10^{-5} ft/s)

Element	u_x	u_y
I	3.0260	−0.9456
II	1.2766	−2.5532
III	7.2813	0.0
IV	3.0260	0.9456
V	1.2766	2.5532

(c)	The formulation for part (c) is identical to that for part (b) except that additional boundary conditions are required, $C_9 = C_{12} = 50$. Final results are given in Table 3.6.

Table 3.6	Concentration with Convection

Node	Concentration (ppm)
1	97.381
2	98.165
3	100.000
4	89.781
5	92.818
6	100.000
7	67.458
8	63.313
9	50.000
10	56.536
11	54.348
12	50.000

(d)	The local stiffness matrix for the chemical reaction term is given in Prob. 3.33. Since all elements have the same dimensions, the local stiffness will be the same for all elements, and that matrix is merely added to the element derived in part (c). The boundary conditions remain the same. The final results for concentration are given in Table 3.7.

Table 3.7	Concentration with Convection
and Chemical Reaction

Node	Concentration (ppm)
1	69.338
2	78.348
3	100.000
4	60.110
5	67.122
6	100.000
7	42.118
8	41.448
9	50.000
10	36.331
11	40.247
12	50.000

3.9. Use Eq. (3.13) and derive the variational function that corresponds to the axially loaded rod of Chap. 2.

Consider the terms within the volume integral first. In one dimension the stress tensor σ_{kj} and the strain tensor ϵ_{kj} correspond to the axial stress σ and axial strain ϵ. Similarly, f_k and u_k are vectors corresponding to body force f and displacement u in one dimension. The constitutive law is Hooke's law in one dimension, or $\sigma = E\epsilon$. The volume integral is written

$$\int_V \left(\frac{1}{2}\, \sigma\epsilon - fu \right) dV$$

Let $dV = dA\, dx$ and $\int_A dA = A$; then the integral can be written

$$\int_0^L \left(\frac{1}{2} E\epsilon^2 - fu \right) A\, dx$$

and substituting Eq. (a) of Prob. 2.9 gives

$$\int_0^L \left[\frac{1}{2} E\left(\frac{du}{dx}\right)^2 - fu \right] A\, dx$$

the desired result. The term $\int_S T_k u_k\, dS$ represents an external surface load and can be interpreted using Fig. 3-8. The term T_k is called a stress vector defining a *surface traction* acting on the area (surface) as shown in the figure. Given the fundamental definition of axial stress $\sigma = P/A$, where P is an applied external force, the surface traction can be interpreted as P/A and the surface integral becomes $\int_A (Pu/A)\, dA = Pu$. In this case u is the displacement at the free end of the rod, and the term Pu can be interpreted as external work.

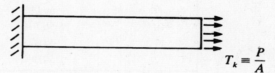

Fig. 3-8   Rod with external load.

3.10. Use the variational function of Eq. (3.13) and write a corresponding function in matrix format that can be used for plane elasticity.

Equation (3.13) is a general statement written in cartesian tensor notation. Problem 3.9 illustrates the one-dimensional problem where each tensor term has been replaced with its one-dimensional counterpart. The two-dimensional function requires a more general approach, but for the reader unfamiliar with tensor notation it is possible to construct the function using matrices and merely use Eq. (3.13) as a model. In fact, using matrices expedites the entire process since the final results are in a proper format for numerical computation. A word of caution is in order. Tensor analysis and matrix theory are two different mathematical concepts that should not be interchanged. Each concept has its own set of rules. Matrix theory is usually used to write equations in a form suitable for computation.

Consider for now only the first term in the volume integral of Eq. (3.13). The analogous matrix statement is

$$\int_V \frac{1}{2}\, [\sigma]^T [\epsilon]\, dV \qquad\qquad (a)$$

Hooke's law in cartesian tensor notation is

$$\sigma_{kl} = C_{klij}\epsilon_{ij}$$

and is completely general. The corresponding statement in matrix notation is

$$\{\sigma\} = [C]\{\epsilon\} \qquad\qquad (b)$$

and each term in Eq. (b) has a form that depends upon the elasticity problem to be solved. The plane elasticity problems of Sec. 3.4 are written as

$$\begin{Bmatrix} \sigma_{xx} \\ \sigma_{yy} \\ \sigma_{xy} \end{Bmatrix} = \begin{bmatrix} C_{11} & C_{12} & 0 \\ C_{12} & C_{22} & 0 \\ 0 & 0 & C_{33} \end{bmatrix} \begin{Bmatrix} \epsilon_{xx} \\ \epsilon_{yy} \\ \epsilon_{xy} \end{Bmatrix} \tag{c}$$

and the [C] matrix corresponds to either plane stress or plane strain. Substituting Eq. (b) into Eq. (a) gives

$$\int_V \frac{1}{2} [\epsilon]^T [C][\epsilon]\, dV \tag{d}$$

where the strains can be written in terms of nodal point displacements using Eqs. (*3.3*). Introduce an operator matrix, as in Prob. 3.4, that relates $[\epsilon]$ to the derivatives of $\{u\}$, the nodal point displacements. The element displacements must be defined in terms of shape functions and nodal point displacements, which requires some organization since there are two unknown displacements at each node. Consider the four-node element of Fig. 3-9 where there are two degrees of freedom per node and a standard numbering scheme is shown. The displacements are u and v, corresponding to the x and y directions, respectively. The matrix of nodal point displacements can be written

$$\{u\} = [u_1 \quad v_1 \quad u_2 \quad v_2 \quad u_3 \quad v_3 \quad u_4 \quad v_4]^T \tag{e}$$

The element displacements are defined in terms of $\{u\}$ using shape functions such as

$$u_e(x, y) = N_1 u_1 + N_2 u_2 + N_3 u_3 + N_4 u_4 \qquad v_e(x, y) = N_1 v_1 + N_2 v_2 + N_3 v_3 + N_4 v_4 \tag{f}$$

In general, the displacements at any x, y location within the element can be written in matrix form as

$$\begin{Bmatrix} u_e \\ v_e \end{Bmatrix} = [N]\{u\} \tag{g}$$

where the format of [N] is dependent upon the numbering sequence used to define nodal displacements on the element. Now, return to Eq. (d) and the operator matrix that was mentioned previously.

Define the operator matrix, using Eqs. (*3.3*) as a model, to relate the strains to the displacements of Eq. (g):

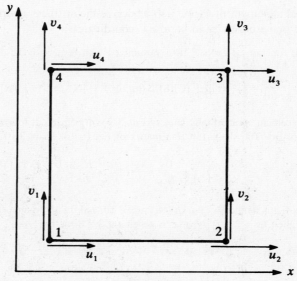

Fig. 3-9

$$\{\epsilon\} = \begin{Bmatrix} \epsilon_{xx} \\ \epsilon_{yy} \\ \epsilon_{xy} \end{Bmatrix} = \begin{bmatrix} \partial/\partial x & 0 \\ 0 & \partial/\partial y \\ \partial/\partial y & \partial/\partial x \end{bmatrix} \begin{Bmatrix} u_e \\ v_e \end{Bmatrix} = [L] \begin{Bmatrix} u_e \\ v_e \end{Bmatrix} \qquad (h)$$

Substitute Eq. (g) into Eq. (h):

$$\{\epsilon\} = [L][N]\{u\} \qquad (i)$$

and recognize that Eq. (i) gives the strains at any x, y location in terms of nodal point displacements. Also, Eq. (i) is a general expression that is applicable for any element since the element is defined by the shape functions. The matrix porduct [L][N] is called the [B] matrix in finite element analysis terminology. Substitute Eq. (i) into Eq. (d) to obtain the final form for the first term of the variational function:

$$\int_V \frac{1}{2}\{u\}^T[N]^T[L]^T[C][L][N]\{u\}\, dV = \int_V \frac{1}{2}\{u\}^T[B]^T[C][B]\{u\}\, dV \qquad (j)$$

The second term of the volume integral of Eq. (3.13) represents the body force and is similar to the heat-source term of Prob. 3.4. It becomes

$$\int_V \{u\}^T[N]^T f\, dV \qquad (k)$$

The surface integral of Eq. (3.13) represents a natural boundary condition and is called a surface traction, a force boundary condition, or a distributed pressure boundary condition. A concentrated force boundary condition applied at a single node in either the x or y direction can be incorporated in the same manner as in Chap. 2. A pressure boundary condition must be distributed to the edge of an element that usually lies on the boundary of the physical problem to be modeled. The surface integral gives information about how to distribute the traction or pressure depending upon the shape function being used. The term T_k, with one subscript, is a vector and as such should have two components, $\{T\} = [T_x \quad T_y]^T$. The surface integral, in matrix form, is as follows. (Additional discussion will be given in a subsequent problem.)

$$\int_S \{u\}^T[N]^T\{T\}\, dS \qquad (l)$$

The final result is obtained by combining Eqs. (j)–(l).

3.11. Use the variational statement of Prob. 3.10 to derive the stiffness matrix for a four-node rectangular finite element for plane elasticity in cartesian coordinates.

The variational function of Prob. 3.10 is minimized with respect to the displacement vector.

$$\frac{\partial J}{\partial \{u\}} = \int_V [N]^T[L]^T[C][L][N]\{u\}\, dV - \int_V [N]^T f\, dV - \int_S [N]^T\{T\}\, dS = 0 \qquad (a)$$

The node and displacement numbering scheme of Fig. 3-9 will be used. It follows that Eq. (e) of Prob. 3.10 is the correct definition for $\{u\}$. The shape function matrix of Eq. (g) of Prob. 3.10 that corresponds to Fig. 3-9 is

$$[N] = \begin{bmatrix} N_1 & 0 & N_2 & 0 & N_3 & 0 & N_4 & 0 \\ 0 & N_1 & 0 & N_2 & 0 & N_3 & 0 & N_4 \end{bmatrix} \qquad (b)$$

The operator matrix [L] is defined by Eq. (h) of Prob. 3.10, and the [B] matrix is the result [L][N], where a 3×2 [L] matrix is multiplied by a 2×8 [N] matrix to give a 3×8 matrix:

$$[B] = \begin{bmatrix} \partial N_1/\partial x & 0 & \partial N_2/\partial x & 0 & \partial N_3/\partial x & 0 & \partial N_4/\partial x & 0 \\ 0 & \partial N_1/\partial y & 0 & \partial N_2/\partial y & 0 & \partial N_3/\partial y & 0 & \partial N_4/\partial y \\ \partial N_1/\partial y & \partial N_1/\partial x & \partial N_2/\partial y & \partial N_2/\partial x & \partial N_3/\partial y & \partial N_3/\partial x & \partial N_4/\partial y & \partial N_4/\partial x \end{bmatrix} \qquad (c)$$

The volume integral is changed to an area integral by assuming a constant thickness in the z direction. The stiffness matrix is computed as

$$[K] = \int_A [B]^T[C][B]t \, dA \qquad (d)$$

Visualize the multiplication indicated by Eq. (d). The [C] matrix is given by Eq. (c) of Prob. 3.10, and the first term of the stiffness matrix is

$$K_{11} = \int_0^a \int_0^b \left(\frac{\partial N_1}{\partial x} C_{11} \frac{\partial N_1}{\partial x} + \frac{\partial N_1}{\partial y} C_{33} \frac{\partial N_1}{\partial y} \right) t \, dx \, dy$$

The derivatives of the shape functions were given in Prob. 3.5 and are substituted to give

$$K_{11} = \int_0^a \int_0^b \left[C_{11} \frac{(b-y)^2}{a^2 b^2} + C_{33} \frac{(a-x)^2}{a^2 b^2} \right] t \, dx \, dy$$

or

$$K_{11} = \left(\frac{C_{11}b}{3a} + \frac{C_{33}a}{3b} \right) t$$

The stiffness matrix is a symmetric 8×8 matrix and follows as Eq. (e).

The second term of Eq. (a) is similar to the heat-source term of Eq. (f) of Prob. 3.5. The surface integral of Eq. (a) will be discussed in a subsequent problem.

$$
\begin{bmatrix}
\frac{C_{11}b}{3a}+\frac{C_{33}a}{3b} & \frac{C_{12}+C_{33}}{4} & -\frac{C_{11}b}{3a}+\frac{C_{33}a}{6b} & \frac{C_{12}-C_{33}}{4} & -\frac{C_{11}b}{6a}-\frac{C_{33}a}{6b} & \frac{-C_{12}-C_{33}}{4} & \frac{C_{11}b}{6a}-\frac{C_{33}a}{3b} & \frac{-C_{12}+C_{33}}{4} \\[2mm]
 & \frac{C_{22}a}{3b}+\frac{C_{33}b}{3a} & \frac{-C_{12}+C_{33}}{4} & \frac{C_{22}a}{6b}-\frac{C_{33}b}{3a} & \frac{-C_{12}-C_{33}}{4} & -\frac{C_{22}a}{6b}-\frac{C_{33}b}{6a} & \frac{C_{12}-C_{33}}{4} & -\frac{C_{22}a}{3b}+\frac{C_{33}b}{6a} \\[2mm]
 & & \frac{C_{11}b}{3a}+\frac{C_{33}a}{3b} & \frac{-C_{12}-C_{33}}{4} & \frac{C_{11}b}{6a}-\frac{C_{33}a}{3b} & \frac{C_{12}-C_{33}}{4} & -\frac{C_{11}b}{6a}-\frac{C_{33}a}{6b} & \frac{C_{12}+C_{33}}{4} \\[2mm]
 & & & \frac{C_{22}a}{3b}+\frac{C_{33}b}{3a} & \frac{-C_{12}+C_{33}}{4} & -\frac{C_{22}a}{3b}+\frac{C_{33}b}{6a} & \frac{C_{12}+C_{33}}{4} & -\frac{C_{22}a}{6b}-\frac{C_{33}b}{6a} \\[2mm]
 & & & & \frac{C_{11}b}{3a}+\frac{C_{33}a}{3b} & \frac{C_{12}+C_{33}}{4} & -\frac{C_{11}b}{3a}+\frac{C_{33}a}{6b} & \frac{C_{12}-C_{33}}{4} \\[2mm]
 & \text{Symmetric} & & & & \frac{C_{22}a}{3b}+\frac{C_{33}b}{3a} & \frac{-C_{12}+C_{33}}{4} & \frac{C_{22}a}{6b}-\frac{C_{33}b}{3a} \\[2mm]
 & & & & & & \frac{C_{11}b}{3a}+\frac{C_{33}a}{3b} & \frac{-C_{12}-C_{33}}{4} \\[2mm]
 & & & & & & & \frac{C_{22}a}{3b}+\frac{C_{33}b}{3a}
\end{bmatrix} \qquad (e)
$$

3.12. Assume the rectangular element of Fig. 3-3 has thickness t and that a uniform pressure loading p_y, in force per unit area, is distributed along the edge connecting nodes 3 and 4. Use the variational function of Probs. 3.10 and 3.11 and determine the distribution of the pressure to nodes 3 and 4.

The surface integral has been minimized as Eq. (a) of Prob. 3.11 and is

$$\int_S [N]^T\{T\} \, dS \qquad (a)$$

Substitute Eq. (b) of Prob. 3.11 into Eq. (a) above to obtain a general expression

$$\int_S \begin{bmatrix} N_1 & 0 \\ 0 & N_1 \\ N_2 & 0 \\ 0 & N_2 \\ N_3 & 0 \\ 0 & N_3 \\ N_4 & 0 \\ 0 & N_4 \end{bmatrix} \begin{Bmatrix} T_x \\ T_y \end{Bmatrix} dS = \int_S \begin{Bmatrix} N_1 T_x \\ N_1 T_y \\ N_2 T_x \\ N_2 T_y \\ N_3 T_x \\ N_3 T_y \\ N_4 T_x \\ N_4 T_y \end{Bmatrix} dS \qquad (b)$$

The element edge connecting nodes 3 and 4 corresponds to the edge $y = b$ for the local element coordinate system. The shape functions of Eq. (b) above are given as Eqs. (e) of Prob. 3.1, and substituting $y = b$ gives $N_1 = N_2 = 0$, $N_3 = x/a$, and $N_4 = (a - x)/a$. The surface tractions are replaced with $T_x = p_x = 0$ and $T_y = p_y$. Finally, dS is replaced with $t\,dx$, where t is the thickness of the element and the integration is along the x axis. Note that if the surface loading is specified in force per unit length, the t term is omitted. Substitute N_3 and N_4 into equation Eq. (b) and integrate with limits 0 to a, and the final load vector is

$$\{0 \quad 0 \quad 0 \quad 0 \quad 0 \quad p_y a/2 \quad 0 \quad p_y a/2\}^T t \qquad (c)$$

Equation (c) indicates that one-half of the pressure loading should be distributed to each node.

3.13. The plate of Fig. 3-10 is 6×10 in and 0.2 in thick and loaded with a tension of 10,000 psi as shown. Compute the node displacements, strains, and stresses using a one-element analysis. Assume $E = 10(10)^6$ psi and solve the problem for $\nu = 0$ and again for $\nu = 0.3$.

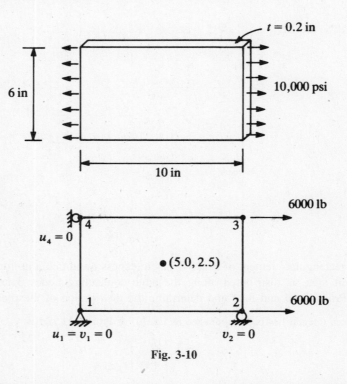

Fig. 3-10

The plate should be analyzed as plane stress. Boundary conditions should allow the plate to deform in both coordinate directions but prevent rigid body translation and rotation. Node 1 is fixed in both directions, node 2 is fixed in the y direction, and node 4 is fixed in the x direction, as shown in Fig. 3-10. For the eight possible node displacements, $u_1 = u_4 = v_1 = v_2 = 0$ are the boundary conditions.

The result obtained in Prob. 3.12 indicates that the pressure loading should be distributed to nodes 2 and 3 as $(10,000 \text{ psi})(6 \text{ in})(0.2 \text{ in})/2 = 6000$ lb. The local stiffness matrix given by Eq. (e) of Prob. 3.11 becomes the global stiffness matrix for this one-element model. The material constants are computed using the plane stress constitutive equations, and for the case $\nu = 0.3$, Eqs. (3.6)–(3.8) give $C_{11} = C_{22} = E/(1 - \nu^2) = 10.9890(10)^6$ psi, $C_{12} = \nu E/(1 - \nu^2) = 3.2967(10)^6$ psi, and $C_{33} = E/2(1 + \nu) = 3.8462(10)^6$ psi. The stiffness matrix after substituting boundary conditions and load vector is, for $\nu = 0.3$,

$$\begin{bmatrix} 1 & 0 & 0 & 0 & 0 & 0 & 0 & 0 \\ 0 & 1 & 0 & 0 & 0 & 0 & 0 & 0 \\ 0 & 0 & 4,334,555 & 0 & -1,037,850 & -137,363 & 0 & 1,785,710 \\ 0 & 0 & 0 & 1 & 0 & 0 & 0 & 0 \\ 0 & 0 & -1,037,850 & 0 & 4,334,555 & 1,785,710 & 0 & -137,363 \\ 0 & 0 & -137,363 & 0 & 1,785,710 & 6,874,240 & 0 & 2,283,270 \\ 0 & 0 & 0 & 0 & 0 & 0 & 1 & 0 \\ 0 & 0 & 1,785,710 & 0 & -137,363 & 2,283,270 & 0 & 6,874,240 \end{bmatrix} \begin{Bmatrix} u_1 \\ v_1 \\ u_2 \\ v_2 \\ u_3 \\ v_3 \\ u_4 \\ v_4 \end{Bmatrix} = \begin{Bmatrix} 0 \\ 0 \\ 6000 \\ 0 \\ 6000 \\ 0 \\ 0 \\ 0 \end{Bmatrix}$$

The displacements are given in Table 3.8.

Table 3.8   Node Displacements (in)

	u_1	v_1	u_2	v_2	u_3	v_3	u_4	v_4
$\nu = 0$	0.0	0.0	0.002	0.0	0.002	0.0	0.0	0.0
$\nu = 0.3$	0.0	0.0	0.002	0.0	0.002	−0.00036	0.0	−0.00036

The solution for displacements is exact because the linear shape functions predict the linear solution for this problem. Also, for $\nu = 0$ the problem is one-dimensional and reduces to the axially loaded member discussed in Chap. 2.

The strains are computed using Eq. (i) of Prob. 3.10. Note that the matrix product $[L][N] = [B]$ is given by Eq. (c) of Prob. 3.11 and the derivatives of the shape functions are given by Eqs. (d) of Prob. 3.5. Choose x and y of Eqs. (d) of Prob. 3.5 as $x = a/2 = 5.0$ and $y = b/2 = 3.0$ to represent the center of the element and compute $\{\epsilon\} = [B]\{u\}$. For $\nu = 0.0$, $\epsilon_{xx} = 0.0002$ and $\epsilon_{yy} = \epsilon_{xy} = 0.0$, and for $\nu = 0.3$, $\epsilon_{xx} = 0.0002$, $\epsilon_{yy} = -0.0006$, and $\epsilon_{xy} = 0.0$. Again the solution is exact for this elementary problem.

The stresses are computed using Eq. (c) of Prob. 3.10, and for both cases $\sigma_{xx} = 2000$ psi and $\sigma_{yy} = \sigma_{xy} = 0.0$.

3.14. Derive a local stiffness matrix for heat transfer using the three-node triangular element defined in Fig. 3-4.

The local stiffness matrix is defined by Eq. (b) of Prob. 3.5:

$$[K] = \int_A [B]^T [k][B] t \, dx \, dy \qquad (a)$$

The temperature is defined in terms of the unknown temperature at each node and is similar to Eq. (b) of Prob. 3.4 except there are only three nodal values:

$$T = [N_1 \quad N_2 \quad N_3]\{T_1 \quad T_2 \quad T_3\}^T = [N]\{T\}^T \qquad (b)$$

The shape functions are given by Eqs. (i) of Prob. 3.2 and, for convenience, are usually written in the following form

$$N_1 = \frac{a_1 + b_1 x + c_1 y}{2A}$$

$$N_2 = \frac{a_2 + b_2 x + c_2 y}{2A} \tag{c}$$

$$N_3 = \frac{a_3 + b_3 x + c_3 y}{2A}$$

where
$$a_1 = x_2 y_3 - x_3 y_2 \qquad b_1 = y_2 - y_3 \qquad c_1 = x_3 - x_2$$

$$a_2 = x_3 y_1 - x_1 y_3 \qquad b_2 = y_3 - y_1 \qquad c_2 = x_1 - x_3$$

$$a_3 = x_1 y_2 - x_2 y_1 \qquad b_3 = y_1 - y_2 \qquad c_3 = x_2 - x_1$$

Define the shape function matrix of Eq. (b), using Eqs. (c) as

$$[\mathrm{N}] = \begin{bmatrix} 1 & x & y \end{bmatrix} \begin{bmatrix} a_1 & a_2 & a_3 \\ b_1 & b_2 & b_3 \\ c_1 & c_2 & c_3 \end{bmatrix} \div 2A$$

The [B] matrix is formed as [L][N] where the operator matrix [L] is given by Eq. (e) of Prob. 3.4 for the heat-transfer problem or any physical problem governed by an equation of the form of Eq. (3.1) when $\beta = 0$:

$$[\mathrm{L}][\mathrm{N}] = \begin{bmatrix} \partial/\partial x \\ \partial/\partial y \end{bmatrix}[\mathrm{N}] = \begin{bmatrix} 0 & 1 & 0 \\ 0 & 0 & 1 \end{bmatrix} \begin{bmatrix} a_1 & a_2 & a_3 \\ b_1 & b_2 & b_3 \\ c_1 & c_2 & c_3 \end{bmatrix} \div 2A$$

or
$$[B] = \begin{bmatrix} b_1 & b_2 & b_3 \\ c_1 & c_2 & c_3 \end{bmatrix} \div 2A$$

The material matrix [k] of Eq. (a) is a 2×2 matrix and is the same as that in Prob. 3.4. The area integration is $\int_A dx\, dy = A$, the area of the triangle, since all terms are constant. Substituting into Eq. (a) gives the symmetric matrix

$$[\mathrm{K}] = \begin{bmatrix} b_1^2 k_x + c_1^2 k_y & b_1 b_2 k_x + c_1 c_2 k_y & b_1 b_3 k_x + c_1 c_3 k_y \\ & b_2^2 k_x + c_2^2 k_y & b_2 b_3 k_x + c_2 c_3 k_y \\ & & b_3^2 k_x + c_3^2 k_y \end{bmatrix} \frac{t}{4A} \tag{d}$$

Equation (d) can be used for numerical computation, however, for computer implementation it is probably more practical to formulate the individual matrices and use a matrix multiplication routine to compute [K].

3.15. Compute the steady-state temperature distribution for the plate of Prob. 3.6: (a) using three triangular elements as shown in Fig. 3-11(a) and (b) using four triangular elements as shown in Fig. 3-11(b).

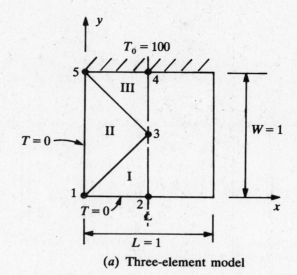

(a) Three-element model

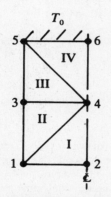

(b) Four-element model   **(c) Alternate four-element model**

Fig. 3-11

(a) Symmetry is used to model the plate as in Prob. 3.6. Element I, formed by nodes 1–3, is used to compute the terms that are substituted into the local stiffness matrix, Eq. (d) of Prob. 3.14. The numbers along the top and side of the local matrix correspond to the row, column location within the gloabl matrix. Note that $k_x = k_y = 1.0$.

Element I:

$$
\begin{array}{lll}
b_1 = -0.5 & c_1 = 0.0 & A = 0.125 \\
b_2 = 0.5 & c_2 = -0.5 & \\
b_3 = 0.0 & c_3 = 0.5 &
\end{array}
\qquad
[K_I] =
\begin{array}{ccc}
1 & 2 & 3 \\
\left[\begin{array}{ccc}
0.5 & -0.5 & 0.0 \\
-0.5 & 1.0 & -0.5 \\
0.0 & -0.5 & 0.5
\end{array}\right] &
\begin{array}{c} 1 \\ 2 \\ 3 \end{array}
\end{array}
$$

The area A can be computed using the result given in Prob. 3.2. Similar computations for elements II and III give the following.

Element II:

$$
\begin{array}{lll}
b_1 = -0.5 & c_1 = -0.5 & A = 0.25 \\
b_2 = 1.0 & c_2 = 0.0 & \\
b_3 = 0.5 & c_3 = -0.5 &
\end{array}
\qquad
[K_{II}] =
\begin{array}{ccc}
1 & 3 & 5 \\
\left[\begin{array}{ccc}
0.5 & -0.5 & 0.0 \\
-0.5 & 1.0 & -0.5 \\
0.0 & -0.5 & 0.5
\end{array}\right] &
\begin{array}{c} 1 \\ 3 \\ 5 \end{array}
\end{array}
$$

Element III:

$$\begin{array}{lll}
b_1 = 0.0 & c_1 = -0.5 & A = 0.125 \\
b_2 = 0.5 & c_2 = 0.5 & \\
b_3 = -0.5 & c_3 = 0.0 &
\end{array}$$

$$[K_{III}] = \begin{bmatrix} 0.5 & -0.5 & 0.0 \\ -0.5 & 1.0 & -0.5 \\ 0.0 & -0.5 & 0.5 \end{bmatrix} \begin{matrix} 3 \\ 4 \\ 5 \end{matrix}$$

$$\begin{matrix} 3 & 4 & 5 \end{matrix}$$

The element stiffness matrices are assembled to form the global stiffness matrix.

$$\begin{bmatrix} 1.0 & -0.5 & -0.5 & 0.0 & 0.0 \\ -0.5 & 1.0 & -0.5 & 0.0 & 0.0 \\ -0.5 & -0.5 & 2.0 & -0.5 & -0.5 \\ 0.0 & 0.0 & -0.5 & 1.0 & -0.5 \\ 0.0 & 0.0 & -0.5 & -0.5 & 1.0 \end{bmatrix} \begin{Bmatrix} T_1 \\ T_2 \\ T_3 \\ T_4 \\ T_5 \end{Bmatrix} = \{0\}$$

Substituting the boundary conditions $T_1 = T_2 = 0$ and $T_4 = T_5 = 100$ reduces the system to one equation with one unknown, and $T_3 = 50.0$ can be computed.

(b) The stiffness matrix for the four-element model is constructed in the same manner; intermediate steps will be omitted. The global matrix before substituting boundary conditions is

$$\begin{bmatrix} 1.0 & -0.5 & 0.0 & -0.5 & 0.0 & 0.0 \\ -0.5 & 1.0 & -0.5 & 0.0 & 0.0 & 0.0 \\ 0.0 & -0.5 & 2.0 & -1.0 & -0.5 & 0.0 \\ -0.5 & 0.0 & -1.0 & 2.0 & 0.0 & -0.5 \\ 0.0 & 0.0 & -0.5 & 0.0 & 1.0 & -0.5 \\ 0.0 & 0.0 & 0.0 & -0.5 & -0.5 & 1.0 \end{bmatrix} \begin{Bmatrix} T_1 \\ T_2 \\ T_3 \\ T_4 \\ T_5 \\ T_6 \end{Bmatrix} = \{0\}$$

After substituting boundary conditions T_3 is the only unknown. It can be computed as $T_3 = 25.0$, which is exact when compared with the results of Prob. 3.6. The question concerning the accuracy of this analysis when compared with the four-element analysis of Prob. 3.6 deserves some comment. The arrangement of the four triangles of this analysis fits the solution surface (the actual temperature distribution) more accurately than the four rectangles of Prob. 3.6. It should not be inferred that three-node triangles are more accurate than four-node rectangles. The interested reader should model the plate using rectangular elements arranged as shown in Fig. 3-11(c) since that arrangement appears to better approximate the solution surface.

3.16. Show that area coordinates are related to shape functions for a three-node triangular finite element and that area coordinates lead to a different, but equivalent, formulation for the shape function.

Consider the triangular element of Fig. 3-12 that is similar to that of Fig. 3-4. The interior point P is not a node but merely an arbitrary reference point. Area coordinates are defined as

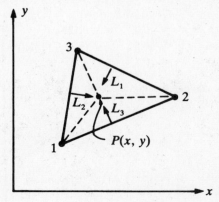

Fig. 3-12

$$L_1 = \frac{\text{area}(2\text{-}3\text{-}P)}{\text{area}(1\text{-}2\text{-}3)} \qquad L_2 = \frac{\text{area}(3\text{-}1\text{-}P)}{\text{area}(1\text{-}2\text{-}3)} \qquad L_3 = \frac{\text{area}(1\text{-}2\text{-}P)}{\text{area}(1\text{-}2\text{-}3)} \qquad (a)$$

The parameters L_1, L_2, and L_3 are defined as area coordinates because each represents a fractional part of the total triangular area and their sum is equal to the total triangular area (Fig. 3-12):

$$\text{Area } (2\text{-}3\text{-}P) + \text{area}(3\text{-}1\text{-}P) + \text{area}(1\text{-}2\text{-}P) = \text{area}(1\text{-}2\text{-}3)$$

and substituting Eqs. (a) gives

$$L_1 + L_2 + L_3 = 1$$

Such coordinates are sometimes called *natural coordinates* and can be visualized as illustrated. The coordinate L_1 emanates from side 2-3 and corresponds to the area defined by side 2-3 and the point P. Note that if L_1 is allowed to increase and reach node point 1, its magnitude will be unity and $L_2 = L_3 = 0$. It appears that L_1 behaves similarly to N_1, the shape function for node 1. In fact, the following relations can be deduced using Fig. 3-12:

$$L_1 = \begin{cases} 1 & \text{at node 1} \\ 0 & \text{at nodes 2 and 3} \end{cases}$$

$$L_2 = \begin{cases} 1 & \text{at node 2} \\ 0 & \text{at nodes 1 and 3} \end{cases}$$

$$L_3 = \begin{cases} 1 & \text{at node 3} \\ 0 & \text{at nodes 1 and 2} \end{cases}$$

Figure 3-4 illustrates the coordinate location of each node. The ratio of area coordinate L_3 relative to the total area is (x, y is the location of point P)

$$L_3 = \frac{\text{area}(1\text{-}2\text{-}P)}{\text{area}(1\text{-}2\text{-}3)} = \frac{1}{2} \det \begin{bmatrix} 1 & 1 & 1 \\ x_1 & x_2 & x \\ y_1 & y_2 & y \end{bmatrix} \div A_{\text{total}}$$

or

$$L_3 = [(x_1 y_2 - x_2 y_1) + x(y_1 - y_2) + y(x_2 - x_1)]/2A_{\text{total}} \qquad (b)$$

Equation (b) is the same as the definition given by Eq. (i) of Prob. 3.2, and it is concluded that $L_3 = N_3$ for the three-node triangular element. Similar computations give equivalent definitions for $L_1 = N_1$ and $L_2 = N_2$.

3.17. Use area coordinates and numerical area integration to derive the stiffness matrix corresponding to a three-node triangular element for the chemical reaction term in Eq. (*3.2*).

Area integration for a three-node triangle can be accomplished using area coordinates and the corresponding integration formula:

$$\int_A L_1^{\alpha} L_2^{\beta} L_3^{\gamma} \, dA = \frac{\alpha! \beta! \gamma!}{(\alpha + \beta + \lambda + 2)!} 2A \qquad (a)$$

The variational form of the stiffness matrix to be formulated is given in Prob. 3.32 (the third term in the function) and after being minimized with respect to {C} appears as

$$\int_A [\text{N}]^T [\text{K}_r][\text{N}]\{C\} t \, dx \, dy \qquad (b)$$

The function is transformed into a finite element model following the method of Prob. 3.4 by defining an operator matrix that acts on [N]. In this case the operator merely defines shape functions and is written as a unit matrix:

$$[\text{L}][\text{N}] = [1][N_1 \quad N_2 \quad N_3] = [N_1 \quad N_2 \quad N_3] \qquad (c)$$

The result is left in the form of Eq. (c) above rather than substituting Eqs. (c) of Prob. 3.14. The reader should compare the various stiffness matrices that have been derived. In every case the [N] matrix defines the element being used, and an operator matrix can be used to define the manner in which the nodal terms appear in the differential equation or variational function. Formally, Eq. (b) is constructed as follows:

$$\int_A \begin{Bmatrix} N_1 \\ N_2 \\ N_3 \end{Bmatrix} [K_r][N_1 \quad N_2 \quad N_3] \, dx \, dy \begin{Bmatrix} C_1 \\ C_2 \\ C_3 \end{Bmatrix} t \tag{d}$$

The matrix multiplications indicated by Eq. (d) result in terms of the type

$$K_{11} = \int_A N_1 K_r N_1 \, dx \, dy$$

The definitions of shape functions given by Eq. (c) of Prob. 3.14 would lead to algebraically complicated mathematics. The shape functions can be replaced by their equivalent definitions using the results of Prob. 3.16 or considering only the first term and Eq. (a):

$$\int_A N_1 K_r N_1 \, dx \, dy = \int_A L_1 K_r L_1 \, dx \, dy = \int_A L_1^2 K_r \, dA = \frac{2!0!0!}{(2+0+0+2)!} K_r 2A = \frac{K_r}{12} 2A$$

Note that the exponents β and γ of Eq. (a) are zero and are included in the computation as such. The area is evaluated as in Prob. 3.15. The complete stiffness matrix for the chemical reaction term becomes

$$\frac{K_r A t}{12} \begin{bmatrix} 2 & 1 & 1 \\ 1 & 2 & 1 \\ 1 & 1 & 2 \end{bmatrix} \tag{e}$$

3.18. Formulate the stiffness matrix in terms of shape functions for the convection terms of Eq. (3.2) using three-node triangular elements and show how area coordinates and area integration can be used to evaluate the final local stiffness matrix.

The convection terms in the variational function given in Prob. 3.32 appear as follows after the minimization is completed.

$$\int_A [N]^T [u][L][N]\{C\} t \, dx \, dy \tag{a}$$

In this instance the operator matrix corresponds to first derivatives as illustrated in Prob. 3.14. The first term $[N]^T$ corresponds to the shape function matrix of the previous problem but must be written to accommodate the two-dimensional formulation. Visualize Eq. (c) of Prob. 3.17 written as

$$[L][N] = \begin{bmatrix} 1 \\ 1 \end{bmatrix} [N_1 \quad N_2 \quad N_3] = \begin{bmatrix} N_1 & N_2 & N_3 \\ N_1 & N_2 & N_3 \end{bmatrix}$$

The stiffness matrix is constructed as follows:

$$\int_A \begin{bmatrix} N_1 & N_1 \\ N_2 & N_2 \\ N_3 & N_3 \end{bmatrix} \begin{bmatrix} u_x & 0 \\ 0 & u_y \end{bmatrix} \begin{bmatrix} \partial N_1/\partial x & \partial N_2/\partial x & \partial N_3/\partial x \\ \partial N_1/\partial y & \partial N_2/\partial y & \partial N_3/\partial y \end{bmatrix} \begin{Bmatrix} C_1 \\ C_2 \\ C_3 \end{Bmatrix} t \, dA \tag{b}$$

The first term of the stiffness matrix is computed as

$$K_{11} = \int_A \left(N_1 u_x \frac{\partial N_1}{\partial x} + N_1 u_y \frac{\partial N_1}{\partial y} \right) t \, dA \tag{c}$$

The shape functions are replaced with their equivalent area coordinates, and numerical integration is used to evaluate Eq. (c). The partial derivatives are not functions of x or y and, using the results of Prob. 3.14, can be evaluated before integrating Eq. (c). The term to be integrated, after replacing shape functions with area coordinates, appears as

$$K_{11} = \int_A \left(L_1 u_x \frac{b_1}{2A} + L_1 u_y \frac{c_1}{2A} \right) t\, dA = \frac{(1!0!0!)2A u_x b_1 t}{(1+0+0+2)!2A} + \frac{(1!0!0!)2A u_y c_1 t}{(1+0+0+2)!2A} = \frac{(u_x b_1 + u_y c_1)t}{6}$$

The local stiffness matrix for convection is 3×3, and the remaining terms are computed in a similar manner. (See Prob. 3.39.)

3.19. Derive the [B] matrix for plane stress analysis using a three-node triangular element.

The derivation follows that of Prob. 3.11. The product [L][N] must be evaluated for a three-node element, and the subsequent derivatives are substituted from Prob. 3.14. (See Prob. 3.40.)

$$[B] = \begin{bmatrix} \partial/\partial x & 0 \\ 0 & \partial/\partial y \\ \partial/\partial y & \partial/\partial x \end{bmatrix} \begin{bmatrix} N_1 & 0 & N_2 & 0 & N_3 & 0 \\ 0 & N_1 & 0 & N_2 & 0 & N_3 \end{bmatrix} = \begin{bmatrix} b_1 & 0 & b_2 & 0 & b_3 & 0 \\ 0 & c_1 & 0 & c_2 & 0 & c_3 \\ c_1 & b_1 & c_2 & b_2 & c_3 & b_3 \end{bmatrix} \div 2A \qquad (a)$$

3.20. Given a uniformly varying pressure loading p_x between nodes 1 and 2 of the triangular element of Fig. 3-13, use the integration formulas to compute the distribution of the load to nodes 1 and 2.

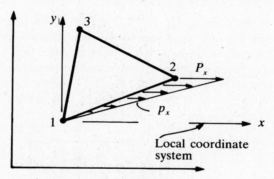

Fig. 3-13

The equation of the pressure loading, assuming a local coordinate system located at node 1, is

$$p_x = P_x \frac{y_1 - y}{y_1 - y_2} = P_x \frac{y_1 - y}{b_3} \qquad (a)$$

In this application of numerical integration the integration occurs along a line rather than over an area. The formula for integration along a line is analogous to Eq. (a) of Prob. 3.17, and the coordinates are referred to as length coordinates:

$$\int_\xi L_1^\alpha L_2^\beta\, d\xi = \frac{\alpha!\beta!}{(\alpha + \beta + 1)!}\, \xi \qquad (b)$$

One-dimensional finite elements can be formulated in terms of length coordinates, but there is usually no advantage when compared with the more traditional cartesian coordinates. Follow the method of analysis used in Eq. (b) of Prob. 3.12 and recall that $\{T_x \quad T_y\}^T$ represents any general traction-type surface load. Along the line connecting nodes 1 and 2 the area coordinate L_3 is zero or in terms of length coordinates only L_1 and L_2 exist and terms corresponding to T_y are zero:

$$\int_s \begin{bmatrix} N_1 & 0 \\ 0 & N_1 \\ N_2 & 0 \\ 0 & N_2 \\ N_3 & 0 \\ 0 & N_3 \end{bmatrix} \begin{Bmatrix} T_x \\ T_y \end{Bmatrix} dS = \int_\xi \begin{Bmatrix} L_1 & P_x(y_1 - y)/b_3 \\ & 0 \\ L_2 & P_x(y_1 - y)/b_3 \\ & 0 \\ & 0 \\ & 0 \end{Bmatrix} t\, dy \qquad (c)$$

In terms of area coordinates and the coordinates of the triangular element,

$$y = L_1 y_1 + L_2 y_2 + L_3 y_3$$

but $L_3 = 0$ between nodes 1 and 2. Substitute y into Eq. (c) and use Eq. (b) to integrate the first term in the matrix:

$$\int_{y_1}^{y_2} L_1 P_x(y_1 - L_1 y_1 - L_2 y_2)\left(\frac{t}{b_3}\right) dy = P_x\left(\frac{1!0!}{2!} - \frac{2!0!}{3!} - \frac{1!1!}{3!}\right)\frac{t}{b_3}(y_2 - y_1) = \frac{P_x t}{6}(y_2 - y_1)$$

A similar computation for the third term in the matrix of Eq. (c) gives the distribution to node 2:

$$\frac{P_x t}{3}(y_2 - y_1)$$

One-third of the distributed pressure should be assigned to node 1, and two-thirds of the pressure should be assigned to node 2.

3.21. Physical equations of the form of Eq. (3.1) have been given several interpretations. In Eq. (f) of Prob. 3.5 it was shown that a constant heat source applied to a body is distributed equally to all four nodes of a rectangular element. In some cases the heat-source term can be considered a point source. Then the distribution of heat generation or dissipation is dependent upon the coordinate location of the source within the element. For the element of Fig. 3-14 assume a point heat source Q is located at the point $x_Q = 4$, $y_Q = 3$, and compute the distribution of heat to each node.

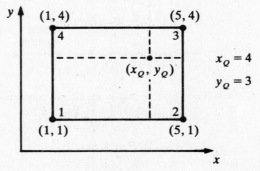

Fig. 3-14

Formally, the integral on the right-hand side of Eq. (a) of Prob. 3.5 is to be evaluated for the point source located at (x_Q, y_Q). The source term Q can be modeled as a *unit impulse function*

$$Q(x, y) = Q(x_Q, y_Q)\, \delta(x - x_Q)\, \delta(y - y_Q) \tag{a}$$

The integral of a function multiplied by an impulse function can be interpreted as the function evaluated at the coordinates defining the location of the function (Wylie, 1960, 340). The shape functions are given in Prob. 3.1 and, similar to Eq. (f) of Prob. 3.5, the distribution of $Q(x_Q, y_Q)$ is

$$\int_A [N]^T Qt\, dx\, dy = \int_A \begin{Bmatrix} N_1 \\ N_2 \\ N_3 \\ N_4 \end{Bmatrix} Q(x_Q, y_Q)\, \delta(x - x_Q)\, \delta(y - y_Q) t\, dx\, dy = \begin{Bmatrix} N_1 \\ N_2 \\ N_3 \\ N_4 \end{Bmatrix} Q(x_Q, y_Q) t$$

The shape functions are evaluated for $a = 4$, $b = 3$, $x_Q = 3$, and $y_Q = 2$, and the final result is

$$\left\{ \begin{array}{c} \frac{1}{12} \\ \frac{3}{12} \\ \frac{6}{12} \\ \frac{2}{12} \end{array} \right\} Q(x_Q, y_Q)$$

The total point source is distributed according to the areas defined in Fig. 3-14. For example, the distribution to node 3 is proportional to the area defined between the point source and node 1.

3.22. Derive the shape functions for the nine-node rectangular element shown in Fig. 3-15.

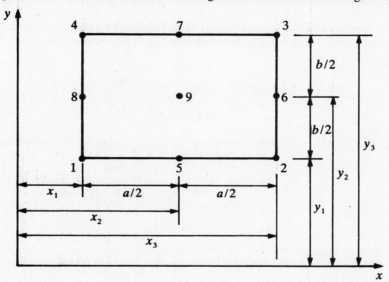

Fig. 3-15

The nine-node element has midside nodes located at the midpoint of each side of the element, and the ninth node located at the center of the element. The node numbering shown is standard, with midside nodes numbered 5 through 8, but the nodes can be numbered in any sequence. In this case the dimensions of the element (the local coordinate system) are assumed as a and b. Other dimensions, such as $2a \times 2b$ could also be used. In later chapters a local coordinate system similar to $-a$ to $+a$ and $-b$ to $+b$ will be introduced.

The shape function can be derived using the results of Prob. 2.31 and Fig. 2-10. The global x coordinates of Fig. 2-10 are reproduced in Fig. 3-15 along with corresponding y coordinates. The results of Prob. 2.30 are repeated for reference and given an extra subscript to indicate the x direction in the global system:

$$N_{1x} = \frac{(x - x_2)(x - x_3)}{(x_1 - x_2)(x_1 - x_3)} \qquad N_{5x} = \frac{(x - x_1)(x - x_3)}{(x_2 - x_1)(x_2 - x_3)} \qquad N_{2x} = \frac{(x - x_1)(x - x_2)}{(x_3 - x_1)(x_3 - x_2)} \qquad (a)$$

In the local system let node 1 correspond to $x_1 = 0$, $x_2 = a/2$, and $x_3 = a$. The shape function, in the x direction only, along nodes 1, 5, and 2 is obtained by substituting into Eqs. (a):

$$N_{1x} = \frac{[x - (a/2)](x - a)}{[0 - (a/2)](0 - a)} = \frac{(2x - a)(x - a)}{a^2} \qquad (b)$$

Similarly, N_{5x} for the nine-node element can be constructed using N_{5x} of Eqs. (a):

$$N_{5x} = \frac{(x - 0)(x - a)}{[(a/2) - 0][(a/2) - a]} = \frac{x(x - a)}{-(a^2/4)} \qquad (c)$$

Visualize that the shape functions of Prob. 2.30 could have been derived along the y axis of the global system using y_1, y_2, and y_3 of Fig. 3-15 and the result would be Eqs. (a) with all x values replaced by y values. A shape function for node 1 of the nine-node element would be identical to Eq. (b) with x and a replaced by y and b, respectively. The two-dimensional shape function for node 1 is the product of the two one-dimensional shape functions for node 1:

$$N_1 = N_{1x}N_{1y} = \frac{(2x - a)(x - a)(2y - b)(y - b)}{a^2b^2} \tag{d}$$

Substituting local coordinates corresponding to node 1 ($x = 0$, $y = 0$) into Eq. (d) will give $N_1 = 1$, and the value of N_1 at all other nodes will be zero.

The y contribution of the two-dimensional shape function for node 5 is constructed using the first Eq. (a) and is the same as that used in Eq. (d) or, corresponding to Fig. 3-15, is written as

$$N_{5y} = \frac{(y - y_9)(y - y_7)}{(y_5 - y_9)(y_5 - y_7)} = \frac{(2y - b)(y - b)}{b^2} \tag{e}$$

Obviously, the nine-node rectangular element is made up of combinations fo quadratic shape functions. It follows that

$$N_5 = N_{5x}N_{5y} = -\frac{4x(x - a)(2b - y)(y - b)}{a^2b^2} \tag{f}$$

A significant result is that two- or three-dimensional shape functions can be constructed from one-dimensional shape functions. The remaining shape functions are obtained in a similar manner and, as an exercise, the reader should verify the results:

$$N_2 = -\frac{x(2x - a)(2y - b)(y - b)}{a^2b^2}$$

$$N_3 = \frac{x(2x - a)y(2y - b)}{a^2b^2}$$

$$N_4 = \frac{(2x - a)(x - a)y(2y - b)}{a^2b^2}$$

$$N_6 = -\frac{4xy(2x - a)(y - b)}{a^2b^2}$$

$$N_7 = -\frac{4xy(x - a)(2y - b)}{a^2b^2}$$

$$N_8 = -\frac{4y(2x - a)(x - a)(y - b)}{a^2b^2}$$

$$N_9 = \frac{16xy(x - a)(y - b)}{a^2b^2}$$

3.23. The rectangular element of Fig. 3-16 is supported at node 2 by a sloping surface such that displacement parallel to the surface is zero, $u_{2\xi} = 0$. Construct the transformation matrix and show the form of the modified local stiffness matrix that accounts for this type of boundary condition.

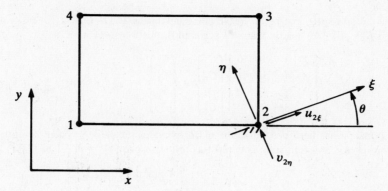

Fig. 3-16

The displacements in the ξ, η system are to be transformed to the x, y system. Therefore, Eq. (3.15) is the proper transformation. The transformation for the local element is

$$[T]^T = \begin{bmatrix} 1 & 0 & 0 & 0 & 0 & 0 & 0 & 0 \\ 0 & 1 & 0 & 0 & 0 & 0 & 0 & 0 \\ 0 & 0 & c & -s & 0 & 0 & 0 & 0 \\ 0 & 0 & s & c & 0 & 0 & 0 & 0 \\ 0 & 0 & 0 & 0 & 1 & 0 & 0 & 0 \\ 0 & 0 & 0 & 0 & 0 & 1 & 0 & 0 \\ 0 & 0 & 0 & 0 & 0 & 0 & 1 & 0 \\ 0 & 0 & 0 & 0 & 0 & 0 & 0 & 1 \end{bmatrix} \qquad (a)$$

where $c = \cos\theta$ and $s = \sin\theta$. The matrix statement that defines the transformation, using Eq. (a), is

$$\{u_{xy}\} = [T]^T\{u_{\xi\eta}\} \qquad (b)$$

But keep in mind that the displacements are unknown at this point and that Eq. (b) is used to transform the stiffness matrix to correspond to the boundary conditions imposed by $u_{2\xi}$ and $v_{2\eta}$. Recall Eq. (j) of Prob. 3.10, the matrix form of the variational function that describes an elasticity problem, in terms of displacements in the x, y system.

$$\int_V \frac{1}{2}\{u_{xy}\}^T[B]^T[C][B]\{u_{xy}\}\, dV \qquad (c)$$

Substitute Eq. (b) into Eq. (c) and note that $\{u_{xy}\}^T = \{u_{\xi\eta}\}^T[T]$:

$$\int_V \frac{1}{2}\{u_{\xi\eta}\}^T[T][B]^T[C][B][T]^T\{u_{\xi\eta}\}\, dV$$

The function is minimized with respect to the new displacement vector, but the definition for the stiffness matrix remains unchanged, $[K] = \int_A [B]^T[C][B]t\, dA$, as given by Eq. ($d$) of Prob. 3.11 since $[T]$ is not a function of the coordinates. The transformed stiffness matrix is

$$[K]_t = [T][K][T]^T \qquad (d)$$

where the subscript t indicates the transformed stiffness matrix. The boundary condition $u_{2\xi} = 0$ can be used in the global formulation for the problem.

3.24. Assume that a local four-node stiffness matrix for the diffusion terms in a mass transport formulation has been inadvertently numbered in the sequence 1, 3, 4, 2 rather than the standard 1, 2, 3, 4 numbering scheme. Derive a transformation matrix that can be used to renumber the element corresponding to the standard scheme. (This problem is somewhat hypothetical but serves to illustrate the transformation using a stiffness matrix with a minimum number of terms.)

In this instance the matrix of nodal point unknowns $\{C_1 \quad C_3 \quad C_4 \quad C_2\}^T$ must be transformed to $\{C_1 \quad C_2 \quad C_3 \quad C_4\}^T$. A matrix that will accomplish this transformation is

$$
\begin{bmatrix} 1 & 0 & 0 & 0 \\ 0 & 0 & 0 & 1 \\ 0 & 1 & 0 & 0 \\ 0 & 0 & 1 & 0 \end{bmatrix} \begin{Bmatrix} C_1 \\ C_3 \\ C_4 \\ C_2 \end{Bmatrix} = \begin{Bmatrix} C_1 \\ C_2 \\ C_3 \\ C_4 \end{Bmatrix} \qquad \text{or} \qquad [T]\{C\}_{\text{old}} = \{C\}_{\text{new}} \tag{a}
$$

The transformation has been defined as the old or previous numbering system being transformed into the new system. The reverse transformation is defined as follows since $[T]^T = [T]^{-1}$ or $[T][T]^T = [I]$:

$$
\{C\}_{\text{old}} = [T]^T \{C\}_{\text{new}} \tag{b}
$$

The nonzero terms in the transformation matrix can be identified using the following rule. Let the row correspond to the new number and the column correspond to the old number, or

$$
T(I_{\text{new}}, J_{\text{old}}) = 1 \qquad \text{all other } T(I, J) = 0 \tag{c}
$$

The nonzero terms in Eq. (a) were obtained as $T(1, 1) = 1$, $T(3, 2) = 1$, $T(2, 4) = 1$, and $T(4, 3) = 1$.

The local stiffness matrix is transformed using the same concept presented in Eq. (d) of Prob. 3.23, $[K]_{\text{new}} = [T][K]_{\text{old}}[T]^T$. As an exercise, the reader can verify that the following matrix multiplications will yield the desired transformation:

$$
\begin{bmatrix} 1 & 0 & 0 & 0 \\ 0 & 0 & 0 & 1 \\ 0 & 1 & 0 & 0 \\ 0 & 0 & 1 & 0 \end{bmatrix} \begin{bmatrix} K_{11} & K_{13} & K_{14} & K_{12} \\ K_{31} & K_{33} & K_{34} & K_{32} \\ K_{41} & K_{43} & K_{44} & K_{42} \\ K_{21} & K_{23} & K_{24} & K_{22} \end{bmatrix} \begin{bmatrix} 1 & 0 & 0 & 0 \\ 0 & 0 & 1 & 0 \\ 0 & 0 & 0 & 1 \\ 0 & 1 & 0 & 0 \end{bmatrix} = \begin{bmatrix} K_{11} & K_{12} & K_{13} & K_{14} \\ K_{21} & K_{22} & K_{23} & K_{24} \\ K_{31} & K_{32} & K_{33} & K_{34} \\ K_{41} & K_{42} & K_{43} & K_{44} \end{bmatrix}
$$

3.25. Deduce the variational function for two-dimensional axisymmetric heat conduction in r, z coordinates and discuss formulation of the corresponding local finite element stiffness matrix.

The variational function is an extension of Eq. (2.31) to include the z coordinate. The governing differential equation corresponds to Eq. (3.16) with partial derivatives with respect to $\theta = 0$ and in this case $\beta = 0$. Recall that dV for an axisymmetric problem is replaced by $2\pi r \, dr \, dz$. The variational function is

$$
J(T) = \int_{r_1}^{r_2} \int_{z_1}^{z_2} \left[k_r \left(\frac{dT}{dr} \right)^2 + k_z \left(\frac{dT}{dz} \right)^2 - 2QT \right] \pi r \, dr \, dz \tag{a}
$$

Note that Eq. (a) above is similar to Eq. (a) of Prob. 3.4 and that Eqs. (e) and (f) of Prob. 3.4 are applicable with x and y replaced by r and z, respectively. The variational function, in matrix form, is

$$
J(T) = \int_{r_1}^{r_2} \int_{z_1}^{z_2} [\{T\}^T [B]^T [k][B]\{T\} - 2\{T\}^T [N]^T Q] \pi r \, dr \, dz \tag{b}
$$

Minimize the function with respect to $\{T\}$, and the final form of the stiffness matrix is

$$
[K] = \int_{r_1}^{r_2} \int_{z_1}^{z_2} [B]^T [k][B] 2\pi r \, dr \, dz \tag{c}
$$

and the heat-source term is

$$
\int_{r_1}^{r_2} \int_{z_1}^{z_2} [N]^T Q 2\pi r \, dr \, dz \tag{d}
$$

The 2π term could have been eliminated from the function, but it is usually left in to avoid confusion when specifying a flux boundary condition.

The shape functions can be obtained using the results from Probs. 3.1 and 3.28 and Fig. 3-17, where the limits on z have been replaced by 0 to b in order to simplify the shape functions. Note that limits on z can be changed to the local element system without changing the final results, but limits on r must be in the global r_1-to-r_2 system since the volume changes relative to $r = 0$:

$$N_1 = \frac{(r - r_2)(z - b)}{Rb} \qquad N_2 = -\frac{(r - r_1)(z - b)}{Rb}$$

$$N_3 = \frac{(r - r_1)z}{Rb} \qquad N_4 = -\frac{(r - r_2)z}{Rb}$$

(e)

where $R = r_2 - r_1$. The construction of the stiffness matrix follows the method outlined in Prob. 3.5. Formulation of the local stiffness will be illustrated using the K_{11} term of Eq. (c):

$$K_{11} = \int_{r_1}^{r_2} \int_0^b \left(\frac{\partial N_1}{\partial r} k_r \frac{\partial N_1}{\partial r} + \frac{\partial N_1}{\partial z} k_z \frac{\partial N_1}{\partial z} \right) 2\pi r \, dr \, dz$$

Substituting the proper derivatives from Eq. (e) gives

$$K_{11} = \int_{r_1}^{r_2} \int_0^b [r(z - b)^2 k_r + r(r - r_2)^2 k_z] \frac{2\pi}{R^2 b^2} \, dr \, dz$$

or
$$K_{11} = 2\pi \left\{ \frac{k_r b^3}{6} (r_2^2 - r_1^2) + \frac{k_z b}{12} [3(r_2^4 - r_1^4) - 8r_2(r_2^3 - r_1^3) + 6r_2^2(r^2 - r_1^2)] \right\}$$

(f)

The result could be reduced somewhat, but the algebra is cumbersome, and Eq. (f) is suitable for computer implementation. In practice, numerical integration is usually used for axisymmetric problems. Deriving a stiffness matrix using terms like Eq. (f) is of value from an instructional viewpoint, however, production computer codes usually use more practical methods. In any case, the reader, as an exercise, can derive the remaining terms for the stiffness matrix.

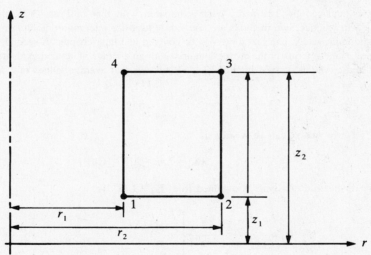

Fig. 3-17   Axisymmetric four-node element.

3.26.   Discuss formulation of the local stiffness matrix for two-dimensional axisymmetric elasticity problems in r, z coordinates using three-node triangular elements.

The derivation of an axisymmetric element for plane elasticity follows the concepts that have been developed in previous problems. The variational statement of the problem is given by Eq. (j) of Prob. 3.10, and it follows that all matrices in that equation must be modified to correspond to a three-node axisymmetric formulation. The strain must be written in terms of an operator matrix [L], a shape function matrix [N], and the displacement vector for a three-node element, or following Eq. (i) of Prob. 3.10, $\{\epsilon\} = [L][N]\{u\}$. There are four strains obtained from Eqs. (3.17) assuming that $v = \partial/\partial\theta = 0$:

$$\begin{Bmatrix} \epsilon_{rr} \\ \epsilon_{\theta\theta} \\ \epsilon_{zz} \\ \epsilon_{rz} \end{Bmatrix} = \begin{Bmatrix} \partial u/\partial r \\ u/r \\ \partial w/\partial z \\ \partial w/\partial r + \partial u/\partial z \end{Bmatrix} = \begin{bmatrix} \partial/\partial r & 0 \\ 1/r & 0 \\ 0 & \partial/\partial z \\ \partial/\partial z & \partial/\partial r \end{bmatrix} \begin{bmatrix} N_1 & 0 & N_2 & 0 & N_3 & 0 \\ 0 & N_1 & 0 & N_2 & 0 & N_3 \end{bmatrix} \begin{Bmatrix} u_1 \\ w_1 \\ u_2 \\ w_2 \\ u_3 \\ w_3 \end{Bmatrix} \qquad (a)$$

The shape functions are given by Eqs. (c) of Prob. 3.14, and u and w are the displacements at each node in the r and z directions, respectively. The [B] matrix is obtained from Eq. (a) and evaluated using the definitions of Prob. 3.14 as follows:

$$[B] = \begin{bmatrix} \partial N_1/\partial r & 0 & \partial N_2/\partial r & 0 & \partial N_3/\partial r & 0 \\ N_1/r & 0 & N_2/r & 0 & N_3/r & 0 \\ 0 & \partial N_1/\partial z & 0 & \partial N_2/\partial z & 0 & \partial N_3/\partial z \\ \partial N_1/\partial z & \partial N_1/\partial r & \partial N_2/\partial z & \partial N_2/\partial r & \partial N_3/\partial z & \partial N_3/\partial r \end{bmatrix} \div 2A$$

or

$$[B] = \begin{bmatrix} b_1 & 0 & b_2 & 0 & b_3 & 0 \\ \dfrac{a_1}{r}+b_1+\dfrac{c_1 z}{r} & 0 & \dfrac{a_2}{r}+b_2+\dfrac{c_2 z}{r} & 0 & \dfrac{a_3}{r}+b_3+\dfrac{c_3 z}{r} & 0 \\ 0 & c_1 & 0 & c_2 & 0 & c_3 \\ c_1 & b_1 & c_2 & b_2 & c_3 & b_3 \end{bmatrix} \div 2A$$

The local stiffness matrix is computed as in previous problems:

$$[K] = \int_V [B]^T [C][B] \, dV$$

where dV is replaced by $2\pi r \, dr \, dz$. When compared with that in Prob. 3.25, the integration is somewhat complicated. In practice, two methods are commonly used for integration. Each term in the stiffness matrix can be integrated numerically, and that topic will be covered in a later chapter. A second, more elementary method is to replace all r and z terms in the stiffness matrix with an average value and evaluate the remaining terms using the a_i, b_i, and c_i values defined in Prob. 3.14. It follows that average values of r and z are simply

$$r_{avg} = \frac{r_1 + r_2 + r_3}{3} \qquad \text{and} \qquad z_{avg} = \frac{z_1 + z_2 + z_3}{3} \qquad (b)$$

The local stiffness matrix can be written as

$$[K] = 2\pi r_{avg} A [B]^T [C][B] \qquad (c)$$

The matrix of material constants is obtained from Eq. (3.18) as

$$[C] = \frac{E}{(1+\nu)(1-2\nu)} \begin{bmatrix} 1-\nu & \nu & \nu & 0 \\ \nu & 1-\nu & \nu & 0 \\ \nu & \nu & 1-\nu & 0 \\ 0 & 0 & 0 & (1-2\nu)/2 \end{bmatrix} \qquad (d)$$

3.27. A plane elasticity problem can be formulated in polar coordinates when dependence upon the z coordinate can be neglected. Assume a four-node element in r, θ coordinates, where θ is in radians, and derive the shape functions and the corresponding [B] matrix.

Shape functions are identical to Eqs. (e) of Prob. 3.25, where θ is in the local system of Fig. 3-18 and varies from 0 to α for all elements:

$$N_1 = \frac{(r-r_2)(\theta-\alpha)}{R\alpha} \qquad N_2 = -\frac{(r-r_1)(\theta-\alpha)}{R\alpha}$$

$$N_3 = \frac{(r-r_1)\theta}{R\alpha} \qquad N_4 = -\frac{(r-r_2)\theta}{R\alpha} \qquad (a)$$

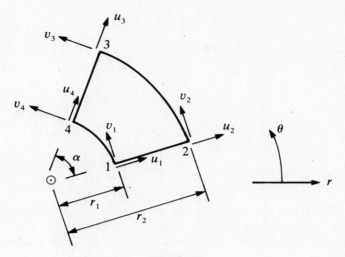

Fig. 3-18

Formulation of the [B] matrix is similar to the method used in Prob. 3.11. In any finite element formulation the [B] matrix can always be written as an operator matrix postmultiplied by the shape function matrix, or [B] = [L][N]. The operator matrix is obtained from Eq. (*3.17*) assuming $w = 0$ and $\partial/\partial z = 0$, where u and v are the displacements in the r and θ directions, respectively. The shape function matrix of Eq. (*b*) of Prob. 3.11 is valid for this formulation since it agrees with the node numbering of Fig. 3-18. The operator matrix is constructed according to the strain-displacement equations $\{\epsilon\} = [L][N]\{u\}$ as outlined in Prob. 3.10:

$$\begin{Bmatrix} \epsilon_{rr} \\ \epsilon_{\theta\theta} \\ \epsilon_{r\theta} \end{Bmatrix} = \begin{bmatrix} \partial/\partial r & 0 \\ 1/r & (1/r)\,\partial/\partial\theta \\ (1/r)\,\partial/\partial\theta & \partial/\partial r - 1/r \end{bmatrix} \begin{bmatrix} N_1 & 0 & N_2 & 0 & N_3 & 0 & N_4 & 0 \\ 0 & N_1 & 0 & N_2 & 0 & N_3 & 0 & N_4 \end{bmatrix} \{u\} \qquad (b)$$

where $\{u\}$ is similar to Eq. (*e*) of Prob. 3.10. The final result is a 3×8 matrix obtained as [L][N] of Eq. (*b*) or

$$\begin{bmatrix} \dfrac{\partial N_1}{\partial r} & 0 & \dfrac{\partial N_2}{\partial r} & 0 & \dfrac{\partial N_3}{\partial r} & 0 & \dfrac{\partial N_4}{\partial r} & 0 \\[2ex] \dfrac{N_1}{r} & \dfrac{1}{r}\dfrac{\partial N_1}{\partial\theta} & \dfrac{N_2}{r} & \dfrac{1}{r}\dfrac{\partial N_2}{\partial\theta} & \dfrac{N_3}{r} & \dfrac{1}{r}\dfrac{\partial N_3}{\partial\theta} & \dfrac{N_4}{r} & \dfrac{1}{r}\dfrac{\partial N_4}{\partial\theta} \\[2ex] \dfrac{1}{r}\dfrac{\partial N_1}{\partial\theta} & \dfrac{\partial N_1}{\partial r} - \dfrac{N_1}{r} & \dfrac{1}{r}\dfrac{\partial N_2}{\partial\theta} & \dfrac{\partial N_2}{\partial r} - \dfrac{N_2}{r} & \dfrac{1}{r}\dfrac{\partial N_3}{\partial\theta} & \dfrac{\partial N_3}{\partial r} - \dfrac{N_3}{r} & \dfrac{1}{r}\dfrac{\partial N_4}{\partial\theta} & \dfrac{\partial N_4}{\partial r} - \dfrac{N_4}{r} \end{bmatrix} \qquad (c)$$

Note that for this element dV becomes $dV = r\,d\theta\,dr$.

Supplementary Problems

3.28.   Given the form

$$\phi = N_1\phi_1 + N_2\phi_2 + N_3\phi_3 + N_4\phi_4$$

derive the shape functions N_i for a rectangular element using general x_i, y_i coordinates as illustrated in Fig. 3-3. Show that these shape functions will reduce to those derived in Prob. 3.1.

3.29. Write Eq. (*b*) of Prob. 3.1 in matrix format.

3.30. Show that $[X]^{-1}$ of Prob. 3.2 is given by

$$\begin{bmatrix} x_2 y_3 - x_3 y_2 & x_3 y_1 - x_1 y_3 & x_1 y_2 - x_2 y_1 \\ y_2 - y_3 & y_3 - y_1 & y_1 - y_2 \\ x_3 - x_2 & x_1 - x_3 & x_2 - x_1 \end{bmatrix} \div 2A$$

where A is the area of the triangle.

3.31. A triangular element has node points located at $(x_1 = 1,\ y_1 = 1)$, $(x_2 = 6,\ y_2 = 1)$, and $(x_3 = 3, y_3 = 4)$. A function has been computed to have nodal point values of $\phi_1 = 900$, $\phi_2 = 600$, and $\phi_3 = 1200$. Use the interpolation function for a three-node triangular element and compute the value of ϕ at $(x = 3, y = 4)$.

3.32. Develop the variational function for mass transport that corresponds to Eq. (*3.2*). Follow the general procedure used in Prob. 3.4 and write the variational function in matrix format.

3.33. Derive the local stiffness matrix for mass transport including the chemical reaction term and the velocity terms using shape functions for a four-node rectangular element.

3.34. Solve the one-dimensional mass transport problem of Prob. 2.18 using a strip of two-dimensional elements.

3.35. Construct the nodal coordinate and connectivity arrays for the dam problem illustrated in Fig. 3-7.

3.36. Solve the steady-state mass transport problem for the space illustrated in Fig. 3-19.

(*a*)   Divide the space into two elements and use symmetry along the axis $y = 0.5$ as shown in Fig. 3-19 and solve for C at node 3.

(*b*)   Divide the space into eight elements as shown and solve for C at all nodes.

(*c*)   Assume the velocity term u_x is zero and solve the problem of steady-state diffusion.

(*d*)   Solve the equation using separation of variables and compare the finite element results with the analytical solution. Assume $D_x = D_y = D = 1.$, $u_x = 1.$, $a = b = 1.$, and $u_y = K_r = 0$. The boundary conditions are shown in Fig. 3-19. Assume $C_0 = 100$ along the edge $x = a$.

3.37. The rectangular element of Fig. 3-3 is loaded with a uniformly varying pressure load as illustrated in Fig. 3-20 and can be described as $T_x = p_x y / b$ and $T_y = 0$ in the element coordinate system. Determine the distribution of the pressure loading to nodes 2 and 3.

3.38. Solutions to problems in plane elasticity are often obtained by solving the partial differential equation formulated in terms of a stress function ϕ. The equation and the corresponding definition for the stresses are

$$\nabla^4 \phi = \frac{\partial^4 \phi}{\partial x^4} + 2 \frac{\partial^4 \phi}{\partial x^2 \partial y^2} + \frac{\partial^4 \phi}{\partial y^4} = 0 \qquad \sigma_{xx} = \frac{\partial^2 \phi}{\partial y^2} \qquad \sigma_{yy} = \frac{\partial^2 \phi}{\partial x^2} \qquad \sigma_{xy} = -\frac{\partial^2 \phi}{\partial x \partial y}$$

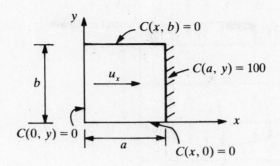

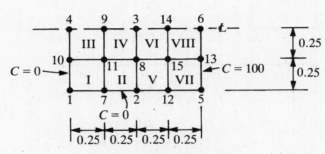

Two-element model

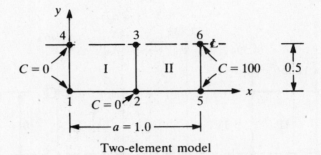

Eight-element model

Fig. 3-19

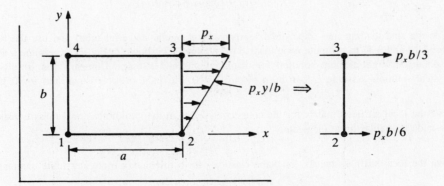

Fig. 3-20

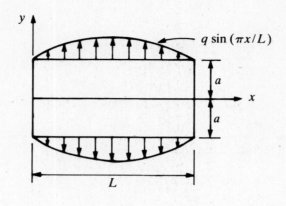

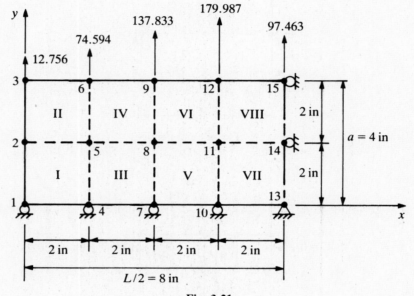

Fig. 3-21

The stress function for the plate of Fig. 3-21 can be obtained using separation of variables as

$$\phi(x, y) = -\frac{q}{\alpha^2} \frac{\alpha y \sinh(\alpha a) \sinh(\alpha y) - [\sinh(\alpha a) + \alpha a \cosh(\alpha a)] \cosh(\alpha y)}{\alpha a + \sinh(\alpha a) \cosh(\alpha a)} \sin(\alpha x) \qquad \alpha = \frac{\pi}{L}$$

Divide the sine loading into trapezoidal segments (rectangular and triangular) and use the results presented in Probs. 3.10 and 3.37 to compute node loads. Use symmetry as shown in Fig. 3-21, assume a unit plate thickness, and obtain an eight-element solution for displacements and stresses. Compare your results for σ_{yy} with the analytical solution. Assume $L = 16$ in, $a = 4$ in, $q = 100$ lb/in, $E = 30(10)^6$ psi, and $\nu = 0.3$.

3.39. Derive the local stiffness matrix for the convection terms in the two-dimensional mass transport equation using three-node triangular finite elements.

3.40. Derive the local stiffness matrix for plane elasticity for a three-node triangular finite element.

3.41. A constant uniform pressure P_x is distributed along the side of a triangular finite element between nodes 1 and 2 in the x-coordinate direction. Determine the distribution of force to each node.

3.42. Assume a uniform body force is applied to a plane elasticity problem modeled using triangular elements. Given that

$$\int_V [N]^T f \, dV = \int_A \begin{Bmatrix} N_1 \\ N_2 \\ N_3 \end{Bmatrix} t \, dA$$

compute the distribution of body force to each node.

3.43. Divide the four-node element of Fig. 3-14 into two triangular elements defined by nodes 1, 2, and 3 and nodes 1, 3, and 4. Compute the distribution of a point source Q located at $x_Q = 4$, $y_Q = 3$, to each node.

3.44. Rework Probs. 3.21 and 3.43 assuming that $x_Q = 4.0$ and $y_Q = 3.25$. Make a comparison of the results for four-node elements versus three-node elements.

3.45. Assume the nine-node element of Fig. 3-15 has a constant flux q applied along the edge defined by nodes 1, 5, and 2. Compute the distribution of flux to each node.

3.46. An eight-node rectangular element has node numbers in sequence with corner nodes numbered 1, 3, 5, 7 and midside nodes numbered 2, 4, 6, 8 with corner nodes appearing first in the solution vector. Derive the transformation matrix that will renumber the element to conform with Prob. 3.22, Fig. 3-15, with corner nodes still appearing first in the solution vector (neglecting the ninth node).

3.47. A three-node triangular element has been derived for an elasticity problem with all x components of displacement appearing first in the solution vector such that $\{u\}^T = \{u_1 \quad u_2 \quad u_3 \quad v_1 \quad v_2 \quad v_3\}^T$. Derive the transformation matrix that will reorder the displacements to conform to the more conventional numbering scheme, $\{u\}^T = \{u_1 \quad v_1 \quad u_2 \quad v_2 \quad u_3 \quad v_3\}^T$.

3.48. Evaluate the [B] matrix for element III of the axisymmetric finite element model shown in Fig. 3-22.

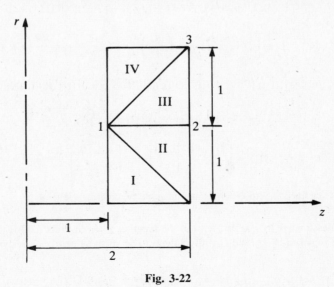

Fig. 3-22

3.49. Assume the shape functions of Prob. 3.27 and derive the form of the individual matrices that form the stiffness matrix for mass transport. Write the stiffness matrix in integral form.

Answers to Supplementary Problems

3.28.

$$N_1 = \frac{(x - x_2)(y - y_4)}{(x_1 - x_2)(y_1 - y_4)} \qquad N_2 = \frac{(x - x_1)(y - y_3)}{(x_2 - x_1)(y_2 - y_3)}$$

$$N_3 = \frac{(x - x_4)(y - y_2)}{(x_3 - x_4)(y_3 - y_2)} \qquad N_4 = \frac{(x - x_3)(y - y_1)}{(x_4 - x_3)(y_4 - y_1)}$$

3.29.

$$\phi = [N_1 \quad N_2 \quad N_3 \quad N_4] \begin{Bmatrix} \phi_1 \\ \phi_2 \\ \phi_3 \\ \phi_4 \end{Bmatrix} \qquad \text{or} \qquad \phi = [N]\{\phi\}$$

3.30. Use the cofactor method to invert $[X]$ (see Chap. 1).

$$[X]^{-1} = \frac{[C]^T}{\det [X]}$$

where $\det [X] = 2A$ and $[C]$ is the cofactor.

$$C_{11} = \begin{bmatrix} x_2 & y_2 \\ x_3 & y_3 \end{bmatrix} = x_2 y_3 - x_3 y_2 \qquad C_{12} = (-1) \begin{bmatrix} 1 & y_2 \\ 1 & y_3 \end{bmatrix} = y_2 - y_3$$

The remaining terms are computed in a similar manner.

3.31. Use Eqs. (g) and (i) of Prob. 3.2.

$$\phi = [21 - (3)(3) - (3)(2)](900/15) + [-1 + (3)(3) - (2)(2)](600/15) + [5 + (5)(2)](1200/15) = 920$$

3.32.

$$J(C) = \int_A \frac{1}{2} \left[D_x \left(\frac{\partial C}{\partial x} \right)^2 + D_y \left(\frac{\partial C}{\partial y} \right)^2 + C u_x \frac{\partial C}{\partial x} + C u_y \frac{\partial C}{\partial y} + K_r C^2 - 2mC \right] t \, dx \, dy$$

$$J(C) = \int_A \left(\frac{1}{2} \{C\}^T [N]^T [L]^T [D][L][N]\{C\} + \frac{1}{2} \{C\}^T [N]^T [u][L][N]\{C\} \right.$$

$$\left. + \frac{1}{2} \{C\}^T [N]^T [K_r][N]\{C\} - \{C\}^T [N]^T m \right) t \, dx \, dy$$

where

$$[D] = \begin{bmatrix} D_x & 0 \\ 0 & D_y \end{bmatrix} \qquad [u] = \begin{bmatrix} u_x & 0 \\ 0 & u_y \end{bmatrix} \qquad [K_r] = [K_r]$$

$[L]$ and $[N]$ are defined in Prob. 3.4 for a four-node rectangular element.

3.33. The local stiffness matrix for the first term of the function given in Prob. 3.32 is the same as Eq. (e) of Prob. 3.5 with k_x and k_y replaced by D_x and D_y, respectively. The stiffness matrix corresponding to the transport velocity terms is

$$\begin{bmatrix} -(u_x b + u_y a)/6 & (2u_x b - u_y a)/12 & (u_x b + u_y a)/12 & (-u_x b + 2u_y a)/12 \\ -(2u_x b + u_y a)/12 & (u_x b - u_y a)/6 & (u_x b + 2u_y a)/12 & (-u_x b + u_y a)/12 \\ -(u_x b + u_y a)/12 & (u_x b - 2u_y a)/12 & (u_x b + u_y a)/6 & (-2u_x b + u_y a)/12 \\ -(u_x b + 2u_y a)/12 & (u_x b - u_y a)/12 & (2u_x b + u_y a)/12 & (-u_x b + u_y a)/6 \end{bmatrix}$$

The stiffness matrix corresponding to the chemical reaction is computed using a matrix equation $\int_A [N]^T [K_r][N] \, dA$. The matrix multiplications appear as

$$\begin{Bmatrix} N_1 \\ N_2 \\ N_3 \\ N_4 \end{Bmatrix} [K_r][N_1 \quad N_2 \quad N_3 \quad N_4]$$

The resulting matrix multiplications and integration give

$$K_r(ab) \begin{bmatrix} \frac{1}{9} & \frac{1}{18} & \frac{1}{36} & \frac{1}{18} \\ \frac{1}{18} & \frac{1}{9} & \frac{1}{18} & \frac{1}{36} \\ \frac{1}{36} & \frac{1}{18} & \frac{1}{9} & \frac{1}{18} \\ \frac{1}{18} & \frac{1}{36} & \frac{1}{18} & \frac{1}{9} \end{bmatrix}$$

3.35. The nodal coordinates and connectivities are given in Tables 3.9 and 3.10, respectively.

Table 3.9

Node	X	Y
1	0.0	0.0
2	0.0	3.0
3	0.0	6.0
4	6.0	0.0
5	6.0	3.0
6	6.0	6.0
7	12.0	0.0
8	12.0	3.0
9	12.0	6.0
10	18.0	0.0
11	18.0	3.0
12	18.0	6.0

Table 3.10

Global element	Local element			
	Node 1	Node 2	Node 3	Node 4
I	1	4	5	3
II	2	5	6	3
III	4	7	8	5
IV	7	10	11	8
V	8	11	12	9

3.36. (a) Finite element, $C_3 = 37.5$ for $u_x = 0.0$, $C_3 = 28.125$ for $u_x = 1.0$. See part (d) for the analytical solution. (b and c) Results are given in Table 3.11 for interior nodes. (d) The separation of variables solution is as follows, and the results are given in Table 3.11.

$$C(x, y) = \frac{4C_0 e^{\alpha x}}{\pi e^{\alpha a}} \sum_{n_{\text{odd}}}^{\infty} \frac{\sinh(\beta x)}{n \sinh(\beta a)} \sin(ky) \qquad \alpha = \frac{u_x}{2D} \qquad k = \frac{n\pi}{b} \qquad \beta = (\alpha^2 + k^2)^{1/2}$$

Note that for $u_x = 0$ this solution is similar to that given in Prob. 3.6.

Table 3.11

Node	Analytical		Finite element	
	Velocity = 0.0	Velocity = 1.0	Velocity = 0.0	Velocity = 1.0
3	25.000	19.116	27.857	20.945
8	18.203	13.924	19.286	14.309
9	9.541	6.398	10.092	6.468
11	6.797	4.558	7.154	4.605
14	54.053	47.231	58.479	50.935
15	43.203	37.779	50.703	44.197

3.37. Use Eq. (*b*) of Prob. 3.12. Let $x = a$ and $N_1 = N_4 = 0$, also $T_y = 0$. The load vector becomes

$$\int_S \{0 \quad 0 \quad N_2 T_x \quad 0 \quad N_3 T_x \quad 0 \quad 0 \quad 0\}^T \, dS$$

The two nonzero terms are evaluated as

$$\int_S N_2 T_x \, dS = \int_0^b \frac{b-y}{b} \, p_x \frac{y}{b} \, t \, dy = \frac{p_x b}{6}$$

$$\int_S N_3 T_x \, dS = \int_0^b \frac{y}{b} \, p_x \frac{y}{b} \, t \, dy = \frac{p_x b}{3}$$

The total force load is $p_x b/2$, of which one-third is applied at node 2 and two-thirds is applied at node 3.

3.38. The analytical solution for σ_{yy} is the same as the stress function multiplied by $-\alpha^2$ since $\sin(\alpha x)$ is the only function of x in the equation. The computed node loading is shown in Fig. 3-21. Results for displacements and stresses are given in Table 3.12.

Table 3.12

Node	u_x (×10⁻⁶)	u_y (×10⁻⁶)	Element	x	y	Finite element			Analytical
						σ_{xx}	σ_{yy}	σ_{xy}	σ_{yy}
1	6.490	0.0	1	1	1	−0.588	23.705	0.931	19.332
2	5.555	0.541	2	1	3	0.588	21.793	1.111	20.054
3	2.506	0.914	3	3	1	−2.727	54.133	1.561	55.053
4	5.931	0.0	4	3	3	2.727	53.958	2.532	57.108
5	5.087	2.643	5	5	1	−5.142	79.574	1.222	82.393
6	2.180	5.153	6	5	3	5.142	80.544	1.736	85.469
7	4.573	0.0	7	7	1	−6.485	93.906	0.425	97.189
8	3.917	4.684	8	7	3	6.485	95.022	0.609	100.89
9	1.555	9.259							
10	2.495	0.0							
11	2.126	6.132							
12	0.810	12.090							
13	0.0	0.0							
14	0.0	6.648							
15	0.0	13.100							

3.39. The computations are similar to those for Prob. 3.18.

$$\begin{bmatrix} u_xb_1 + u_yc_1 & u_xb_2 + u_yc_2 & u_xb_3 + u_yc_3 \\ u_xb_1 + u_yc_1 & u_xb_2 + u_yc_2 & u_xb_3 + u_yc_3 \\ u_xb_1 + u_yc_1 & u_xb_2 + u_yc_2 & u_xb_3 + u_yc_3 \end{bmatrix} \frac{t}{6}$$

3.40. Use Eq. (d) of Prob. 3.11 and [B] given in Prob. 3.19.

$$\begin{bmatrix} b_1^2C_{11} + c_1^2C_{33} & b_1c_1C_{12} + b_1c_1C_{33} & b_1b_2C_{11} + c_1c_2C_{33} & b_1c_1C_{12} + b_2c_1C_{33} & b_1b_3C_{11} + c_1c_2C_{33} & b_1c_3C_{12} + b_3c_1C_{33} \\ & c_1^2C_{11} + b_1^2C_{33} & b_2c_1C_{12} + b_1c_2C_{33} & c_1c_2C_{11} + b_1b_2C_{33} & b_3c_1C_{12} + b_1c_3C_{33} & c_1c_3C_{11} + b_1b_3C_{33} \\ & & b_2^2C_{11} + c_2^2C_{33} & b_2c_2C_{12} + b_2c_2C_{33} & b_2b_3C_{11} + c_2c_3C_{33} & b_2c_3C_{12} + b_3c_2C_{33} \\ & \text{Symmetric} & & c_2^2C_{11} + b_2^2C_{33} & b_3c_2C_{12} + b_2c_3C_{33} & c_2c_3C_{11} + b_2b_3C_{33} \\ & & & & b_3^2C_{11} + c_3^2C_{33} & b_3c_3C_{12} + b_3c_3C_{33} \\ & & & & & c_3^2C_{11} + b_3^2C_{33} \end{bmatrix} t \div 2A$$

3.41. Both nodes are loaded with one-half of the total pressure $(P_xt/2)(y_2 - y_1)$.

3.42. Substitute area coordinates for shape functions and integrate the result numerically using area integration. The local force matrix is $(fAt/3)\{1 \quad 1 \quad 1\}^T$, or one-third of the body force is distributed to each node.

3.43. The point source lies within the element defined by nodes 1, 2, and 3, and consequently the element defined by nodes 1, 3, and 4 will receive zero distribution. Use the shape functions defined in Prob. 3.14 and follow the method of analysis given in Prob. 3.21. Note that $x_Q = 4$ and $y_Q = 3$.

$$N_1 = \frac{15 - 3x_Q}{12} = \frac{3}{12}$$

$$N_2 = \frac{1 + 3x_Q - 4y_Q}{12} = \frac{1}{12}$$

$$N_3 = \frac{-4 + 4y_Q}{12} = \frac{8}{12}$$

The distribution is according to the value of the shape functions.

3.44. The point (4.00, 3.25) lies on the diagonal between nodes 1 and 2. It follows that each of the triangular elements receives one-half of the point source. For the triangular element defined by nodes 1, 2, and 3 the shape functions of Prob. 3.43 can be used to compute the distribution to each node $N_1 = \frac{3}{12}$, $N_2 = 0$, and $N_3 = \frac{9}{12}$. The shape functions for the element defined by nodes 1, 3, and 4 are

$$N_1 = \frac{16 - 4x_Q}{12} = \frac{3}{12}$$

$$N_3 = \frac{-3 + 3x_Q}{12} = \frac{9}{12}$$

$$N_4 = \frac{-1 - 3x_Q + 4y_Q}{12} = 0$$

Therefore, one-fourth of the source is applied at node 1, and three-fourths is applied at node 3 with zero distributed to nodes 2 and 4.

Repeating the analysis given in Prob. 3.21 for $x_Q = 3.00$ and $y_Q = 2.25$ (in the local element coordinate system) gives the ratio $\{6.75/12 \quad 2.25/12 \quad 0.75/12 \quad 2.25/12\}^T Q(x_Q, y_Q)$. Comparing the results in this case shows that the rectangular element distributes the source to all nodes, whereas the triangular formulation omits two of the nodes.

3.45. The shape functions are given in Prob. 3.22. The analysis is formulated as $\int_0^a \{N_1 \quad N_2 \quad 0 \quad 0 \quad N_5 \quad 0 \quad 0 \quad 0 \quad 0\}^T q \, dx$, and the integration is along the line defined by nodes 1, 5, and 2 with $y = 0$.

Substituting shape functions and integrating gives $\{aq/6 \quad aq/6 \quad 0 \quad 0 \quad 2aq/3 \quad 0 \quad 0 \quad 0\}^T$, or one-sixth of the flux is applied to each of the corner nodes and two-thirds of the flux is applied to the midside node.

3.46.

$$
\begin{bmatrix}
1 & 0 & 0 & 0 & 0 & 0 & 0 & 0 \\
0 & 0 & 0 & 0 & 1 & 0 & 0 & 0 \\
0 & 1 & 0 & 0 & 0 & 0 & 0 & 0 \\
0 & 0 & 0 & 0 & 0 & 1 & 0 & 0 \\
0 & 0 & 1 & 0 & 0 & 0 & 0 & 0 \\
0 & 0 & 0 & 0 & 0 & 0 & 1 & 0 \\
0 & 0 & 0 & 1 & 0 & 0 & 0 & 0 \\
0 & 0 & 0 & 0 & 0 & 0 & 0 & 1
\end{bmatrix}
\begin{Bmatrix}
C_1 \\ C_3 \\ C_5 \\ C_7 \\ C_2 \\ C_4 \\ C_6 \\ C_8
\end{Bmatrix}
=
\begin{Bmatrix}
C_1 \\ C_2 \\ C_3 \\ C_4 \\ C_5 \\ C_6 \\ C_7 \\ C_8
\end{Bmatrix}
$$

3.47.

$$
\begin{bmatrix}
1 & 0 & 0 & 0 & 0 & 0 \\
0 & 0 & 0 & 1 & 0 & 0 \\
0 & 1 & 0 & 0 & 0 & 0 \\
0 & 0 & 0 & 0 & 1 & 0 \\
0 & 0 & 1 & 0 & 0 & 0 \\
0 & 0 & 0 & 0 & 0 & 1
\end{bmatrix}
\begin{Bmatrix}
u_1 \\ u_2 \\ u_3 \\ v_1 \\ v_2 \\ v_3
\end{Bmatrix}
=
\begin{Bmatrix}
u_1 \\ v_1 \\ u_2 \\ v_2 \\ u_3 \\ v_3
\end{Bmatrix}
$$

3.48. Element III is described using nodes 1, 2, and 3; $a_1 = 2$, $a_2 = 0$, $a_3 = -1$, $b_1 = -1$, $b_2 = 1$, $b_3 = 0$, $c_1 = 0$, $c_2 = -1$, $c_3 = 1$, $r_{\text{avg}} = (1 + 2 + 2)/3 = \frac{5}{3}$, $z_{\text{avg}} = (1 + 1 + 2)/3 = \frac{4}{3}$, $A = \frac{1}{2}$.

$$
[B] =
\begin{bmatrix}
-1 & 1 & 1 & 0 & 0 & 0 \\
\frac{1}{5} & 0 & \frac{1}{5} & 0 & \frac{1}{5} & 0 \\
0 & 0 & 0 & -1 & 0 & 1 \\
0 & -1 & -1 & 1 & 1 & 0
\end{bmatrix}
\frac{1}{2(\frac{1}{2})}
$$

3.49.

$$
[N] =
\begin{bmatrix}
N_1 & N_2 & N_3 & N_4 \\
N_1 & N_2 & N_3 & N_4
\end{bmatrix}
$$

$$
[B] =
\begin{bmatrix}
\partial/\partial r \\
(1/r)\,\partial/\partial\theta
\end{bmatrix}
[N_1 \quad N_2 \quad N_3 \quad N_4]
$$

$$
=
\begin{bmatrix}
\partial N_1/\partial r & \partial N_2/\partial r & \partial N_3/\partial r & \partial N_4/\partial r \\
(1/r)\,\partial N_1/\partial\theta & (1/r)\,\partial N_2/\partial\theta & (1/r)\,\partial N_3/\partial\theta & (1/r)\,\partial N_4/\partial\theta
\end{bmatrix}
$$

$$
\int_A ([B]^T[D][B]\{C\} + [N]^T[u][B]\{C\} + [N]^T[k_r][N]\{C\})tr\,dr\,d\theta
$$

where $[D]$, $[u]$, and $[k_r]$ have definitions similar to those in Prob. 3.32.

Chapter 4

Beam and Frame Finite Elements

4.1. INTRODUCTION

Beam finite elements are related directly to *matrix analysis of structures*. Additionally, the term *displacement method* or *stiffness method* is used within the field of study known as theory of structures. These methods of structural analysis are, historically, the forerunner of the finite element method. This chapter will serve as an introduction to beam finite elements. The emphasis is on the connection between the beam finite element and elementary beam theory as it is presented in any standard textbook on *mechanics of materials*. The reader who is well versed in matrix analysis of structures can proceed to the next chapter. However, the reader who is unfamiliar with beam analysis will find this chapter to be an introduction to the study of beam finite elements.

The primary difference between the beam finite element and the finite elements of the previous chapters is the order of the governing differential equation. Beam theory is based upon a fourth-order ordinary differential equation, whereas the differential equations of the previous chapters were second order. An elementary comparison can be established: the second-order ordinary differential equations of Chap. 2 require two boundary conditions to describe a physical situation, while the fourth-order equation requires four boundary conditions. The situation is confusing because the use of the boundary conditions is sometimes hidden within a theory that has primary application for statically indeterminate beams and frames.

In textbooks that discuss the beam finite element it is customary to begin with the axially loaded bar element and develop a finite element theory that is applicable to truss analysis. In this chapter the transversely loaded beam will be given major emphasis, and for truss analysis the reader is referred to textbooks that cover the topic in depth. The bar element was derived in Chap. 2 and will be included in the formulation for beams with axial loading later in this chapter. An introduction to free vibration of beams is given in Chap. 7.

4.2. GOVERNING DIFFERENTIAL EQUATION

The governing differential equation for beam deflection in two dimensions is derived in textbooks on mechanics of materials. The beam is shown in Fig. 4-1, and the governing equation is written in terms of deflection of the beam $v(x)$ as

$$EI \frac{d^4 v(x)}{dx^4} = w(x) \tag{4.1}$$

The modulus of elasticity E and second moment (or moment of inertia) of the cross-sectional area I can be functions of x, the axial coordinate, but in the finite element formulation EI is assumed to be constant within any given element. The external loading $w(x)$ applied to the beam is assumed to be positive when directed upward in this formulation. In some derivations of Eq. (4.1) the y axis is assumed positive downward, but for the purpose of this discussion it does not matter which direction is taken. Equation (4.1) is sometimes referred to as describing a small deflection theory for beam analysis. The limitation of the theory is based upon the slope of the beam rather than the deflection. Small deflection is somewhat of a misnomer since the question of "small compared to what" is not describable mathematically. The first derivative of the deflection is called the slope θ and, as illustrated in Fig. 4-1, is small if $\tan \theta \simeq \sin \theta \simeq dv/dx$. It follows that in elementary beam theory the assumption that the slope $dv/dx = \theta$ is of primary importance. The longitudinal axis of the beam before loading and deflection corresponds to the x axis of the coordinate system. The deflected axis of the beam is called the *elastic curve* of the deflected beam. The slope can be further defined as the angle between the original undeflected axis of the beam and the deflected elastic curve. The assumptions that constitute elementary beam theory lead to a *linear* theory for beam analysis. The theory is quite accurate as long as the basic assumptions are not violated. Each derivative of the deflection has a particular name, and the following relationships are important for the analysis of beams.

$$v(x) \qquad\qquad \text{(deflection)}$$

$$\frac{dv(x)}{dx} = \theta(x) \qquad\qquad \text{(slope)}$$

$$EI \frac{d^2v(x)}{d^2x} = M(x) \qquad\qquad \begin{array}{l}\text{(bending moment that is related directly}\\ \text{to the curvature in beam theory)}\end{array} \qquad (4.2)$$

$$EI \frac{d^3v(x)}{d^3x} = \frac{dM(x)}{dx} = V(x) \qquad \text{(transverse shear)}$$

$$EI \frac{d^4v(x)}{d^4x} = \frac{dV(x)}{dx} = w(x) \qquad [\text{load; see Eq. } (4.1)]$$

The differential equations of Eq. (4.2) are based upon a classical *beam sign convention*, and the beam finite element is based upon the so-called *joint sign convention*. The beam sign convention is discussed in Prob. 4.1 and is necessary for successfully developing the relationships for deriving the beam finite element.

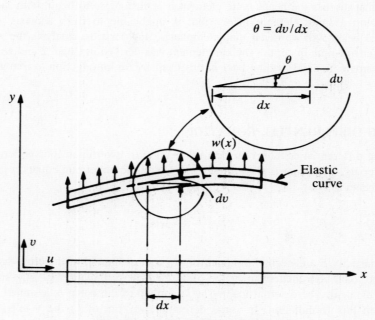

Fig. 4-1 Transversely loaded beam.

4.3.   THE DISPLACEMENT METHOD FOR BEAM ANALYSIS

The displacement method is often referred to as the stiffness method and predates the use of computers in structural analysis. The *three-moment equation* and the *slope-deflection method* were forerunners of the displacement method. In fact, the computer successfully revived these methods of analysis because it

Table 4-1   Equivalent Joint Loadings

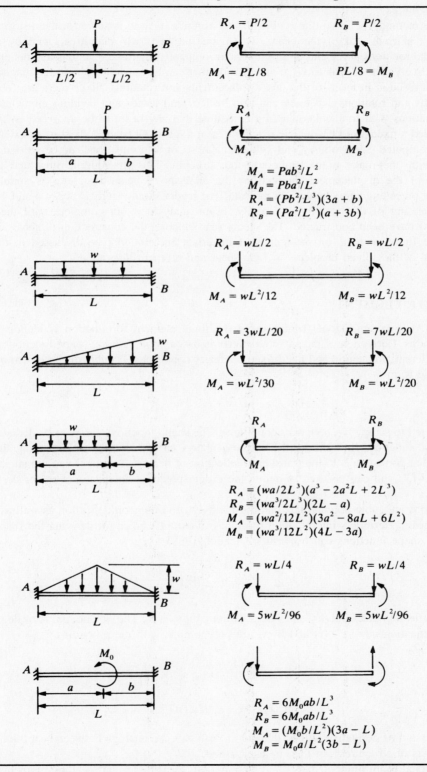

became practical to solve large numbers of simultaneous equations and the modern stiffness method eventually evolved.

The stiffness method, for a two-dimensional beam structure, is used to solve for the displacements and rotations (slopes) at each end of the beam. These are the first two quantities given by Eqs. (4.2). Displacements and rotations at the end of a beam are accompanied by force reactions and bending moment reactions, respectively. The idea is to write an equation that can be solved to obtain the magnitude of the force or moment reaction in terms of the end displacements and rotations. Since there are four unknown end displacements and rotations and there can also be four end forces and bending moments, the result will be four equations with four unknowns. First, assume that any beam to be analyzed is fixed at both ends. This is called a fixed-fixed beam. The reactions for a fixed-fixed beam subject to an external applied loading can be computed (see Prob. 4.3 and Table 4.1). A set of four equations can be written for each of the individual beams that make up the total continuous structure. This procedure is illustrated in Prob. 4.4. The application of the displacement method for the analysis of beam structures is somewhat more involved than the preceding paragraph might indicate. The reader should study Probs. 4.4 and 5.5 to obtain a better understanding of this technique. Usually, beam analysis is not complete until the shear and moment diagrams have been constructed. The theory that supports the construction of shear and moment diagrams can be found in texts on mechanics of materials and will not be discussed in this chapter. However, several of the solved problems include shear and moment diagrams.

4.4. BEAM FINITE ELEMENTS

The analysis of two-dimensional beams using the finite element formulation is identical to matrix analysis of structures. Derivation of the stiffness matrix is based upon defining shape functions that satisfy the governing differential equation and fixed-fixed boundary conditions. A cubic displacement function can be assumed in the form

$$v(x) = a_0 + a_1 x + a_2 x^2 + a_3 x^3 \qquad (4.3)$$

and the a_i computed to satisfy the boundary conditions. The shape functions can then be deduced from the form of the cubic function. In addition, the stiffness matrix can be derived by satisfying the potential energy function that corresponds to the transverse deflection of beams. A general variational function was defined by Eq. (3.13) and can be used to define the proper potential energy function for deriving beam finite elements.

Shape functions are cubic polynomials that must satisfy conditions on deflection as well as rotation at each end of the beam. Hermite's interpolation formula is a natural choice for deriving the shape function. This formula for shape functions can be written (Scheid, 1988)

$$N_i = \sum_{i=0}^{n} U_i(x) v_i + \sum_{i=0}^{n} W_i(x) \frac{dv_i}{dx} \qquad (4.4)$$

where v_i is the deflection at x_i and dv_i/dx is the rotation (slope) at x_i. The polynomial is of degree $2n + 1$, and for a cubic equation $n = 1$. $U_i(x)$ and $W_i(x)$ are polynomials with the properties

$$U_i(x) = \left[1 - 2 \frac{dL(x_i)}{dx} (x - x_i) \right] [L_i(x)]^2 \qquad (4.5)$$

$$W_i(x) = (x - x_i)[L_i(x)]^2 \qquad (4.6)$$

The function $L_i(x)$ and its derivative will be defined in Prob. 4.9. An analogy between beam finite elements and matrix analysis of structures can then be established.

Two-dimensional beam finite elements that include only shear forces and bending moments cannot be used effectively for frame-type problems without the addition of axial forces. The finite element model for axial force acting on a rod or beam was developed in Chap. 2. The basic beam finite element can be modified to include axial force by merely adding two degrees of freedom—the axial deformation at either end of the beam—and is derived in Prob. 4.11. Coordinate transformations as matrix transformations for the combined action of shear forces and axial forces must be understood before applying beam finite element analysis to frames.

4.5. MATRIX TRANSFORMATIONS

Matrix transformations that describe a boundary condition in plane elasticity problems were discussed in Chap. 3. The vector transformations of Eqs. (3.14) and (3.15) are used to transform beam finite elements from a local ξ, η system to a global x, y system. The application given in Prob. 3.23 described the representation of a displacement boundary condition that remained in the local ξ, η system. The application for frame-type problems is somewhat the reverse situation. Problem 4.12 describes the coordinate transformation for a beam element formulated in the local ξ, η system and transformed to the global x, y system so that the boundary conditions can be specified in the x, y system.

Solved Problems

4.1.   The beam shown in Fig. 4-2 (with a positive joint rotation applied at $x = 0$) is important in the derivation of the equations for the stiffness matrix for beam finite elements. It is also useful when interpreting the results of the final beam analysis. Use the differential equation defined by Eq. (4.1) to solve for shear and moment reactions for the fixed-fixed beam of Fig. 4-2.

In this case the differential equation is written using the shorthand notation

$$EIv^{\mathrm{IV}} = w = 0 \qquad \text{or} \qquad v^{\mathrm{IV}} = 0 \tag{a}$$

The beam of Fig. 4-2 is statically indeterminant. There are four unknown external reactions, and there are only two equations available from statics. The fact that there are two additional unknown reactions requires two boundary conditions to solve the differential equations. Integrating Eq. (a),

$$v^{\mathrm{III}} = C_1 \tag{b}$$

A boundary condition at $x = 0$ is that the shear is the unknown reaction R_1. Using Eq. (4-2) it follows that $EIv^{\mathrm{III}}(0) = V = R_1$. Actually, in this application the shear V is constant for $0 \leq x \leq L$. Substituting into Eq. (b) gives

$$v^{\mathrm{III}} = \frac{R_1}{EI} \tag{c}$$

Integrating again gives

$$v^{\mathrm{II}} = R_1 \frac{x}{EI} + C_2 \tag{d}$$

The boundary condition, using Fig. 4-2, is $EIv^{\mathrm{II}}(0) = -M_1$. Substituting into Eq. (d) and solving for C_2 gives

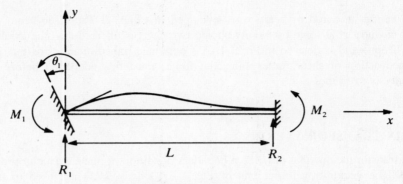

(a) Applied rotation θ_1 at $x = 0$

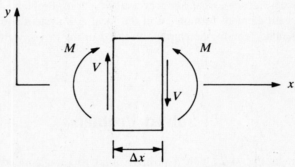

(b) Sign convention for shear and moment

Fig. 4-2

$$v^{II} = R_1\,\frac{x}{EI} - \frac{M_1}{EI} \qquad\qquad (e)$$

A discussion of sign conventions is in order for the reader who is not familiar with beam theory. The beam sign convention was probably originally developed for successful application of the last three equations given by Eqs. (4-2), that is, the relation among applied load w, transverse shear V, and bending moment M. The sign convention for beam theory [the sign convention for solving Eq. (4.1)] is illustrated in Fig. 4-2(b). The bending moments applied on either side of the beam element are assumed positive and act to cause compression at the top of the beam. Positive shear is a force that is assumed to act downward on the positive side of the beam element. Similarly, positive shear acts upward on the negative side of the beam element. This sign convention is foolproof as long as the positive x axis is directed to the right. The y axis can be directed either up or down. The reader must keep in mind that for the y axis directed upward an applied load directed downward would be negative when using Eq. (4.1). Therefore, the moment M_1 that is applied at $x = 0$ in Fig. 4-2 is negative when used as a boundary condition in Eq. (d).

The joint sign convention is used for the matrix analysis of structures and for beam finite elements. It is assumed that moments that rotate a joint or support in a *counterclockwise direction* are positive and that positive forces acting on the joint act *upward* in the positive y direction. In this and future beam problems it is desirable to assume that forces and moments are positive according to the joint sign convention.

Integrating Eq. (e) twice gives an equation for the slope and an equation for the deflection:

$$v^I = \frac{R_1 x^2}{2EI} - \frac{M_1 x}{EI} + C_3 \qquad\qquad (f)$$

$$v = \frac{R_1 x^3}{6EI} - \frac{M_1 x^2}{2EI} + C_3 x + C_4 \qquad\qquad (g)$$

The most convenient boundary conditions to use for computing C_3 and C_4 are those that occur at $x = 0$,

$v^{1}(0) = \theta_1$, the assumed joint rotation, and $v(0) = 0$. Substituting into Eqs. (f) and (g) and solving for C_3 and C_4 gives $C_3 = \theta_1$ and $C_4 = 0$. The conditions at the boundary $x = L$, $v^{1}(L) = 0$, and $v(L) = 0$ can be used with Eqs. (f) and (g) to give two equations that can be solved for R_1 and M_1. It follows that

$$R_1 = \frac{6EI\theta_1}{L^2} \quad \text{and} \quad M_1 = \frac{4EI\theta_1}{L} \tag{h}$$

There are two additional equations that can be used to solve for R_2 and M_2. The reader who is familiar with statics will recognize that $\Sigma F_y = 0$ and $\Sigma M_z = 0$. Otherwise, one can merely substitute $EIv^{III}(L) = V = R_2$ into Eq. (c) and $EIv^{II}(L) = M_2$ into Eq. (e) and compute

$$R_2 = -\frac{6EI\theta_1}{L^2} \quad \text{and} \quad M_2 = \frac{2EI\theta_1}{L} \tag{i}$$

For completeness the equations of statics are (see Fig. 4-2)

$$\Sigma F_y = R_1 + R_2 = 0 \quad \text{and} \quad \Sigma M_z(0) = M_1 + M_2 + R_2 L = 0$$

4.2. The beam shown in Fig. 4-3 with a positive joint displacement applied at $x = 0$ is important in the derivation of the equations for the stiffness matrix for beam finite elements. Use the differential equation defined by Eq. (4.1) to solve for shear and moment reactions for the fixed-fixed beam of Fig. 4-3. Note that reaction R_2 is assumed downward.

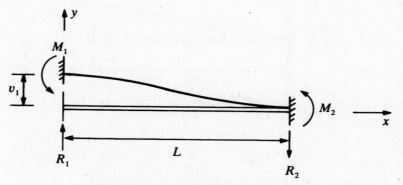

Fig. 4-3　Applied displacement at $x = 0$.

Beginning with Eq. (4.1) and integrating twice results in

$$v^{IV} = 0$$

$$v^{III} = C_1 \tag{a}$$

$$v^{II} = C_1 x + C_2 \tag{b}$$

The appropriate boundary conditions are $EIv^{III}(0) = R_1$ and $EIv^{II}(0) = -M_1$. Substituting into Eqs. (a) and (b) and solving for the constants gives

$$v^{II} = R_1 \frac{x}{EI} - \frac{M_1}{EI} \tag{c}$$

Integrating again gives

$$v^{1} = \frac{R_1 x^2}{2EI} - \frac{M_1 x}{EI} + C_3 \tag{d}$$

$$v = \frac{R_1 x^3}{6EI} - \frac{M_1 x^2}{2EI} + C_3 x + C_4 \qquad (e)$$

The boundary conditions are $v^{\text{I}}(0) = 0$ and $v(0) = v_1$, the assumed joint displacement. The constants are computed as $C_3 = 0$ and $C_4 = v_1$. The remaining boundary conditions $v^{\text{I}}(L) = 0$ and $v(L) = 0$ can be used with Eqs. (d) and (e) to obtain two equations that can be solved for R_1 and M_1. It follows that

$$R_1 = \frac{12EIv_1}{L^3} \qquad \text{and} \qquad M_1 = \frac{6EIv_1}{L^2} \qquad (f)$$

The equations of statics can be used to compute M_2 and R_2:

$$R_2 = \frac{12EIv_1}{L^3} \qquad \text{and} \qquad M_2 = \frac{6EIv_1}{L^2} \qquad (g)$$

These results are summarized in Table 4.1.

4.3. The fixed-fixed beam shown in Fig. 4-4 has a continuous loading that can be described as $w(x) = -wx/L$. Compute the shear and moment reactions.

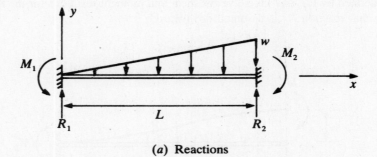

(a) Reactions

$$R_1 = 3wL/20 \qquad R_2 = 7wL/20$$

$$M_1 = wL^2/30 \qquad M_2 = wL^2/20$$

(b) Equivalent joint loading

Fig. 4-4

The fourth-order equation, Eq. (4.1), will be used as

$$EIv^{\text{IV}} = -\frac{wx}{L}$$

Integrating the governing equation twice results in

$$EIv^{\text{III}} = -\frac{wx^2}{2L} + C_1 \qquad (a)$$

$$EIv^{\text{II}} = -\frac{wx^3}{6L} + C_1 x + C_2 \qquad (b)$$

Use the boundary conditions $EIv^{III}(0) = R_1$ and $EIv^{II}(0) = -M_1$ to compute the constants of integration as $C_1 = R_1$ and $C_2 = -M_1$. Substitute into Eqs. (a) and (b) and integrate twice more:

$$EIv^I = -\frac{wx^4}{24L} + \frac{R_1 x^2}{2} - M_1 x + C_3 \qquad\qquad (c)$$

$$EIv = -\frac{wx^5}{120L} + \frac{R_1 x^3}{6} - \frac{M_1 x^2}{2} + C_3 x + C_4 \qquad\qquad (d)$$

The boundary conditions $v^I(0) = 0$ and $v(0) = 0$ can be used to evaluate $C_3 = C_4 = 0$. In addition, the boundary conditions $v^I(L) = 0$ and $v(L) = 0$ can be used to construct two equations to compute R_1 and M_1:

$$-\frac{wL^3}{24} + \frac{R_1 L^2}{2} - M_1 L = 0 \qquad \text{and} \qquad -\frac{wL^4}{120} + \frac{R_1 L^3}{6} - \frac{M_1 L^2}{2} = 0$$

Solving gives $M_1 = wL^2/30$ and $R_1 = 3wL/20$. The equations of statics can be used to compute $M_2 = wL^2/20$ and $R_2 = 7wL/20$. The results shown in Fig. 4-4 are the reactions caused by the external load. When these results are used in later applications in this chapter, the *reactions will be reversed and considered equivalent joint loadings to replace the external beam loading.*

4.4. The beam and the notation of Fig. 4-5(a) can be used to identify the end displacements and rotations for a beam subjected to the end forces and bending moments shown in Fig. 4-5(b). Construct a set of four equations that relate each force or moment action of Fig. 4-5(b) to the displacements and rotations of Fig. 4-5(a).

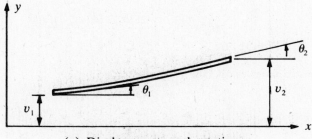

(a) Displacements and rotations

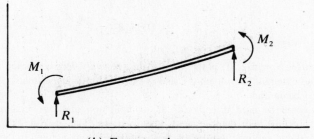

(b) Forces and moments

Fig. 4-5

It is desirable, first, to write an equation that relates R_1 to each of the displacements and rotations of Fig. 4-5(a), or

$$R_1 = f(v_1, \theta_1, v_2, \theta_2) \qquad\qquad (a)$$

Assume, before assigning boundary conditions to the beam problem, that both supports are fixed. The function

$f(v_1)$ of Eq. (a) is evaluated in terms of a displacement applied at the left end of the beam. The analysis was carried out in Prob. 4.2, and that result can be used. Similarly, R_1 is affected by a rotation $f(\theta_1)$ at the left end of the beam, and that result is given by Prob. 4.1. The remaining functions are computed in Probs. 4.15 and 4.16. The function of Eq. (a) is written as

$$R_1 = \frac{12EIv_1}{L^3} + \frac{6EI\theta_1}{L^2} - \frac{12EIv_2}{L^3} + \frac{6EI\theta_2}{L^2} \tag{b}$$

Similarly,

$$M_1 = f(v_1, \theta_1, v_2, \theta_2)$$

and the functions are again evaluated using the results given by Probs. 4.1, 4.2, 4.15, and 4.16:

$$M_1 = \frac{6EIv_1}{L^2} + \frac{4EI\theta_1}{L} - \frac{6EIv_2}{L^2} + \frac{2EI\theta_2}{L} \tag{c}$$

The reactions are evaluated at the right end of the beam in a similar manner:

$$R_2 = -\frac{12EIv_1}{L^3} - \frac{6EI\theta_1}{L^2} + \frac{12EIv_2}{L^3} - \frac{6EI\theta_2}{L^2} \tag{d}$$

$$M_2 = \frac{6EIv_1}{L^2} + \frac{2EI\theta_1}{L} - \frac{6EIv_2}{L^2} + \frac{4EI\theta_2}{L} \tag{e}$$

A more compact matrix statement of Eqs. (b)–(e) is

$$\begin{Bmatrix} R_1 \\ M_1 \\ R_2 \\ M_2 \end{Bmatrix} = \frac{EI}{L^3} \begin{bmatrix} 12 & 6L & -12 & 6L \\ 6L & 4L^2 & -6L & 2L^2 \\ -12 & -6L & 12 & -6L \\ 6L & 2L^2 & -6L & 4L^2 \end{bmatrix} \begin{Bmatrix} v_1 \\ \theta_1 \\ v_2 \\ \theta_2 \end{Bmatrix} \tag{f}$$

or
$$\{f\} = [K]\{v\} \tag{g}$$

where [K] is the stiffness matrix, {f} is the *force* matrix of *equivalent* computed shear and moment end actions (dependent upon the external beam loading), and {v} is the *displacement* matrix of the unknown displacements and rotations of the beam ends.

Equation (g) could be used directly for the computation of beam reactions if all beam loadings were specified as corresponding joint displacements and rotations. The analysis is carried out in two parts. First, equivalent fixed-fixed end actions representing the beam loading are applied to the ends (joints) of the beam, and that becomes the left-hand side of Eq. (f). Second, the actual reactions are computed as the sum of the equivalent joint forces and moments plus the forces and moments caused by the joint displacements and rotations. Problem 4.5 illustrates this procedure for a simple beam.

4.5.   The beam of Fig. 4-6 is statically determinate, and computations for the reactions R_A and R_B are elementary using the equations of statics. However, the beam will serve the purpose of illustrating the displacement method. Use the displacement method to compute the reactions for the beam of Fig. 4-6.

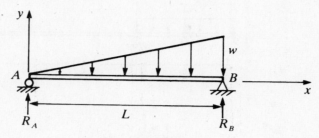

Fig. 4-6

Assume the beam is fixed-fixed and compute equivalent joint loadings to represent the uniformly varying load. Equivalent joint loadings are given by the results computed in Prob. 4.3 or can be obtained from Table 4.1. Substitute into Eq. (f) of Prob. 4.4 to compute joint rotations.

$$\begin{Bmatrix} -3wL/20 \\ -wL^2/30 \\ -7wL/20 \\ wL^2/20 \end{Bmatrix} = \frac{EI}{L^3} \begin{bmatrix} 12 & 6L & -12 & 6L \\ 6L & 4L^2 & -6L & 2L^2 \\ -12 & -6L & 12 & -6L \\ 6L & 2L^2 & -6L & 4L^2 \end{bmatrix} \begin{Bmatrix} v_1 \\ \theta_1 \\ v_2 \\ \theta_2 \end{Bmatrix} \qquad (a)$$

Equation (a) is an exact model of the fixed-fixed beam, and if solved for all joint displacements and rotations, they will compute as zero. Apply the boundary conditions $v(0) = v(L) = 0$ to make the fixed-fixed beam appear as the simple beam that is being analyzed. The given boundary conditions are deleted from Eq. (a) to give the matrix equation

$$\begin{Bmatrix} -wL^2/30 \\ wL^2/20 \end{Bmatrix} = \frac{EI}{L^3} \begin{bmatrix} 4L^2 & 2L^2 \\ 2L^2 & 4L^2 \end{bmatrix} \begin{Bmatrix} \theta_1 \\ \theta_2 \end{Bmatrix} \qquad (b)$$

Solving Eq. (b) gives the rotations at each joint:

$$\theta_1 = -\frac{7wL^3}{360EI} \qquad \text{and} \qquad \theta_2 = \frac{wL^3}{45EI} \qquad (c)$$

The final reactions are computed as the reactions caused by the joint rotations plus the applied joint loadings that replaced the applied beam loading:

$$\begin{Bmatrix} R_A \\ M_A \\ R_B \\ M_B \end{Bmatrix} = \frac{EI}{L^3} \begin{bmatrix} 12 & 6L & -12 & 6L \\ 6L & 4L^2 & -6L & 2L^2 \\ -12 & -6L & 12 & -6L \\ 6L & 2L^2 & -6L & 4L^2 \end{bmatrix} \begin{Bmatrix} 0 \\ -7wL^3/36EI \\ 0 \\ wL^3/45EI \end{Bmatrix} + \begin{Bmatrix} 3wL/20 \\ wL^2/30 \\ 7wL/20 \\ -wL^2/20 \end{Bmatrix} \qquad (d)$$

The matrix equation is evaluated to give the final results:

$$R_A = \frac{wL}{6} \qquad R_B = \frac{wL}{3} \qquad M_A = M_B = 0$$

4.6. The two-span beam of Fig. 4-7(a) is fixed at both ends and supported between the ends with a simple support that allows rotation. Compute the rotation at the simple support and reactions at all supports. Construct the corresponding shear and moment diagrams.

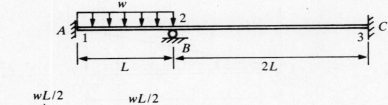

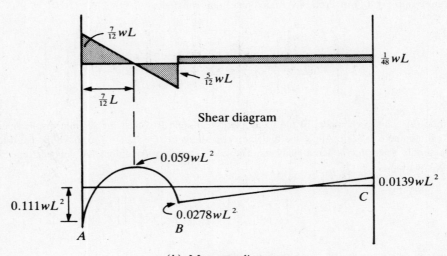

(a) Equivalent joint loading, beam 1-2

(b) Moment diagram

Fig. 4-7

Two local elements will be used for this structure and combined to form a global stiffness matrix. The beams have a common rotation and displacement at support 2, and the connectivity of the two beam elements will occur at that support. The local stiffness matrices will be constructed first using Eq. (f) of Prob. 4.4 as a model. The equivalent joint loads for the uniform load of beam 1-2 are obtained from Table 4.1.

Beam 1-2:

$$\begin{Bmatrix} -wL/2 \\ -wL^2/12 \\ -wL/2 \\ wL^2/12 \end{Bmatrix} = \frac{EI}{L^3} \begin{bmatrix} 12 & 6L & -12 & 6L \\ 6L & 4L^2 & -6L & 2L^2 \\ -12 & -6L & 12 & -6L \\ 6L & 2L^2 & -6L & 4L^2 \end{bmatrix} \begin{Bmatrix} v_1 \\ \theta_1 \\ v_2 \\ \theta_2 \end{Bmatrix} \qquad (a)$$

Beam 2-3:

$$\begin{Bmatrix} 0 \\ 0 \\ 0 \\ 0 \end{Bmatrix} = \frac{EI}{(2L)^3} \begin{bmatrix} 12 & 6(2L) & -12 & 6(2L) \\ 6(2L) & 4(2L)^2 & -6(2L) & 2(2L)^2 \\ -12 & -6(2L) & 12 & -6(2L) \\ 6(2L) & 2(2L)^2 & -6(2L) & 4(2L)^2 \end{bmatrix} \begin{Bmatrix} v_2 \\ \theta_2 \\ v_3 \\ \theta_3 \end{Bmatrix} \qquad (b)$$

The two local stiffness matrices are combined through v_2 and θ_2 to give a 6×6 global matrix. Each local stiffness matrix is defined in the matrix of Eq. (c), and the connectivity is illustrated by the matrix elements grouped inside the center box.

$$
\frac{EI}{L^3}
\begin{bmatrix}
12 & 6L & -12 & 6L & 0 & 0 \\
6L & 4L^2 & -6L & 2L^2 & 0 & 0 \\
-12 & -6L & 12+12/8 & -6L+12L/8 & -12/8 & 12L/8 \\
6L & 2L^2 & -6L+12L/8 & 4L^2+16L^2/8 & 12L/8 & 8L^2/8 \\
0 & 0 & -12/8 & -12/8 & 12/8 & -12/8 \\
0 & 0 & 12L/8 & 8L^2/8 & -12L/8 & 16L^2/8
\end{bmatrix}
\begin{Bmatrix}
v_1 \\ \theta_1 \\ v_2 \\ \theta_2 \\ v_3 \\ \theta_3
\end{Bmatrix}
=
\begin{Bmatrix}
-wL/2 \\ -wL^2/12 \\ -wL/2 \\ wL^2/12 \\ 0 \\ 0
\end{Bmatrix}
\tag{c}
$$

The boundary conditions for the beam of Fig. 4-7 are $v_1 = v_2 = v_3 = \theta_1 = \theta_3 = 0$. The solution reduces to one equation and one unknown θ_2. Visualize that all rows and columns of Eq. (c) corresponding to the zero boundary conditions are removed, and the result is

$$
\frac{EI}{L^3}(4L^2 + 2L^2)\theta_2 = \frac{wL^2}{12} \qquad \text{or} \qquad \theta_2 = \frac{wL^3}{72EI}
\tag{d}
$$

The results for each span must be computed individually using the local stiffness matrix for that span. The reactions for span 1-2 are V_A, V_B, M_A, and M_B. The stiffness matrix is the same as Eq. (a), and the computation appears as

$$
\begin{Bmatrix}
V_A \\ M_A \\ V_{B1} \\ M_B
\end{Bmatrix}
=
\frac{EI}{L^3}
\begin{bmatrix}
12 & 6L & -12 & 6L \\
6L & 4L^2 & -6L & 2L^2 \\
-12 & -6L & 12 & -6L \\
6L & 2L^2 & -6L & 4L^2
\end{bmatrix}
\begin{Bmatrix}
0 \\ 0 \\ 0 \\ wL^3/72EI
\end{Bmatrix}
+
\begin{Bmatrix}
wL/2 \\ wL^2/12 \\ wL/2 \\ -wL^2/12
\end{Bmatrix}
\tag{e}
$$

Multiplication gives

$$
V_A = R_A = \frac{7wL}{12} \qquad V_{B1} = \frac{5wL}{12} \qquad M_A = \frac{wL^2}{9} \qquad M_B = -\frac{wL^2}{36}
$$

The notation V_{B1} requires some additional explanation. The total reaction at support B is composed of the end shear at point 2 of beam 1-2 plus the end shear at point 2 of beam 2-3. Then, $R_B = V_{B1} + V_{B2}$.

Similarly, the stiffness matrix of Eq. (b) is used to compute the shears and moments for beam 2-3.

$$
\begin{Bmatrix}
V_{B2} \\ M_B \\ V_C \\ M_C
\end{Bmatrix}
=
\frac{EI}{L^3}
\begin{bmatrix}
1.5 & 1.5L & -1.5 & 1.5L \\
1.5L & 2L^2 & -1.5L & L^2 \\
-1.5 & -1.5L & 1.5 & -1.5L \\
1.5L & L^2 & -1.5L & 2L^2
\end{bmatrix}
\begin{Bmatrix}
0 \\ wL^3/72EI \\ 0 \\ 0
\end{Bmatrix}
+
\begin{Bmatrix}
0 \\ 0 \\ 0 \\ 0
\end{Bmatrix}
\tag{f}
$$

$$
V_{B2} = \frac{wL}{48} \qquad V_C = R_C = -\frac{wL}{48} \qquad M_B = \frac{wL^2}{36} \qquad M_C = \frac{wL^2}{72} \qquad R_B = V_{B1} + V_{B2}
$$

Internal supports such as joint C should always have the same numerical value of bending moment but with opposite signs, thus indicating equilibrium at the joint. The bending moments obtained from the stiffness analysis must be converted from the joint sign convention to the beam sign convention to construct shear and moment diagrams. Review Fig. 4-2 for the beam sign convention and note that M_A is counterclockwise using the joint sign convention and is a negative moment using the beam sign convention. Similarly, M_B is a negative moment and M_C is a positive moment using the beam sign convention. Shear and moment diagrams are shown in Fig. 4-7(b).

4.7. The two-span beam structure of Fig. 4-8(a) is free to rotate at supports A and B and is fixed at joint C. Compute the rotations at supports A and B and the reactions at all supports. Construct shear and moment diagrams.

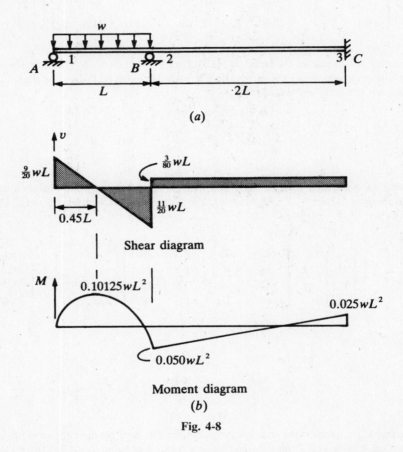

(a)

Shear diagram

Moment diagram

(b)

Fig. 4-8

The structure is identical to the one that was analyzed in Prob. 4.6 except that the support at A is free to rotate. There are two unknowns, θ_1 and θ_2. The stiffness matrices are identical to Eqs. (a)–(c) of Prob. 4.6, and the boundary conditions are $v_1 = v_2 = v_3 = \theta_3 = 0$. Equation (c) of Prob. 4.6 reduces to two equations and two unknowns:

$$\frac{EI}{L^3}\begin{bmatrix} 4L^2 & 2L^2 \\ 2L^2 & 4L^2 \end{bmatrix}\begin{Bmatrix} \theta_1 \\ \theta_2 \end{Bmatrix} = \begin{Bmatrix} -wL^2/12 \\ wL^2/12 \end{Bmatrix} \qquad (a)$$

Solving Eq. (a) gives $\theta_1 = -wL^3/30EI$ and $\theta_2 = wL^3/40EI$. Joint 1 rotates clockwise, and joint 2 rotates counterclockwise.

Use the stiffness matrix of Eq. (a) given in Prob. 4.6 to formulate the matrix equation that can be solved for the reactions at the ends of beam 1-2:

$$\begin{Bmatrix} V_A \\ M_A \\ V_{B1} \\ M_B \end{Bmatrix} = \frac{EI}{L^3} \begin{bmatrix} 12 & 6L & -12 & 6L \\ 6L & 4L^2 & -6L & 2L^2 \\ -12 & -6L & 12 & -6L \\ 6L & 2L^2 & -6L & 4L^2 \end{bmatrix} \begin{Bmatrix} 0 \\ -wL^3/30EI \\ 0 \\ wL^3/40EI \end{Bmatrix} + \begin{Bmatrix} wL/2 \\ wL^2/12 \\ wL/2 \\ -wL^2/12 \end{Bmatrix} \qquad (b)$$

The reactions are computed as

$$V_A = R_A = \frac{9wL}{20} \qquad V_{B1} = \frac{11wL}{20} \qquad M_A = 0 \qquad M_B = -\frac{wL^2}{20}$$

Similarly, the stiffness matrix of Eq. (b) of Prob. 4.6 is used to compute the reactions for beam 1-2:

$$\begin{Bmatrix} V_{B2} \\ M_B \\ V_C \\ M_C \end{Bmatrix} = \frac{EI}{L^3} \begin{bmatrix} 1.5 & 1.5L & -1.5 & 1.5L \\ 1.5L & 2L^2 & -1.5L & L^2 \\ -1.5 & -1.5L & 1.5 & -1.5L \\ 1.5L & L^2 & -1.5L & 2L^2 \end{bmatrix} \begin{Bmatrix} 0 \\ wL^3/40EI \\ 0 \\ 0 \end{Bmatrix} + \begin{Bmatrix} 0 \\ 0 \\ 0 \\ 0 \end{Bmatrix} \qquad (c)$$

$$V_{B2} = \frac{3wL}{80} \qquad V_C = R_C = -\frac{3wL}{80} \qquad M_B = \frac{wL^2}{20} \qquad M_C = \frac{wL^2}{40} \qquad R_B = V_{B1} + V_{B2}$$

The shear and moment diagrams are shown in Fig. 4-8(b).

4.8. The beam structure of Fig. 4-9(a) is sometimes called a propped cantilever beam. Compute the rotations θ_1 and θ_2 and the deflection v_1. Then compute the reactions at B and C. Construct the shear and moment diagrams.

The beam structure is similar to that of Prob. 4.6 but has different boundary conditions. The span from A to

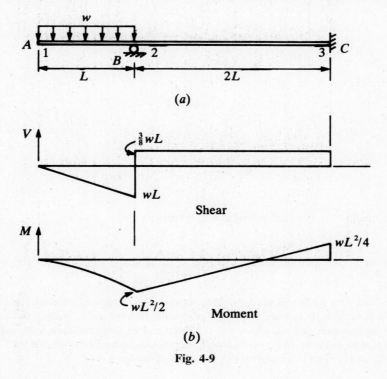

Fig. 4-9

B is free to rotate and deflect and can be analyzed by merely specifying the proper boundary conditions for the global matrix formulation given in Eq. (c) of Prob. 4.6. In the theory of matrix analysis of structures the overhanging span, A to B, can be replaced by an equivalent shear force and moment acting at B. The method illustrated here corresponds to the method used in computer applications; merely specify the proper boundary conditions.

The matrix formulation of Eq. (c) of Prob. 4.6 will be used with boundary conditions $v_2 = v_3 = \theta_3 = 0$. Delete the rows and columns that correspond to the zero boundary conditions and the result is the following three equations.

$$\frac{EI}{L^3}\begin{bmatrix} 12 & 6L & 6L \\ 6L & 4L^2 & 2L^2 \\ 6L & 2L^2 & 4L^2 \end{bmatrix}\begin{Bmatrix} v_1 \\ \theta_1 \\ \theta_2 \end{Bmatrix} = \begin{Bmatrix} -wL/2 \\ -wL^2/12 \\ wL^2/12 \end{Bmatrix} \tag{a}$$

Solving Eq. (a) gives $\theta_1 = 5wL^3/12EI$, $\theta_2 = wL^3/4EI$, and $v_1 = -3wL^4/8EI$. Both joints rotate counterclockwise, and the deflection is downward. The final results for reactions are computed using the local stiffness matrices of Prob. 4.6.

$$\begin{Bmatrix} V_A \\ M_A \\ V_{B1} \\ M_B \end{Bmatrix} = \frac{EI}{L^3}\begin{bmatrix} 12 & 6L & -12 & 6L \\ 6L & 4L^2 & -6L & 2L^2 \\ -12 & -6L & 12 & -6L \\ 6L & 2L^2 & -6L & 4L^2 \end{bmatrix}\begin{Bmatrix} -3wL^4/8EI \\ 5wL^3/12EI \\ 0 \\ wL^3/4EI \end{Bmatrix} + \begin{Bmatrix} wL/2 \\ wL^2/12 \\ wL/2 \\ -wL^2/12 \end{Bmatrix} \tag{b}$$

and

$$\begin{Bmatrix} V_{B2} \\ M_B \\ V_C \\ M_C \end{Bmatrix} = \frac{EI}{L^3}\begin{bmatrix} 1.5 & 1.5L & -1.5 & 1.5L \\ 1.5L & 2L^2 & -1.5L & L^2 \\ -1.5 & -1.5L & 1.5 & -1.5L \\ 1.5L & L^2 & -1.5L & 2L^2 \end{bmatrix}\begin{Bmatrix} 0 \\ wL^3/4EI \\ 0 \\ 0 \end{Bmatrix} + \begin{Bmatrix} 0 \\ 0 \\ 0 \\ 0 \end{Bmatrix} \tag{c}$$

Solving Eqs. (b) and (c) results in

$$V_A = R_A = 0 \qquad V_{B1} = wL \qquad V_{B2} = \frac{3wL}{8} \qquad V_C = R_C = -\frac{3wL}{8} \qquad M_A = 0 \qquad M_B = \frac{wL^2}{2}$$

$$M_C = \frac{wL^2}{4} \qquad R_B = V_{B1} + V_{B2} = \frac{11wL}{8}$$

Shear and moment diagrams are shown in Fig. 4-9(b).

4.9. Use Hermite's interpolation formula to derive cubic shape functions for the transverse deflection of beams.

Shape functions for beams must reflect the behavior of the possible boundary conditions on deflection and rotation at each joint. It follows that just any one-dimensional cubic shape function may not be adequate. Hermite's interpolation formula allows the deflection and its first derivative to be satisfied at each joint. The shape function N_1 is shown in Fig. 4-10(a) with boundary conditions $v(0) = 1$ and $v(L) = v'(0) = v'(L) = 0$. The functions L_i of Eqs. (4.5) and (4.6) are defined as follows for $n = 1$:

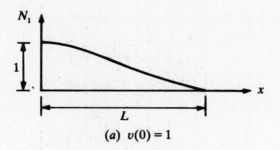

(a) $v(0) = 1$

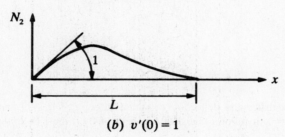

(b) $v'(0) = 1$

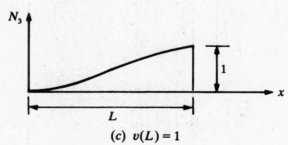

(c) $v(L) = 1$

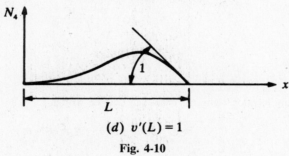

(d) $v'(L) = 1$

Fig. 4-10

$$L_0(x) = \frac{x - x_L}{x_0 - x_L} \qquad L_L(x) = \frac{x - x_0}{x_L - x_0} \tag{a}$$

where $x_0 = 0$ and $x_L = L$. Substituting into Eq. (a) gives

$$L_0(x) = \frac{x - L}{-L} \qquad L_L(x) = \frac{x}{L} \qquad \frac{dL_0(x)}{dx} = \frac{-1}{L} \qquad \frac{dL_L(x)}{dx} = \frac{1}{L} \tag{b}$$

The boundary conditions require that only one term of Eq. (4.4) must be evaluated. The shape function N_1 has $v(0) = 1$ as the only nonzero boundary condition, and that requires that only $U_i(x) = U_0(x)$ of Eq. (4.5) be evaluated:

$$U_0(x) = \left[1 - 2\left(\frac{-1}{L}\right)x \right]\left[\frac{(x - L)^2}{L^2} \right]$$

Substituting into Eq. (4.4) with $v(0) = 1$ and rearranging gives the shape function

$$N_1 = 1 - 3\frac{x^2}{L^2} + 2\frac{x^3}{L^3} \qquad (c)$$

Shape function N_2 satisfies the slope condition at $x = 0$, $v'(0) = 1$, and the remaining conditions are $v(0) = v(L) = v'(L) = 0$. It follows that only one term of $W_i(x)$ must be evaluated in Eq. (4.4):

$$W_0(x) = (x)\left[\frac{(x-L)^2}{L^2}\right]$$

and Eq. (4.4) gives

$$N_2 = x\left(1 - 2\frac{x}{L} + \frac{x^2}{L^2}\right) \qquad (d)$$

Shape function N_3 satisfies the deflection condition at $x = L$, $v(L) = 1$, and the remaining conditions are $v(0) = v'(0) = v'(L) = 0$. It follows that only one term of $U_i(x)$ must be evaluated in Eq. (4.4):

$$U_L(x) = \left[1 - 2\left(\frac{1}{L}\right)(x - L)\right]\left(\frac{x^2}{L^2}\right)$$

Evaluating Eq. (4.4) gives

$$N_3 = 3\frac{x^2}{L^2} - 2\frac{x^3}{L^3} \qquad (e)$$

Similarly, N_4 must satisfy $v'(L) = 1$ and $v(0) = v(L) = v'(0) = 0$. Equation (4.6) gives

$$W_L(x) = \frac{(x - L)x^2}{L^2}$$

and

$$N_4 = x\left(\frac{x^2}{L^2} - \frac{x}{L}\right) \qquad (f)$$

The deflection at any location along the beam can now be written

$$v(x) = N_1 v_1 + N_2 \theta_1 + N_3 v_2 + N_4 \theta_2$$

or

$$v(x) = [N]\{v\} \qquad (g)$$

where $[N] = [N_1 \quad N_2 \quad N_3 \quad N_4]$ and $\{v\} = [v_1 \quad \theta_1 \quad v_2 \quad \theta_2]^T$ (h)

4.10. Use the potential energy function defined by Eq. (3.13) and the results of Prob. 4.9 to derive the stiffness matrix for beam finite elements.

The variational function, Eq. (3.13), is written as a volume integral and a surface integral. The volume is the volume of the beam that can be assumed to have a constant cross section within any finite element, and the surface corresponds to any surface where an external loading is applied. In the case of a two-dimensional beam the loading is the external applied loading and can be any load condition, some of which are illustrated in Table 4.1. The surface traction term T_K of Eq. (3.13) is replaced by $w(x)$, which represents the external transverse loading. In the previous discussion concerning the stiffness method, the external transverse loading was applied as an equivalent joint loading. The potential energy of the external loading will give a similar result. In addition, the body force term of Eq. (3.13) will be neglected. Equation (3.13) can be written

$$J(v) = \int_A \int_0^L \frac{1}{2}\, \sigma\epsilon \, dx \, dA - \int_0^L wv \, dx \qquad\qquad (a)$$

The first term of Eq. (a) is called the strain energy and is of interest because it is used to derive the stiffness matrix. The second term is the potential of the external loading. The strain can be defined in terms of displacement of the elastic curve of the beam. A more familiar approach is to define the bending stress in terms of bending moment. The stress caused by bending moment is derived in mechanics of materials as (see Fig. 4-1)

$$\sigma(x) = \frac{M(x)y}{I} \qquad\qquad (b)$$

and for elementary beam theory, Hooke's law, Eq. (b), and the second of Eqs. (4.2) can be combined to give

$$\epsilon(x) = \frac{\sigma(x)}{E} = \frac{M(x)y}{EI} = y\, \frac{d^2v(x)}{dx^2} \qquad\qquad (c)$$

Equation (a) can be written

$$J(v) = \int_A \int_0^L \frac{y^2}{2} \left(\frac{d^2v(x)}{dx^2} \right)^2 E \, dA \, dx - \int_0^L wv \, dx$$

Recall that $\int_A y^2 \, dA = I$, and a more compact form is

$$J(v) = \frac{1}{2} \int_0^L EI(v'')^2 \, dx - \int_0^L wv \, dx \qquad\qquad (d)$$

The second derivative in Eq. (d) is defined in terms of the shape functions and joint (nodal point) displacements given by Eq. (g) of Prob. 4.9. It follows that

$$v'' = [N'']\{v\}$$

and substituting into Eq. (d) gives the potential energy in matrix form:

$$J(v) = \frac{1}{2} \int_0^L \{v\}^T [N'']^T [EI][N'']\{v\} \, dx - \int_0^L \{v\}^T [N]^T w \, dx \qquad\qquad (e)$$

Minimize $J(V)$ with respect to $\{v\}$ and set the result equal to zero to obtain the matrix equation that defines the local beam finite element:

$$\int_0^L [N'']^T [EI][N'']\{v\} \, dx - \int_0^L [N]^T w \, dx = 0 \qquad\qquad (f)$$

Substituting Eqs. (c)–(f) of Prob. 4.9 and their second derivatives results in the matrix equation

$$\int_0^L \begin{bmatrix} \dfrac{6}{L^2} + \dfrac{12x}{L^3} \\[2mm] -\dfrac{4}{L} + \dfrac{6x}{L^2} \\[2mm] \dfrac{6}{L^2} - \dfrac{12x}{L^3} \\[2mm] -\dfrac{2}{L} + \dfrac{6x}{L^2} \end{bmatrix} [EI] \begin{bmatrix} \dfrac{6}{L^2} + \dfrac{12x}{L^3} & -\dfrac{4}{L} + \dfrac{6x}{L^2} & \dfrac{6}{L^2} - \dfrac{12x}{L^3} & -\dfrac{2}{L} + \dfrac{6x}{L^2} \end{bmatrix} \begin{Bmatrix} v_1 \\ \theta_1 \\ v_2 \\ \theta_2 \end{Bmatrix} dx$$

$$= \int_0^L \begin{Bmatrix} 1 - 3\dfrac{x^2}{L^2} + 2\dfrac{x^3}{L^3} \\[3mm] x - 2\dfrac{x^2}{L} + \dfrac{x^3}{L^2} \\[3mm] 3\dfrac{x^2}{L^2} - 2\dfrac{x^3}{L^3} \\[3mm] \dfrac{x^3}{L^2} - \dfrac{x^2}{L} \end{Bmatrix} w\, dx \qquad (g)$$

Performing the matrix multiplications and integrating will give the stiffness matrix, which can be compared with Eq. (f) of Prob. 4.4. The matrix of equivalent joint loading corresponds to the uniformly loaded fixed-fixed beam of Table 4.1:

$$\frac{EI}{L^3} \begin{bmatrix} 12 & 6L & -12 & 6L \\ 6L & 4L^2 & -6L & 2L^2 \\ -12 & -6L & 12 & -6L \\ 6L & 2L^2 & -6L & 4L^2 \end{bmatrix} \begin{Bmatrix} v_1 \\ \theta_1 \\ v_2 \\ \theta_2 \end{Bmatrix} = \begin{Bmatrix} wL/2 \\ wL^2/2 \\ wL/2 \\ -wL^2/2 \end{Bmatrix} \qquad (h)$$

Note that the y axis is upward in the coordinate system being used and that the uniform load in this derivation acts upward. Thus equivalent joint loading on the right-hand side of Eq. (g) is opposite in sign from the results shown in Table 4.1.

4.11. Derive the local finite element stiffness matrix for a beam with combined transverse loading and axial force.

The stiffness matrix for axial force acting on a rod or beam was derived as an example of elementary elasticity in Prob. 2.9 using the variational function, and the derivation was repeated in Prob. 2.21 using a direct approach. In this application the body force will be omitted and replaced by axial forces N_1 and N_2 acting at joints (nodes) 1 and 2. The node loadings and corresponding displacements are shown in Fig. 4-11. The stiffness matrix for axial forces from Prob. 2.9 is repeated here:

$$\frac{AE}{L} \begin{bmatrix} 1 & -1 \\ -1 & 1 \end{bmatrix} \begin{Bmatrix} u_1 \\ u_2 \end{Bmatrix} = \begin{Bmatrix} N_1 \\ N_2 \end{Bmatrix} \qquad (a)$$

where u_1 and u_2 are the node displacements. The complete local stiffness matrix is obtained by combining Eq. (a) above and Eq. (f) of Prob. 4.4:

$$\begin{Bmatrix} N_1 \\ R_1 \\ M_1 \\ N_2 \\ R_2 \\ M_2 \end{Bmatrix} = \begin{bmatrix} C_1 & 0 & 0 & -C_1 & 0 & 0 \\ 0 & 12C_2 & 6C_2L & 0 & -12C_2 & 6C_2L \\ 0 & 6C_2L & 4C_2L^2 & 0 & -6C_2L & 2C_2L^2 \\ -C_1 & 0 & 0 & C_1 & 0 & 0 \\ 0 & -12C_2 & -6C_2L & 0 & 12C_2 & -6C_2L \\ 0 & 6C_2L & 2C_2L^2 & 0 & -6C_2L & 4C_2L^2 \end{bmatrix} \begin{Bmatrix} u_1 \\ v_1 \\ \theta_1 \\ u_2 \\ v_2 \\ \theta_2 \end{Bmatrix} \qquad (b)$$

where $\qquad\qquad C_1 = AE/L \qquad$ and $\qquad C_2 = EI/L^3$. $\hfill (c)$

The axial and transverse deformations are uncoupled for beams when the local axis of the beam coincides with the global axis.

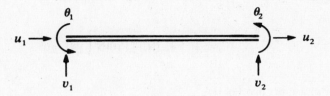

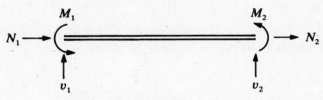

Fig. 4-11

4.12. Derive the transformation matrix and corresponding stiffness matrix for a beam oriented in a local ξ, η coordinate system and referenced to the global x, y coordinate system.

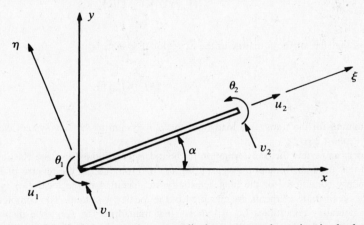

Fig. 4-12 Positive axial displacement, transverse displacement, and rotation in the local ξ, η system.

The beam is shown in Fig. 4-12, and the vector transformations are given by Eqs. (*3.14*) and (*3.15*) and illustrated in Fig. 3-2. The transformations are given in matrix form as

$$\{v\}_{\xi\eta} = [T]\{v\}_{xy} \hfill (a)$$

$$\{v\}_{xy} = [T]^T\{v\}_{\xi\eta} \hfill (b)$$

where $$[T] = \begin{bmatrix} \cos\alpha & \sin\alpha \\ -\sin\alpha & \cos\alpha \end{bmatrix} \hfill (c)$$

In this application the transformation is from the local ξ, η system to the global x, y system. The transpose of $\{v\}_{\xi\eta}$ is

$$\{v\}_{\xi\eta}^{T} = \{v\}_{xy}^{T}[T]^{T} \tag{d}$$

The first term of Eq. (e) of Prob. 4.10 can be used to derive the transformation. Assume that the matrices of Prob. 4.10 correspond to the local ξ, η system, and the energy term appears as

$$\frac{1}{2}\int_{0}^{L}\{v\}_{\xi\eta}^{T}[N'']_{\xi\eta}^{T}[EI][N'']_{\xi\eta}\{v\}_{\xi\eta}\,dx \tag{e}$$

Substitute Eqs. (a) and (d) into Eq. (e).

$$\frac{1}{2}\int_{0}^{L}\{v\}_{xy}^{T}[T]^{T}[N'']_{\xi\eta}^{T}[EI][N'']_{\xi\eta}[T]\{v\}_{xy}\,dx \tag{f}$$

The local (ξ, η) stiffness matrix is

$$[K]_{\xi\eta} = \int_{0}^{L}[N'']_{\xi\eta}^{T}[EI][N'']_{\xi\eta}\,dx$$

and the expression inside the integral can be written

$$\{v\}_{xy}^{T}[T]^{T}[K]_{\xi\eta}[T]\{v\}_{xy}$$

and it follows that the stiffness matrix in the x, y global system is

$$[K]_{xy} = [T]^{T}[K]_{\xi\eta}[T] \tag{g}$$

The transformation for the transverse load from the ξ, η system to the x, y system can be obtained in a similar manner (see Prob. 4.26).

The reader can review the transformation for boundary conditions for plane elasticity discussed in Prob. 3.23 and will find that the transformation given by Eq. (d) of Prob. 3.23 is the reverse of Eq. (g) given above. Note that the boundary conditions for the plane elasticity are specified in the local ξ, η system, while the boundary conditions for beam finite elements are specified in the local x, y system. The complete transformation for node displacement boundary conditions, Eq. (a) above, that transforms the ξ, η node displacements to the x, y node displacements, using the displacement vector defined by Eq. (b) of Prob. 4.11, is

$$\{v\}_{\xi\eta} = \begin{Bmatrix} u_{1\xi\eta} \\ v_{1\xi\eta} \\ \theta_{1\xi\eta} \\ u_{2\xi\eta} \\ v_{2\xi\eta} \\ \theta_{2\xi\eta} \end{Bmatrix} = \begin{bmatrix} c & s & 0 & 0 & 0 & 0 \\ -s & c & 0 & 0 & 0 & 0 \\ 0 & 0 & 1 & 0 & 0 & 0 \\ 0 & 0 & 0 & c & s & 0 \\ 0 & 0 & 0 & -s & c & 0 \\ 0 & 0 & 0 & 0 & 0 & 1 \end{bmatrix} \begin{Bmatrix} u_1 \\ v_1 \\ \theta_1 \\ u_2 \\ v_2 \\ \theta_2 \end{Bmatrix} = [T]\{v_{xy}\} \tag{h}$$

where $c = \cos\alpha$ and $s = \sin\alpha$.

After substituting into Eq. (g) and using the stiffness matrix of Eq. (b) given in Prob. 4.11, the stiffness matrix in the global system becomes (see Logan, 1986)

$$[K]_{xy} = \frac{E}{L} \begin{bmatrix} Ac^2 + \dfrac{12I}{L^2} s^2 & \left(A - \dfrac{12I}{L^2}\right)cs & -\dfrac{6I}{L} s & -\left(Ac^2 + \dfrac{12I}{L^2} s^2\right) & -\left(A - \dfrac{12I}{L^2}\right)cs & -\dfrac{6I}{L} s \\ & As^2 + \dfrac{12I}{L^2} c^2 & \dfrac{6I}{L} c & -\left(A - \dfrac{12I}{L^2}\right)cs & -\left(As^2 + \dfrac{12I}{L^2} c^2\right) & \dfrac{6I}{L} c \\ & & 4I & \dfrac{6I}{L} s & -\dfrac{6I}{L} c & 2I \\ & & & Ac^2 + \dfrac{12I}{L^2} s^2 & \left(A - \dfrac{12I}{L^2}\right)cs & \dfrac{6I}{L} s \\ & \text{Symmetric} & & & As^2 + \dfrac{12I}{L^2} c^2 & -\dfrac{6I}{L} c \\ & & & & & 4I \end{bmatrix}$$

$$(i)$$

4.13. The frame of Fig. 4-13 is fixed at node 1 and free to translate in the x direction at node 3. Compute the displacements and reactions at all nodes.

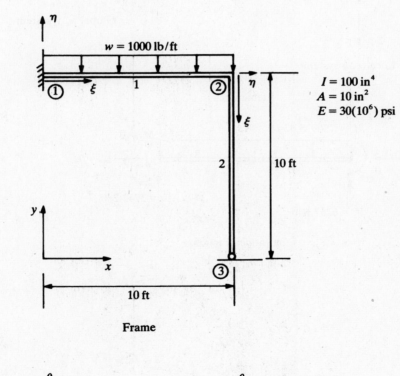

Frame

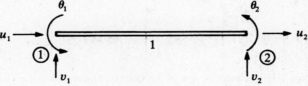

Element 1, x, y and ξ, η coincide.

Fig. 4-13

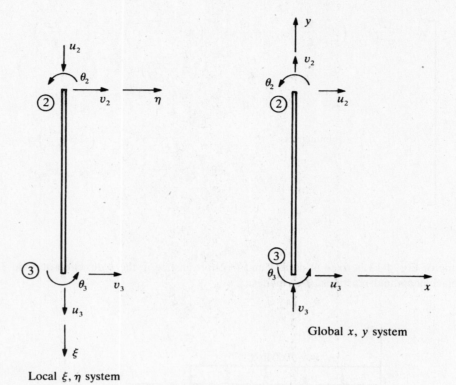

Local ξ, η system

Global x, y system

Element 2

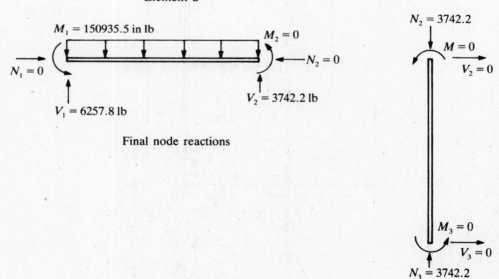

Final node reactions

Fig. 4-13   (*Continued*)

The analysis of this problem will illustrate the use of the matrix transformation derived in Eq. (*i*) of Prob. 4.12. There is considerable detail included in the analysis to illustrate how deformation boundary conditions specified in the local system are transformed and then specified in the global system. Let $I = 100$ in^4, $A = 10$ in^2, and $E = 30(10)^6$ psi for both structural members (elements). The stiffness matrix and corresponding displacement matrix will be evaluated for both elements.

Element 1 (nodes 1–2):

The element axis is assumed to be directed from node 1 to node 2 and coincides with the global x axis; by inspection, $\alpha = 0°$. Then,

$$c = \cos 0° = 1 \qquad s = \sin 0° = 0$$

Also, for both elements

$$\frac{12I}{L^2} = \frac{12(100)}{(120)^2} = 0.083333$$

$$\frac{6I}{L} = \frac{6(100)}{120} = 5$$

$$\frac{E}{L} = \frac{30(10)^6}{120} = 250{,}000$$

Substituting into Eq. (i) of Prob. 4.12 gives the stiffness matrix

$$250{,}000 \begin{bmatrix} 10 & 0 & 0 & -10 & 0 & 0 \\ 0 & 0.083333 & 5 & 0 & -0.083333 & 5 \\ 0 & 5 & 400 & 0 & -5 & 200 \\ -10 & 0 & 0 & 10 & 0 & 0 \\ 0 & -0.083333 & -5 & 0 & 0.083333 & -5 \\ 0 & 5 & 200 & 0 & -5 & 400 \end{bmatrix} \begin{Bmatrix} u_1 \\ v_1 \\ \theta_1 \\ u_2 \\ v_2 \\ \theta_2 \end{Bmatrix} \qquad (a)$$

Element 2 (nodes 2–3):

The element axis is assumed to be directed from node 2 to node 3; by inspection, the angle between the local axis and the global x axis is $\alpha = 270°$. Then,

$$c = \cos 270° = 0 \qquad s = \sin 270° = -1$$

Again, substituting into Eq. (i) of Prob. 4.12 gives the stiffness matrix

$$250{,}000 \begin{bmatrix} 0.08333 & 0 & 5 & -0.08333 & 0 & 5 \\ 0 & 10 & 0 & 0 & -10 & 0 \\ 5 & 0 & 400 & -5 & 0 & 200 \\ -0.08333 & 0 & -5 & 0.08333 & 0 & -5 \\ 0 & -10 & 0 & 0 & 10 & 0 \\ 5 & 0 & 200 & -5 & 0 & 400 \end{bmatrix} \begin{Bmatrix} u_2 \\ v_2 \\ \theta_2 \\ u_3 \\ v_3 \\ \theta_3 \end{Bmatrix} \qquad (b)$$

The global stiffness matrix is obtained by combining Eqs. (a) and (b) through the connectivity at node 2.

$$250{,}000 \begin{bmatrix} 10 & 0 & 0 & -10 & 0 & 0 & 0 & 0 & 0 \\ 0 & 0.083333 & 5 & 0 & -0.083333 & 5 & 0 & 0 & 0 \\ 0 & 5 & 400 & 0 & -5 & 200 & 0 & 0 & 0 \\ -10 & 0 & 0 & 10.08333 & 0 & 5 & -0.08333 & 0 & 5 \\ 0 & -0.083333 & -5 & 0 & 10.083333 & -5 & 0 & -10 & 0 \\ 0 & 5 & 200 & 5 & -5 & 800 & -5 & 0 & 200 \\ 0 & 0 & 0 & -0.083333 & 0 & -5 & 0.083333 & 0 & -5 \\ 0 & 0 & 0 & 0 & -10 & 0 & 0 & 10 & 0 \\ 0 & 0 & 0 & 5 & 0 & 200 & -5 & 0 & 400 \end{bmatrix}$$

$$\times \begin{Bmatrix} u_1 \\ v_1 \\ \theta_1 \\ u_2 \\ v_2 \\ \theta_2 \\ u_3 \\ v_3 \\ \theta_3 \end{Bmatrix} = \begin{Bmatrix} 0 \\ -5000 \\ -1(10)^5 \\ 0 \\ -5000 \\ 1(10)^5 \\ 0 \\ 0 \\ 0 \end{Bmatrix} \qquad (c)$$

The right-hand side of Eq. (c) represents the applied joint loading that replaces the uniform load for beam element 1-2, $V_1 = V_2 = -wL/2 = -5,000$, and $M_1 = -wL^2/2 = -100,000 = -M_2$. The boundary conditions are $u_1 = v_1 = \theta_1 = v_3 = 0$. The zero boundary condition at node 3 corresponds to zero displacement in the global y direction described by v_3 even though the displacement is axial in the local ξ, η system. After coordinate transformation all boundary conditions must be referenced to the x, y system, or displacements u and v always correspond to x and y, respectively. Note that in two dimensions the rotation θ is represented by a vector normal to the plane of the structure and does not require transformation.

The final matrix equation to be solved is obtained by striking out the rows and columns that correspond to the zero boundary conditions:

$$250,000 \begin{bmatrix} 10.083333 & 0 & 5 & -0.083333 & 5 \\ 0 & 10.083333 & -5 & 0 & 0 \\ 5 & -5 & 800 & -5 & 200 \\ -0.083333 & 0 & -5 & 0.083333 & -5 \\ 5 & 0 & 200 & -5 & 400 \end{bmatrix} \begin{Bmatrix} u_2 \\ v_2 \\ \theta_2 \\ u_3 \\ \theta_3 \end{Bmatrix} = \begin{Bmatrix} 0 \\ -5000 \\ 1(10)^5 \\ 0 \\ 0 \end{Bmatrix} \qquad (d)$$

Solving Eq. (d) gives the results for node displacements:

$$\begin{Bmatrix} u_2 \\ v_2 \\ \theta_2 \\ u_3 \\ \theta_3 \end{Bmatrix} = \begin{Bmatrix} -2.91038(10)^{-11} \\ -1.49688(10)^{-3} \\ 9.81293(10)^{-4} \\ 0.117755 \\ 9.81293(10)^{-4} \end{Bmatrix} \qquad (e)$$

Substituting the displacements into Eq. (a) gives the results for element 1:

$$250,000 \begin{bmatrix} 10 & 0 & 0 & -10 & 0 & 0 \\ 0 & 0.083333 & 5 & 0 & -0.083333 & 5 \\ 0 & 5 & 400 & 0 & -5 & 200 \\ -10 & 0 & 0 & 10 & 0 & 0 \\ 0 & -0.083333 & -5 & 0 & 0.083333 & -5 \\ 0 & 5 & 200 & 0 & -5 & 400 \end{bmatrix} \begin{Bmatrix} 0 \\ 0 \\ 0 \\ -2.91038(10)^{-11} \\ -1.49688(10)^{-3} \\ 9.81293(10)^{-4} \end{Bmatrix} \qquad (f)$$

The joint loading must be added to the results given by Eq. (f), or

$$\begin{Bmatrix} 0 \\ 1257.8 \\ 50,935.5 \\ 0 \\ -1257.8 \\ 100,000 \end{Bmatrix} + \begin{Bmatrix} 0 \\ 5000 \\ 100,000 \\ 0 \\ 5000 \\ -100,000 \end{Bmatrix} = \begin{Bmatrix} 0 \\ 6257.8 \\ 150,935.5 \\ 0 \\ 3742.2 \\ 0 \end{Bmatrix} = \begin{Bmatrix} N_1 \\ V_1 \\ M_1 \\ N_2 \\ V_2 \\ M_2 \end{Bmatrix} \qquad (g)$$

The results are shown in Fig. 4-13. Substituting the displacements into Eq. (b) gives the results for element 2:

$$250,000 \begin{bmatrix} 0.08333 & 0 & 5 & -0.08333 & 0 & 5 \\ 0 & 10 & 0 & 0 & -10 & 0 \\ 5 & 0 & 400 & -5 & 0 & 200 \\ -0.08333 & 0 & -5 & 0.08333 & 0 & -5 \\ 0 & -10 & 0 & 0 & 10 & 0 \\ 5 & 0 & 200 & -5 & 0 & 400 \end{bmatrix} \begin{Bmatrix} -2.91038(10)^{-11} \\ -1.49688(10)^{-3} \\ 9.81293(10)^{-4} \\ 0.117755 \\ 0 \\ 9.81293(10)^{-4} \end{Bmatrix} \qquad (h)$$

The joint loading is zero for element 2 since there is no external loading. The multiplication indicated by Eq. (h) gives the final results in the global system and formally appears as

$$\begin{Bmatrix} 0 \\ -3742.2 \\ 0 \\ 0 \\ 3742.2 \\ 0 \end{Bmatrix} + \begin{Bmatrix} 0 \\ 0 \\ 0 \\ 0 \\ 0 \\ 0 \end{Bmatrix} = \begin{Bmatrix} 0 \\ -3742.2 \\ 0 \\ 0 \\ 3742.2 \\ 0 \end{Bmatrix} = \begin{Bmatrix} N_2 \\ V_2 \\ M_2 \\ N_3 \\ V_3 \\ M_3 \end{Bmatrix} \tag{i}$$

The results, Eq. (*i*), should be transformed back to the local system for final interpretation. Equation (*a*) of Prob. 4.12 gives the proper vector transformation and can be written as $\{F\}_{\xi\eta} = [T]\{F\}_{xy}$, where $\{F\}$ represents the final node reactions. The matrix form of the transformation is given by Eq. (*h*) of Prob. 4.12. The final results, in the local system, are computed as

$$\begin{Bmatrix} N_{2\xi\eta} \\ V_{2\xi\eta} \\ M_{2\xi\eta} \\ N_{3\xi\eta} \\ V_{3\xi\eta} \\ M_{3\xi\eta} \end{Bmatrix} = \begin{bmatrix} 0 & -1 & 0 & 0 & 0 & 0 \\ 1 & 0 & 0 & 0 & 0 & 0 \\ 0 & 0 & 1 & 0 & 0 & 0 \\ 0 & 0 & 0 & 0 & -1 & 0 \\ 0 & 0 & 0 & 1 & 0 & 0 \\ 0 & 0 & 0 & 0 & 0 & 1 \end{bmatrix} \begin{Bmatrix} 0 \\ -3742.2 \\ 0 \\ 0 \\ 3742.2 \\ 0 \end{Bmatrix} = \begin{Bmatrix} 3742.2 \\ 0 \\ 0 \\ -3742.2 \\ 0 \\ 0 \end{Bmatrix} \tag{j}$$

The final results are shown in Fig. 4-13. The reader should note that the local axis for element 2 is directed from node 2 to node 3 and that the results given by Eq. (*j*) indicate that the column is in tension.

4.14. The frame of Fig. 4-14 is supported at joints 1 and 3 with pin-type supports that are free to rotate but not to translate. The length and structural properties are the same as in Prob. 4.13. Compute the reactions at each support and internal actions at joint 2.

Element 1-2 (nodes 1–2):

The beam element is identical to element 1-2 of Prob. 4.13, and the stiffness matrix is the same as Eq. (*a*) of Prob. 4.13.

Element 2 (nodes 2–3):

The element axis is assumed to be directed from node 2 to node 3 and, by inspection, the angle between the local axis and the global x axis is $\alpha = 300°$.

$$c = \cos 300° = 0.5 \qquad s = \sin 300° = -0.86603$$

Substituting into Eq. (*i*) of Prob. 4.12 gives the symmetric stiffness matrix

$$250{,}000 \begin{bmatrix} 2.5625 & -4.2941 & 4.3301 & -2.5625 & 4.2941 & 4.3301 \\ & 7.5208 & 2.5 & 4.2941 & -7.5208 & 2.5 \\ & & 400 & -4.3301 & -2.5 & 200 \\ & & & 2.5625 & -4.2941 & -4.3301 \\ & & & & 7.5208 & -2.5 \\ & & & & & 400 \end{bmatrix} \begin{Bmatrix} u_2 \\ v_2 \\ \theta_2 \\ u_3 \\ v_3 \\ \theta_3 \end{Bmatrix} \tag{a}$$

The element stiffness matrices are combined through the connectivity at node 2. Substituting the boundary conditions $u_1 = v_1 = u_3 = v_3 = 0$ gives the final global stiffness matrix to be solved for translations and rotations:

$$250{,}000 \begin{bmatrix} 400 & 0 & -5 & 200 & 0 \\ & 12.5625 & -4.2941 & 4.3301 & 4.3301 \\ & & 7.6042 & -2.5 & 2.5 \\ & \text{Symmetric} & & 800 & 200 \\ & & & & 400 \end{bmatrix} \begin{Bmatrix} \theta_1 \\ u_2 \\ v_2 \\ \theta_2 \\ \theta_3 \end{Bmatrix} = \begin{Bmatrix} -1(10)^5 \\ 0 \\ -5000 \\ 1(10)^5 \\ 0 \end{Bmatrix} \tag{b}$$

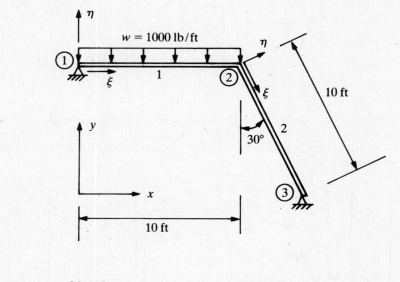

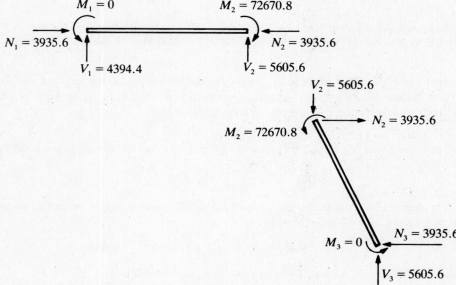

Fig. 4-14 Node reactions in the x, y system.

Solving Eq. (b) gives the results for node displacements:

$$\begin{Bmatrix} \theta_1 \\ u_2 \\ v_2 \\ \theta_2 \\ \theta_3 \end{Bmatrix} = \begin{Bmatrix} -1.54936(10)^{-3} \\ -1.57457(10)^{-3} \\ -4.06005(10)^{-3} \\ 9.97222(10)^{-4} \\ -4.56194(10)^{-4} \end{Bmatrix} \qquad (c)$$

Substituting the displacements into Eq. (a) of Prob. 4.12 and adding the equivalent joint loading gives the results for element 1:

$$250{,}000 \begin{bmatrix} 10 & 0 & 0 & -10 & 0 & 0 \\ 0 & 0.083333 & 5 & 0 & -0.083333 & 5 \\ 0 & 5 & 400 & 0 & -5 & 200 \\ -10 & 0 & 0 & 10 & 0 & 0 \\ 0 & -0.083333 & -5 & 0 & 0.083333 & -5 \\ 0 & 5 & 200 & 0 & -5 & 400 \end{bmatrix} \begin{Bmatrix} 0 \\ 0 \\ -1.54936(10)^{-3} \\ -1.57457(10)^{-3} \\ -4.06005(10)^{-3} \\ 9.97222(10)^{-4} \end{Bmatrix}$$

$$= \begin{Bmatrix} 3935.6 \\ -605.6 \\ -100{,}000 \\ -3935.6 \\ 605.6 \\ 27{,}329.2 \end{Bmatrix} + \begin{Bmatrix} 0 \\ 5000 \\ 100{,}000 \\ 0 \\ 5000 \\ -100{,}000 \end{Bmatrix} = \begin{Bmatrix} 3935.6 \\ 4394.6 \\ 0 \\ -3935.6 \\ 5605.6 \\ -72{,}670.8 \end{Bmatrix} = \begin{Bmatrix} N_1 \\ V_1 \\ M_1 \\ N_2 \\ V_2 \\ M_2 \end{Bmatrix} \quad (d)$$

Similarly, computations using Eq. (a) above give the final reactions at nodes 2 and 3 of element 2 in the global coordinate system:

$$250{,}000 \begin{bmatrix} 2.5625 & -4.2941 & 4.3301 & -2.5625 & 4.2941 & 4.3301 \\ & 7.5208 & 2.5 & 4.2941 & -7.5208 & 2.5 \\ & & 400 & -4.3301 & -2.5 & 200 \\ & & & 2.5625 & -4.2941 & -4.3301 \\ & \text{Symmetric} & & & 7.5208 & -2.5 \\ & & & & & 400 \end{bmatrix} \begin{Bmatrix} -1.57457(10)^{-3} \\ -4.06005(10)^{-3} \\ 9.97222(10)^{-4} \\ 0 \\ 0 \\ -4.56194(10)^{-4} \end{Bmatrix}$$

$$= \begin{Bmatrix} 3935.6 \\ -5605.6 \\ 72{,}670.8 \\ -3935.6 \\ 5605.6 \\ 0 \end{Bmatrix} + \begin{Bmatrix} 0 \\ 0 \\ 0 \\ 0 \\ 0 \\ 0 \end{Bmatrix} = \begin{Bmatrix} 3935.6 \\ -5605.6 \\ 2670.8 \\ -3935.6 \\ 5605.6 \\ 0 \end{Bmatrix} = \begin{Bmatrix} N_2 \\ V_2 \\ M_2 \\ N_3 \\ V_3 \\ M_3 \end{Bmatrix} \quad (e)$$

The node reactions can be converted to the local system using the transformation $\{F\}_{\xi\eta} = [T]\{F\}_{xy}$, where $[T]$ is given by Eq. (h) of Prob. 4.12.

Supplementary Problems

4.15. Use the differential equation defined by Eq. (4.1) to solve for shear and moment reactions for the fixed-fixed beam of Fig. 4-15.

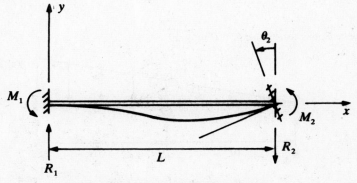

Fig. 4-15 Applied rotation θ_2 at $x = L$.

4.16. Use the differential equation defined by Eq. (*4.1*) to solve for shear and moment reactions for the fixed-fixed beam of Fig. 4-16.

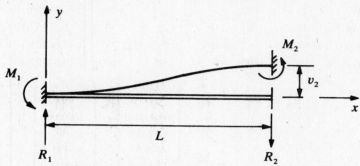

Fig. 4-16   Applied displacement at $x = L$.

4.17. Verify the fixed-fixed beam equivalent loadings given by Table 4.1.

4.18. Compute the reactions for the beam of Fig. 4-17.

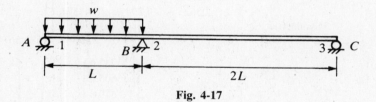

Fig. 4-17

4.19. Compute the reactions for the beam of Fig. 4-18.

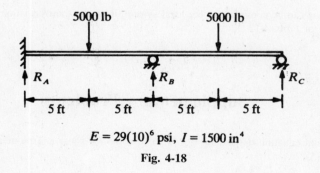

$E = 29(10)^6$ psi, $I = 1500$ in^4

Fig. 4-18

4.20. Compute the reactions for the beam of Fig. 4-19.

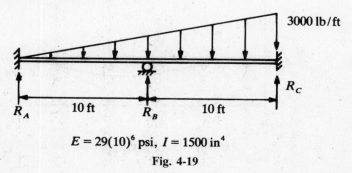

$E = 29(10)^6$ psi, $I = 1500$ in^4

Fig. 4-19

4.21. Compute the reactions for the beam of Fig. 4-20.

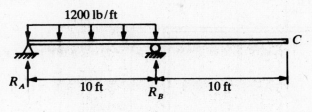

$$E = 29(10)^6 \text{ psi}, \ I = 1500 \text{ in}^4$$

Fig. 4-20

4.22. Compute the reactions for the beam of Fig. 4-21.

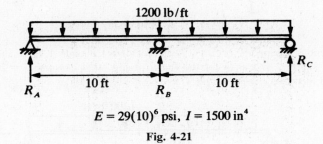

$$E = 29(10)^6 \text{ psi}, \ I = 1500 \text{ in}^4$$

Fig. 4-21

4.23. Show that the equivalent joint loadings computed in Prob. 4.3 and tabulated in Table 4.1 for a uniformly varying external load can be obtained from the potential energy expression given by Eq. (f) of Prob. 4.10.

4.24. Derive the shape functions for a beam finite element of length L assuming a cubic polynomial in the form

$$v(x) = a_0 + a_1 x + a_2 x^2 + a_3 x^3$$

by satisfying the boundary conditions $v(0) = 0$, $v(L) = 0$, $v'(0) = 0$, and $v'(L) = 0$.

4.25. The fixed-fixed beam of Fig. 4-22 has an axial force applied at $2L/3$. Use the finite element derived in Prob. 4.11 to compute the axial reactions.

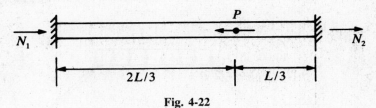

Fig. 4-22

4.26. Derive the transformation from local ξ, η coordinates to global x, y coordinates for transverse beam loading.

4.27. Repeat Prob. 4.13 assuming additional boundary conditions of $u_2 = v_2 = 0$.

4.28. Compute the reactions at supports A and D and joint rotations at B and C for the frame of Fig. 4-23. Assume $E = 29(10)^6$ psi, $I = 1800$ in^4, and $A = 20$ in^2 for all members.

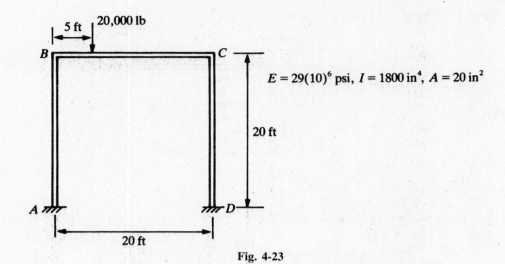

$E = 29(10)^6$ psi, $I = 1800$ in^4, $A = 20$ in^2

Fig. 4-23

4.29.   Compute the reactions at supports A, C, and D and rotations for joints A, B, C, and D along with the horizontal displacements of supports C and D for the frame of Fig. 4-24. Assume $E = 29(10)^6$ psi, $I = 1800$ in^4, and $A = 20$ in^2 for all members.

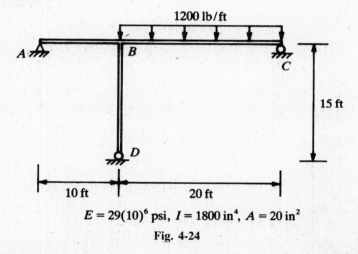

$E = 29(10)^6$ psi, $I = 1800$ in^4, $A = 20$ in^2

Fig. 4-24

4.30.   Compute reactions, joint rotations, and the horizontal displacement of joint A for the frame of Fig. 4-25. Assume $E = 29(10)^6$ psi, $I = 1800$ in^4, and $A = 20$ in^2 for all members.

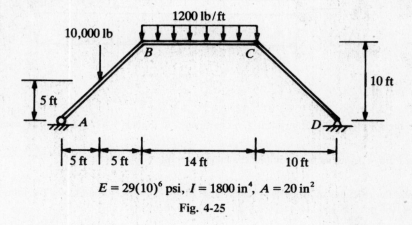

$E = 29(10)^6$ psi, $I = 1800$ in^4, $A = 20$ in^2

Fig. 4-25

Answers to Supplementary Problems

4.15. $M_1 = 2EI\theta_2/L$, $M_2 = 4EI\theta_2/L$, $R_1 = 6EI\theta_2/L^2$, $R_2 = 6EI\theta_2/L^2$.

4.16. $M_1 = M_2 = 6EIv_2/L^2$, $R_1 = R_2 = 12EIv_2/L^3$.

4.18. $M_A = M_C = 0$, $M_B = -3wL^2/72$ (acting on member A-B), $R_A = 33wL/72$, $R_B = 81wL/144$, $R_C = -3wL/144$, $\theta_A = -5wL^3/144EI$, $\theta_B = wL^3/36EI$, $\theta_C = -wL^3/72EI$, $v_A = v_B = v_C = 0$.

4.19. $M_A = 5360$ ft·lb, $M_B = -8040$ (acting on member A-B), $R_A = 2230$ lb, $R_B = 6070$ lb, $R_C = 1700$ lb, $\theta_A = 0$, $\theta_B = -1.48(10)^{-5}$, $\theta_C = 5.91(10)^{-5}$.

4.20. $M_A = 2500$ ft·lb, $M_B = -12,500$ (acting on member A-B), $M_C = -22,500$, $R_A = 1500$ lb, $R_B = 15,000$, $R_C = 13,500$, $\theta_B = -4.14(10)^{-5}$.

4.21. $R_A = 6000$ lb, $R_B = 6000$, $M_B = 0$, $\theta_A = -1.655(10)^{-3}$, $\theta_B = \theta_C = 1.655(10)^{-3}$, $v_C = 0.1986$ in.

4.22. $M_A = M_C = 0$, $M_B = -15,000$ (acting on member A-B), $R_A = R_C = 4500$ lb, $R_B = 15,000$, $\theta_A = -\theta_C = -8.28(10)^{-5}$, $\theta_B = 0$.

4.23.   Compute $\int_0^L [N]^T(-wx/L)\,dx$, where $[N]^T$ is given by Eq. (g) of Prob. 4.10.

4.24.   Substitute boundary conditions:

$$v(0) = v_1 = a_0$$
$$v(L) = v_2 = a_0 + a_1 L + a_2 L^2 + a_3 L^3$$
$$v'(0) = \theta_1 = a_1$$
$$v'(L) = \theta_2 = a_1 + 2a_2 L + 3a_3 L^2$$

Solving for the coefficients gives

$$a_0 = v_1 \qquad a_1 = \theta_1 \qquad a_2 = -\frac{3(v_1 - v_2)}{L^2} - \frac{2\theta_1 + \theta_2}{L} \qquad a_3 = \frac{2(v_1 - v_2)}{L^3} + \frac{\theta_1 + \theta_2}{L^2}$$

Rearranging and writing $v(x)$ in matrix form $v(x) = [N]\{v\}$ gives $[N]$ identical to Prob. 4.9.

4.25.   The axial and transverse deformations are uncoupled in this application of beam finite elements, and Eq. (a) of Prob. 4.11 is sufficient to obtain a solution. The local elements are

$$\frac{3AE}{2L}\begin{bmatrix} 1 & -1 \\ -1 & 1 \end{bmatrix}\begin{Bmatrix} u_1 \\ u_2 \end{Bmatrix} = \begin{Bmatrix} 0 \\ -P \end{Bmatrix} \qquad \frac{3AE}{L}\begin{bmatrix} 1 & -1 \\ -1 & 1 \end{bmatrix}\begin{Bmatrix} u_2 \\ u_3 \end{Bmatrix} = \begin{Bmatrix} 0 \\ 0 \end{Bmatrix}$$

The global finite element matrix becomes

$$\begin{bmatrix} \frac{3}{2} & -\frac{3}{2} & 0 \\ -\frac{3}{2} & \frac{9}{2} & -3 \\ 0 & -3 & 3 \end{bmatrix}\begin{Bmatrix} u_1 \\ u_2 \\ u_3 \end{Bmatrix} = \begin{Bmatrix} 0 \\ -P \\ 0 \end{Bmatrix}$$

with boundary conditions $u_1 = u_3 = 0$. The matrix equation reduces to one equation that can be solved for $u_2 = -2PL/9AE$. Substituting u_2 into the local matrices gives the results $N_1 = P/3$ and $N_3 = 2P/3$, both acting to the right.

4.26. Refer to Eq. (*e*) of Prob. 4.10. The transverse load term is $\int_0^L \{v\}^T [N]^T w \, dx$. Substitute Eq. (*e*) of Prob. 4.12, $\int_0^L \{v\}_{xy}^T [T]^T [N]^T w \, dx$, and the final transformation is $[T]^T \int_0^L [N]^T w \, dx$.

4.27. The additional boundary conditions assume no axial deformation for the column and correspond to the standard mechanics of materials analysis. The results are $M_1 = 150,000$ in·lb, $V_1 = 6250$ lb, and $V_3 = 3750$ lb.

4.28. $M_A = -9990$ ft·lb, $M_B = 27,340$ (acting on member B-C), $M_C = -22,540$ (acting on member B-C), $M_D = 14,780$; vertical reactions $V_A = 15,240$ lb, $V_D = 4760$; horizontal reactions $H_A = H_D = 1870$ lb; $\theta_A = \theta_D = 0$, $\theta_B = -4.787(10)^{-4}$, $\theta_C = 2.141(10)^{-4}$.

4.29. $M_A = M_C = M_D = 0$, $M_B = 37,520$ ft·lb (acting on member B-C); vertical reactions $V_A = -3750$ lb, $V_C = 10,120$, $V_D = 17,630$; $\theta_A = 1.296(10)^{-4}$, $\theta_B = \theta_D = -3.906(10)^{-4}$, $\theta_C = 7.812(10)^{-4}$; horizontal displacement at D, $u_D = -0.0703$ in, $u_C = 0$.

4.30. $M_A = M_D = 0$, $M_B = -108,960$ ft·lb (acting on member B-C), $M_C = 94,400$ (acting on member B-C), $V_A = 17,370$ lb, $V_D = 9450$, $\theta_A = -4.187(10)^{-3}$, $\theta_B = -2.246(10)^{-3}$, $\theta_C = 2.438(10)^{-3}$, $\theta_D = 4.279(10)^{-3}$, $u_A = -0.90$ in.

Chapter 5

Variational Principles, Galerkin Approximation, and Partial Differential Equations

5.1. INTRODUCTION

The fundamental numerical method of analysis utilizing finite elements has been developed in previous chapters with elementary elements and formulations of problems and basic equations of mathematical physics and engineering. In this chapter a more fundamental development of variational principles will be presented. The reader who is familiar with the mathematical structure of variational principles but unfamiliar with applications to finite element analysis will find detailed examples in terms of applications. For the reader who is unfamiliar with the mathematics of functional analysis and vector spaces, the examples are intended to translate mathematical concepts into more meaningful applied situations. In either case several references are given in the bibliography that can be consulted for additional reading.

The Galerkin method of numerical analysis is introduced and is emphasized in this chapter. Examples illustrating the mathematical method will aid the reader in understanding fundamental concepts referred to as *weighted residual methods*. Detailed examples are intended to illustrate application of the Galerkin method for deriving local finite element models. The Galerkin method is a more powerful tool for finite element analysis than the variational method because almost any physical problem described by one or more differential equations can be modeled using this technique. Variational principles do not exist for all physical problems, however, mathematical results proving that a variational formulation either does or does not exist can tell the analyst more about the fundamental mathematical structure of a problem than the Galerkin method. These issues are of an advanced nature and beyond the scope of this book.

The finite element method will be extended to coupled partial differential equations in this chapter. The equations of elasticity of Chap. 3 are an example of a coupled problem since the finite elements were derived for displacements in two independent directions. There are a variety of such problems in mathematical physics. There should be sufficient examples in this chapter to enable the reader to derive a finite element for any situation. Initial-value problems will be discussed here, but their numerical solution will be addressed in Chap. 7. Subscript tensor notation will be used, and the reader unfamiliar with the notation will find a brief outline in Chap. 1.

5.2. VARIATIONAL PRINCIPLES

The more abstract mathematical representation of a variational principle is given by

$$J(u) = [u, Au]_\Omega - 2[u, f]_\Omega \tag{5.1}$$

where A is a self-adjoint positive definite operator, f is a vector of specified quantities within the region Ω, and u is a vector of unknown quantities. The notation $[\ ,\ ]_\Omega$ denotes an inner product and can be considered as having a property similar to the inner product or scalar product of vector analysis. The terms within the brackets, when multiplied together, give a scalar result. Also, within the context of Eq. (5.1), the notation can be interpreted as an integral taken over the volume of the domain Ω.

The operator A and vectors u and f define the boundary-value problem

$$Au = f \text{ on } \Omega \tag{5.2}$$

A fundamental concept of variational calculus is that the value of u that minimizes Eq. (5.1) is the solution of Eq. (5.2). There are several conditions imposed upon these rather elementary ideas that are related to functional analysis and the theory of vector spaces. In this book the emphasis is placed upon an interpretation of the abstract concepts and the construction of variational principles.

Equation (5.2) is not complete without a set of boundary conditions. The boundary conditions can be written

$$Cu = g \text{ on } \partial\Omega \tag{5.3}$$

where C is an operator, u is again the vector of unknowns, and g is a set of given quantities (boundary conditions). Equation (5.1) can be extended to

$$J(u) = [u, Au]_\Omega - 2[u, f]_\Omega + [u, Cu]_{\partial\Omega} - 2[u, g]_{\partial\Omega} \tag{5.4}$$

and includes the boundary conditions within the definition of the variational principle.

The fundamental representation given by Eq. (5.1) is said to be valid for homogeneous boundary conditions, while Eq. (5.4) is valid for nonhomogeneous boundary conditions. The first term in brackets in Eq. (5.1) is usually written in the more general form $[u, Av]$, where u and v are defined within a space, called a *vector space*, and the term is called a *bilinear mapping* and must satisfy certain mathematical requirements (see Reddy, 1986). The operator A is said to be symmetric if

$$[u, Av]_\Omega = [v, Au]_\Omega = [Au, v]_\Omega \tag{5.5}$$

A symmetric operator defined by Eq. (5.5) does *not* imply a symmetric matrix within the context of matrix analysis. The adjoint of the operator A is denoted as A^* and is defined within the following context. If

$$[u, Av]_\Omega = [A^*u, v]_\Omega = [v, A^*u]_\Omega \tag{5.6a}$$

for homogeneous boundary conditions or

$$[u, Av]_\Omega = [v, A^*u]_\Omega + D_{\partial\Omega}(v, u) \tag{5.6b}$$

then A^* is the adjoint of A, where u and v are defined within the same vector space and Ω indicates the domain of the space. The term $D_{\partial\Omega}(u, v)$ represents possible boundary conditions. If $A = A^*$, the operator A is said to be self-adjoint.

Symmetry defined by Eq. (5.5) and applied to Eq. (5.1) implies the simple relation $[u, Au] = [u, Au]$. However, when u represents more than one field variable, the interpretation is

$$[u_i, A_{ij}u_j] = [u_j, A_{ji}u_j] \qquad \text{(no sum on repeated indices)} \tag{5.7}$$

The concept of self-adjoint operators is dependent upon the form of the bilinear mapping. The usual form associated with an ordinary differential equation such as those of Chap. 2 is

$$[p, q] = \int_\Omega p(x)q(x)\,dx \tag{5.8}$$

Time-dependent problems may require a different definition of the bilinear mapping. Equation (5.8) can be extended to include time t as

$$[p, q] = \int_\Omega \int_0^t p(x, t)q(x, t)\,dx\,dt \tag{5.9}$$

and the spatial dimension can represent one, two, or three dimensions. Equations (5.5)–(5.7) are dependent upon the definition of the bilinear mapping. Differential equations that have a first-order derivative (or any other odd derivative) in time are not self-adjoint with respect to the bilinear form of Eq. (5.9). Reddy (1975) and Oden and Reddy (1976) have shown that equations of the heat-conduction type (initial-value problems with first-order derivatives in time) are self-adjoint with respect to the bilinear mapping

$$[p, q] = \int_\Omega \int_0^t p(x, t)q(x, t - \tau)\, dx\, d\tau \tag{5.10}$$

Initial-value problems of the heat-conduction type are characterized by an operator [in the context of Eq. (5.2)] with the time derivative on the diagonal of the operator, as well as by initial conditions.

A physical situation that can be classified as an initial-value problem with first-order time derivatives that cannot be placed on the diagonal of the operator or cannot be placed symmetrically within the operator will not be self-adjoint with respect to the bilinear mapping of Eq. (5.10). An example is the equations that describe time-dependent linear coupled thermoelasticity. A convolution type of bilinear mapping was first used by Gurtin (1963) for the study of linear viscoelasticity, and the corresponding variational principle has since been termed a *Gurtin-type variational principle*. The convolution has been used extensively for situations other than viscoelasticity [see Oden and Reddy (1976) and references therein]. The convolution bilinear mapping is defined using the Laplace transform as

$$L^{-1}[p', q'] = \int_\Omega \int_0^t p(x, \tau)q(x, t - \tau)\, dx\, d\tau = [p * q] \tag{5.11}$$

where $L^{-1}\{p'(s)\} = p(x, t)$ and $L^{-1}\{q'(s)\} = q(x, t)$.

The variational principle derived using Eq. (5.1) or (5.4) is a function of all the field variables. While it is a proper variational function, it is not suitable for finite element analysis. Additional variational principles can be derived from the basic or general function and are called *extended variational principles*. The extended principle is obtained by assuming that one or more of the governing equations and boundary conditions are identically satisfied (see Prob. 5.5).

5.3. GALERKIN APPROXIMATION

The Galerkin method of approximate analysis is classified as a *method of weighted residuals*. The analysis is based upon assuming an approximate solution for a differential equation. Since the assumption is an approximation, the differential equation will not be satisfied and there will be an error in the solution. The error (residual) is then optimized with respect to some parameter, and the optimization procedure is called a weighted residual method. Given a differential equation, such as the heat-conduction equation of Chap. 2, a possible assumed solution T_R is

$$T_R = a_0 + a_1 x + a_2 x^2 + a_3 x^3 + \cdots \tag{5.12}$$

where the a_i are unknown constants. The assumed solution must satisfy the boundary conditions, and it follows that Eq. (5.12) must have at least one more unknown constant than there are boundary conditions. This requirement is easily satisfied with finite element analysis. The exact solution of the heat conduction equation is defined as T, and the approximate solution is T_R. The error or residual R is the difference between the two, or

$$R = T - T_R \tag{5.13}$$

The method of weighted residuals requires that the unknowns of Eq. (5.12) be computed using the criterion

$$\int_\Omega w_i(x)R(x; a_i)\, dV = 0 \tag{5.14}$$

where there is a one-to-one correspondence between each $w_i(x)$ and $R(x; a_i)$ and Ω represents the domain of the problem. The Galerkin method requires that each w_i be the function multiplied by the corresponding a_i of Eq. (5.12). When using the Galerkin method in finite element analysis, the assumed functions in Eq. (5.12) are the shape functions. It will be illustrated that the Galerkin method and the variational formulation lead to an identical finite element formulation. Obviously, the Galerkin method is quite powerful since it can be applied to physical problems that do not have an alternative variational formulation.

Once the reader has mastered the Galerkin technique for deriving finite element models of differential equations, it will appear that the Galerkin method is superior to the variational method. This text will not attempt to establish that either is superior, but when both methods are available to the numerical analyst, they offer a check on each other. There are relative merits to be claimed for each method. For instance, derivation of the transformation matrices of Chap. 3 (see Prob. 3.23) is more definitive when starting with the variational function. On the other hand, problems with two or more degrees of freedom per finite element node might be modeled using a different shape function for each degree of freedom (see Prob. 5.14), and that derivation is more definitive when the Galerkin method is used.

5.4. COUPLED PARTIAL DIFFERENTIAL EQUATIONS

The elasticity problems of Chap. 3 are examples of coupled partial differential equations. There are two unknown displacements, and they are dependent upon each other. In the corresponding finite element formulation there are two degrees of freedom per node that must be solved simultaneously. The majority of topics in engineering and physics that must be studied numerically are of the coupled variety. A coupled theory of mass diffusion is analyzed in Probs. 5.14–5.16. Steady-state thermoelasticity is discussed in Probs. 5.17 and 5.18, where heat effects and mechanical displacements effects are uncoupled because of the steady-state assumption. Displacements are dependent upon temperature, but temperature is independent of displacement or traction boundary loading. However, for simplicity the problem is formulated as a coupled problem, and unknown temperatures and displacements are computed using a single finite element formulation. The fully coupled thermoelasticity problem is time-dependent and is discussed in Prob. 5.8. A theory that governs coupled electrical and deformation effects in materials is called piezoelectricity. The theory is fully coupled and is discussed as Prob. 5.19.

5.5. INITIAL-VALUE PROBLEMS

There are two basic types of initial-value problems. Unsteady heat conduction is one example of a problem that contains only first-order derivatives with respect to time in the formulation. Dynamic problems with second-order time derivatives constitute the second type, and an example is the equations governing dynamic elasticity. The theory of thermoelasticity, depending upon the formulation, can be elastostatic but coupled to first-order time derivatives in the energy part of the formulation. Dynamic thermoelasticity can govern problems of free vibration or wave propagation with heat-conduction effects. The elementary case would be steady-state thermoelasticity with the heat-conduction effect uncoupled from the elastic displacements. Steady-state thermoelasticity, of course, cannot be an initial-value problem.

Solved Problems

5.1. Assume an abstract set of equations

$$A_{11}u_1 + A_{12}u_2 = f_1$$

$$A_{21}u_1 + A_{22}u_2 = f_2$$

and relate these equations to Eqs. (5.1) and (5.2).

The operator of Eq. (5.2) is

$$A = \begin{bmatrix} A_{11} & A_{12} \\ A_{21} & A_{22} \end{bmatrix}$$

with $u = \{u_1 \quad u_2\}^T$ and $f = \{f_1 \quad f_2\}^T$. The inner product of Eq. (5.1) can be written as

$$J(u) = \{u_1 \quad u_2\} \begin{bmatrix} A_{11} & A_{12} \\ A_{21} & A_{22} \end{bmatrix} \begin{Bmatrix} u_1 \\ u_2 \end{Bmatrix} - 2\{u_1 \quad u_2\} \begin{Bmatrix} f_1 \\ f_2 \end{Bmatrix}$$

The third matrix in the first term on the right-hand side is premultiplied by the second matrix and can be written as

$$J(u) = \{u_1 \quad u_2\} \begin{bmatrix} A_{11}u_1 + A_{12}u_2 \\ A_{21}u_1 + A_{22}u_2 \end{bmatrix} - 2\{u_1 \quad u_2\} \begin{Bmatrix} f_1 \\ f_2 \end{Bmatrix}$$

The multiplication is completed similarly to matrix multiplication but written in the notation of Eq. (5.1) as

$$J(u) = [u_1, A_{11}u_1] + [u_1, A_{12}u_2] + [u_2, A_{21}u_1] + [u_2, A_{22}u_2] - 2[u_1, f_1] - 2[u_2, f_2] \qquad (a)$$

5.2. Discuss symmetry in terms of the results obtained in Prob. 5.1.

Symmetry of the operator A, using Eq. (5.7) and Eq. (a) of Prob. 5.1, means that $[u_1, A_{12}u_2] = [u_2, A_{21}u_1]$. Substituting into Eq. (a) of Prob. 5.1 leads to the elimination of A_{21}, or

$$J(u) = [u_1, A_{11}u_1] + 2[u_1, A_{12}u_2] + [u_2, A_{22}u_2] - 2[u_1, f_1] - 2[u_2, f_2]$$

Similarly, $[u_1, A_{12}u_2]$ can be eliminated.

5.3. Derive a variational principle for one-dimensional steady heat conduction in the format of the abstract funcion of Prob. 5.2.

The equations of one-dimensional steady-state heat conduction are given as Eqs. (2.9) and (2.10) and can be written as

$$\frac{dq}{dx} = Q \qquad \text{and} \qquad q + k\frac{dT}{dx} = 0$$

The equations can be written in the format of Eq. (5.2) by defining

$$A = \begin{bmatrix} 0 & d/dx \\ -d/dx & -1/k \end{bmatrix} \qquad (a)$$

with $u = \{T \quad q\}^T$ and $f = \{Q \quad 0\}^T$. The variational function is constructed following the procedure of Prob. 5.1:

$$J(u) = \{T \quad q\} \begin{bmatrix} 0 & d/dx \\ -d/dx & -1/k \end{bmatrix} \begin{Bmatrix} T \\ q \end{Bmatrix} - 2\{T \quad q\} \begin{Bmatrix} Q \\ 0 \end{Bmatrix}$$

and after multiplication gives

$$J(u) = \left[T, \frac{dq}{dx} \right]_\Omega - \left[q, \frac{dT}{dx} \right]_\Omega - \left[q, \frac{q}{k} \right]_\Omega - 2[T, Q]_\Omega \qquad (b)$$

Symmetry, as discussed in Prob. 5.2, implies

$$\left[T, \frac{dq}{dx} \right]_\Omega = -\left[q, \frac{dT}{dx} \right]_\Omega \qquad (c)$$

Assume homogeneous boundary conditions for now, and Eq. (c) relates to Eq. (5.6a) with $T \equiv u(x)$, $q \equiv v(x)$, and $A \equiv d/dx$. Assume the function is bounded with the domain Ω replaced by the one-dimensional limits 0 to L. It follows that

$$\left[T, \frac{dq}{dx} \right]_\Omega \Rightarrow \int_\Omega \left[T, \frac{dq}{dx} \right] dx = \int_0^L T(x) \frac{dq(x)}{dx} dx$$

Integration by parts gives

$$\left[T, \frac{dq}{dx} \right]_\Omega = Tq \Big|_0^L - \int_0^L q \frac{dT}{dx} dx = -\left[q, \frac{dT}{dx} \right]_\Omega \qquad (d)$$

Homogeneous boundary conditions at 0 and L make the boundary term zero in Eq. (d). Equation (d) gives the same result as Eq. (c), and the problem is self-adjoint in the sense of the mapping defined by Eq. (5.8) or (5.9) when there is no time dependence. Note that the operator A defined by Eq. (a) is self-adjoint. Equation (d) shows that the first-order derivative by itself is not self-adjoint.

A more general concept employs the divergence theorem (the Green-Gauss theorem) and illustrates the inclusion of boundary conditions on q. The divergence theorem was discussed in Chap. 1, and in the notation used here gives the following result when applied to the term $[T, dq/dx]$:

$$\left[T, \frac{dq}{dx} \right]_\Omega = -\left[\frac{dT}{dx}, q \right]_\Omega + [T, qn]_{\partial\Omega q} \qquad (e)$$

where n is an outward normal to the surface defined by $\partial\Omega q$ (the surface where a boundary flux is applied). Equation (b) can now be written in the form

$$J(u) = -2 \left[q, \frac{dT}{dx} \right]_\Omega - \left[q, \frac{q}{k} \right]_\Omega - 2[T, Q]_\Omega + [T, qn]_{\partial\Omega q} \qquad (f)$$

5.4. Show that the functional derived in Prob. 5.3 is the variational principle corresponding to steady heat conduction.

The variables for the variational principle derived in Prob. 5.3 are $u = \{T \quad q\}$. It follows that $\delta J(u)/\delta(u) = 0$ or, taken independently, $\delta J(u)/\delta T = 0$ and $\delta J(u)/\delta q = 0$ should yield the governing equations. Equation (f) of Prob. 5.3 can be written in the analogous integral form

$$J(u) = \int_\Omega \left(-2q \frac{dT}{dx} - \frac{q^2}{k} - 2TQ \right) dx + \int_{\partial\Omega q} Tqn \, dx \qquad (a)$$

and

$$\frac{\delta J(u)}{\delta q} = \int_\Omega \left(-2 \frac{dT}{dx} - 2 \frac{q}{k} \right) \delta q \, dx = 0 \qquad (b)$$

The term within the parentheses is equivalent to

$$k \frac{dT}{dx} + q = 0$$

The variation with respect to T of the functional given by Eq. (a) can be obtained directly from that equation or the divergence theorem can be used to eliminate the dT/dx term. Recall Eq. (e) of Prob. 5.3 and substitute into Eq. (a) above to obtain

$$J(u) = \int_\Omega \left(2T \frac{dq}{dx} - \frac{q^2}{k} - 2TQ \right) dx \qquad (c)$$

and

$$\frac{\delta J(u)}{\delta T} = \int_\Omega \left(2 \frac{dT}{dx} - 2Q \right) \delta T \, dx = 0 \qquad (d)$$

Again, the term within the parentheses is the governing equation

$$\frac{dq}{dx} - Q = 0$$

5.5. Show that the variational function of Chap. 2, Eq. (2.23), that was used to derive the finite element model can be obtained as an extended variational principle of the variational principle given by Eq. (f) of Prob. 5.3.

The variational function of Chap. 2 is a special case of Eq. (f) of Prob. 5.3 and is often called an extended variational principle. The unknown functions are temperature T and heat flux q, and it is desirable to eliminate q. Assume the governing equation $q = -k(dT/dx)$ is identically satisfied and substitute into the first and second terms on the right-hand side of Eq. (f):

$$J(u) = 2k\left[\frac{dT}{dx}, \frac{dT}{dx}\right]_\Omega - k\left[\frac{dT}{dx}, \frac{dT}{dx}\right]_\Omega - 2[T, Q]_\Omega + [T, qn]_{\partial\Omega_q}$$

or
$$J(u) = k\left[\frac{dT}{dx}, \frac{dT}{dx}\right]_\Omega - 2[T, Q]_\Omega + [T, qn]_{\partial\Omega_q} \qquad (a)$$

Equation (a) is equivalent to the integral form given by Eq. (2.23). The variation with respect to T will give the governing differential equation. Also, construction of the variational principle using the methods of this chapter gives information concerning the boundary conditions. Use of the divergence theorem introduces the boundary condition on flux. Of course, there are boundary conditions on T, and the inclusion of that type of boundary condition will be discussed in a later problem.

5.6. Show that the equations of Chap. 2 that define mass transport with convection are not self-adjoint with respect to the bilinear mapping of Eq. (5.8).

Equations (2.16) and (2.17) are used to formulate an equation similar to Eq. (5.1):

$$J(u) = \{C \quad j\}\begin{bmatrix} K_r + u\,d/dx & d/dx \\ -d/dx & -1/D \end{bmatrix}\begin{Bmatrix} C \\ j \end{Bmatrix} - 2\{C \quad j\}\begin{Bmatrix} m \\ 0 \end{Bmatrix} \qquad (a)$$

It follows that

$$J(u) = \left[C, K_r C + u\frac{dC}{dx}\right]_\Omega + \left[C, \frac{dj}{dx}\right]_\Omega - \left[j, \frac{dC}{dx}\right]_\Omega - \left[j, \frac{j}{D}\right]_\Omega - 2[C, m]_\Omega \qquad (b)$$

As in Eq. (e) of Prob. 5.3 it can be shown that symmetry is satisfied for the off-diagonal terms:

$$\left[C, \frac{dj}{dx}\right]_\Omega = -\left[\frac{dC}{dx}, j\right]_\Omega + [C, jn]_{\partial\Omega_j}$$

The first term within brackets in Eq. (b) does not satisfy the fundamental criterion and can be written in two parts as

$$[C, K_r C]_\Omega + \left[C, u\frac{dC}{dx}\right]_\Omega \qquad (c)$$

Then, the first term of Eq. (c) satisfies Eq. (5.7). However, the second term can be integrated by parts, as in Eq. (d) of Prob. 5.3, to show that the term is not self-adjoint for homogeneous boundary conditions:

$$\left[C, u\frac{dC}{dx}\right]_\Omega = CuC \Big|_0^L - \int_0^L \left(u\frac{dC}{dx}\right)C\,dx = -\left[u\frac{dC}{dx}, C\right]_\Omega = -\left[C, u\frac{dC}{dx}\right]_\Omega \qquad (d)$$

Equation (d) illustrates that the term $Cu(dC/dx)$ in Eq. (2.24) will always result in zero if an attempt is made to use the Rayleigh-Ritz method. Hence Eq. (2.24) was termed a pseudofunction. However, in view of the fact that Eq. (2.24) was used to successfully derive a local finite element and because it will be shown (Prob. 5.13) that the local finite element was correct, it follows that Eq. (2.24) is self-adjoint with respect to some bilinear mapping other than Eq. (5.8). (Review Probs. 2.15 and 2.16.)

5.7.   The equations of static elasticity can be written in subscript tensor notation as

$$\sigma_{ij},_j + f_i = 0 \qquad \sigma_{ij} = \sigma_{ji} \qquad \text{for } i \neq j \tag{a}$$

$$\epsilon_{ij} = \tfrac{1}{2}(u_i,_j + u_j,_i) \tag{b}$$

$$\sigma_{ij} = C_{ijkl}\epsilon_{kl} \tag{c}$$

where σ_{ij}, ϵ_{ij}, and u_i are the stresses, strains, and displacements, respectively. Boundary conditions are specified on displacement and surface traction as

$$u_i = \bar{u}_i \text{ on } \partial u \qquad \text{and} \qquad t_i = \sigma_{ij}n_j = \bar{t}_i \text{ on } \partial t \tag{d}$$

Use Eq. (5.4) to derive a general variational principle that includes boundary conditions. Obtain an extended variational principle in terms of displacement that is suitable for finite element analysis.

Define the vector u of Eq. (5.4) as

$$\{u\} = \{u_m, \epsilon_{ij}, \sigma_{ij}; u_i, t_i\} \tag{e}$$

The f and g terms of Eq. (5.4) are written as one array

$$\{f; g\} = \{f_m, 0, 0; \bar{t}_i, -\bar{u}_i\} \tag{f}$$

A semicolon separates the quantities defined within the region from those defined on the boundary. The operator equation, corresponding to Eq. (5.4), is constructed as

$$J(u) = \{u\} \begin{bmatrix} 0 & 0 & -L & & \\ 0 & C_{ijkl}\delta_{ik}\delta_{jl} & -1 & & \\ L & -1 & 0 & & \\ & & & 0 & 1 \\ & & & -1 & 0 \end{bmatrix} \begin{Bmatrix} u_m \\ \epsilon_{ij} \\ \sigma_{ij} \\ u_i \\ t_i \end{Bmatrix} - 2\{u\} \begin{Bmatrix} f_m \\ 0 \\ 0 \\ \bar{t}_i \\ -\bar{u}_i \end{Bmatrix} \tag{g}$$

where $L = \tfrac{1}{2}(\delta_{im} \partial/\partial x_j + \delta_{jm} \partial/\partial x_i)$. The variational principle is obtained by performing the matrix multiplications:

$$J_1(u) = \{-[u_m, \sigma_{mj},_j] + [\epsilon_{ij}, C_{ijkl}\epsilon_{kl}] - [\epsilon_{ij}, \sigma_{ij}] + [\sigma_{ij}, \tfrac{1}{2}(u_i,_j + u_j,_i)]$$

$$- [\sigma_{ij}, \epsilon_{ij}] - 2[u_m, f_m]\}_\Omega$$

$$+ [u_i, t_i]_{\partial t} - [t_i, u_i]_{\partial u} - 2[u_i, \bar{t}_i]_{\partial t} + 2[t_i, \bar{u}_i]_{\partial u} \tag{h}$$

The first term is rewritten, using the Green-Gauss theorem, as

$$-[u_m, \sigma_{mj},_j]_\Omega = [u_m,_j, \sigma_{mj}]_\Omega - [u_m, t_m]_{\partial t} \tag{i}$$

and combined with the fourth term of Eq. (h) and the first boundary term to give

$$J_2(u) = \{2[\sigma_{ij}, \tfrac{1}{2}(u_i,_j + u_j,_i)] + [\epsilon_{ij}, C_{ijkl}\epsilon_{kl}] - 2[\sigma_{ij}, \epsilon_{ij}] - 2[u_m, f_m]\}_\Omega$$

$$- 2[u_i, \bar{t}_i]_{\partial t} + [t_i, 2\bar{u}_i - u_i]_{\partial u} \tag{j}$$

The extended variational principle is obtained from Eq. (j) by assuming that Eq. (b), Eq. (c), and the first Eq. (d) are identically satisfied and can be eliminated from Eq. (j):

$$J_3(u_i) = [\epsilon_{ij}, C_{ijkl}\epsilon_{kl}]_\Omega - 2[u_i, f_i]_\Omega - 2[u_i, \bar{t}_i]_{\partial t} + [t_i, u_i]_{\partial u} \tag{k}$$

The last term can be omitted since u_i is specified on the boundary ∂u as a constant or zero and the variation with respect to u_i would be zero. Equation (k) can be compared with Eq. (3.13) and the formulation developed in Chap. 3 for two-dimensional elasticity. Identical results can be obtained using Eq. (k).

5.8. The equations of thermoelasticity with time-dependent heat conduction are given below in a general form using notation that is sometimes encountered in continuum mechanics (see Nickell and Sackman, 1968). Derive the variational principle that corresponds to these equations.

Elasticity:

$$\sigma_{ij},_j + f_i = 0 \qquad \sigma_{ij} = \sigma_{ji} \qquad \text{for } i \neq j \tag{a}$$

$$\epsilon_{ij} = \tfrac{1}{2}(u_i,_j + u_j,_i) \tag{b}$$

$$\sigma_{ij} = C_{ijkl}\epsilon_{kl} - \beta\theta\delta_{ij} \tag{c}$$

Heat conduction:

$$q_i,_i + \rho T_0 \frac{\partial\eta}{\partial t} = r \tag{d}$$

$$q_i = -k_{ij}\theta,_j \qquad \theta = T - T_0 \tag{e}$$

$$\rho T_0\eta = c\theta - \beta\epsilon_{ij}\delta_{ij} \tag{f}$$

With boundary conditions and initial conditions

$$u_i = \bar{u}_i \text{ on } \partial u \qquad t_i = \sigma_{ij}n_j = \bar{t}_i \text{ on } \partial t \tag{g}$$

$$\theta = \bar{\theta} \text{ on } \partial\theta \qquad Q = q_i n_i = \bar{Q} \text{ on } \partial Q \tag{h}$$

$$u_i(t=0) = 0 \qquad \text{and} \qquad \theta(t=0) = T_0 \tag{i}$$

where Eq. (i) specifies that the deformation process starts from rest and the initial temperature is T_0 and θ is defined in Eq. (e). Entropy is denoted by η, and all other terms have been defined previously.
    Substitute Eq. (f) into Eq. (d):

$$c\frac{\partial\theta}{\partial t} - \beta\delta_{ij}\frac{\partial\epsilon_{ij}}{\partial t} + q_i,_i = r \tag{j}$$

The governing equations can be written in the operator format given by Eq. (5.2), and it can be verified that the symmetry required by Eq. (5.1) or (5.4) does not exist because of the time derivatives in Eq. (j). A variational principle can be derived using the convolution bilinear mapping of Eq. (5.11). The time derivatives in Eq. (j) are eliminated by taking the Laplace transform, substituting initial conditions, and inverting to give a modified form of the equation:

$$c[\theta - T_0] - \beta\delta_{ij}[\epsilon_{ij} - \epsilon_{ij}(0)] - g*q_i,_i = g*r \tag{k}$$

where $\epsilon_{ij}(0) = 0$ by the first Eq. (i) and $g = g(t) = 1$. The modified equation contains the initial condition on temperature. The field variables are arranged as

$$\{u\} = \{u_k, \epsilon_{ij}, \sigma_{ij}, q_i, \theta; u_i, t_i, \theta, Q\}$$

with $$\{f; g\} = \{f_k, 0, 0, 0, g*r + cT_0; \bar{t}_i, -\bar{u}_i, g*\bar{Q}, -g*\bar{\theta}\}$$

The operator A is constructed using Eqs. (a)–(c), the first (e), (g), (h), and (k):

$$A = \begin{bmatrix} 0 & 0 & -L & 0 & 0 \\ 0 & C_{ijkl}\delta_{ik}\delta_{jl} & -1 & 0 & -\beta\delta_{ij} \\ L & -1 & 0 & 0 & 0 \\ 0 & 0 & 0 & -g^*k_{ij}\delta_{ij} & -g^*\,\partial/\partial x_i \\ 0 & -\beta\delta_{ij} & 0 & g^*\,\partial/\partial x_i & c \end{bmatrix}$$

$$\begin{bmatrix} 0 & 1 & 0 & 0 \\ -1 & 0 & 0 & 0 \\ 0 & 0 & 0 & g^* \\ 0 & 0 & -g^* & 0 \end{bmatrix}$$

where $L = \frac{1}{2}(\delta_{ik}\,\partial/\partial x_j + \delta_{jk}\,\partial/\partial x_i)$. The variational principle is obtained as in previous problems and can be written

$$
\begin{aligned}
J_1(u) = &\{-[u_m{}^*\sigma_{mj,j}] + [\epsilon_{ij}{}^*C_{ijkl}\epsilon_{kl}] - [\epsilon_{ij}{}^*\sigma_{ij}] - [\epsilon_{ij}{}^*\beta\theta\delta_{ij}] \\
&+ [\sigma_{ij}{}^*\tfrac{1}{2}(u_{i,j}+u_{j,i})] - [\sigma_{ij}{}^*\epsilon_{ij}] - [g^*q_i{}^*k_{ij}^{-1}q_j] - [g^*q_i{}^*\theta_{,i}] \\
&- [\beta\theta\delta_{ij}{}^*\epsilon_{ij}] + [g^*\theta^*q_{i,i}] + [\theta^*c\theta] - 2[u_m{}^*f_m] - 2\theta^*[g^*r + cT_0]\}_\Omega \\
&+ [u_i{}^*t_i]_{\partial t} - [t_i{}^*u_i]_{\partial u} + [g^*\theta^*Q]_{\partial Q} - [g^*Q^*\theta]_{\partial\theta} \\
&- 2[u_i{}^*\bar{t}_i]_{\partial t} + 2[t_i{}^*\bar{u}_i]_{\partial u} - 2[g^*\theta^*\bar{Q}]_{\partial Q} + 2[g^*Q^*\bar{\theta}]_{\partial\theta}
\end{aligned}
\tag{l}
$$

5.9. Obtain an extended variational principle for coupled thermoelasticity that is suitable for finite element analysis.

Refer to Prob. 5.8 and assume that Eq. (b), Eq. (e), the first Eq. (g), and the first Eq. (h) are identically satisfied. The Green-Gauss theorem is used to modify terms containing $\sigma_{mj,j}$ and $q_{i,i}$ as

$$-[u_m{}^*\sigma_{mj,j}]_\Omega = [u_{m,j}{}^*\sigma_{mj}]_\Omega - [u_m{}^*t_m]_{\partial t}$$

$$[g^*\theta^*q_{i,i}]_\Omega = -[g^*\theta_i{}^*q_i]_\Omega + [g^*\theta^*Q]_{\partial Q}$$

Substitute into Eq. (l) of Prob. 2.8:

$$
\begin{aligned}
J_2(u_i, \theta) = &\{[\epsilon_{ij}{}^*C_{ijkl}{}^*\epsilon_{kl}] + [g^*k_{ii}\theta^*\theta] + [\theta^*c\theta] - 2[\epsilon_{ii}{}^*\beta\theta] \\
&- 2[u_m f_m] - 2[g^*\theta^*r] - 2[\theta^*cT_0]\}_\Omega - 2[g^*u_i{}^*\bar{t}_i]_{\partial t} - 2[g^*\theta^*\bar{Q}]_{\partial Q}
\end{aligned}
\tag{a}
$$

5.10. The one-dimensional steady-state heat-conduction equation is

$$\frac{d^2T}{dx^2} = \frac{Q}{k} \tag{a}$$

Assume boundary conditions $T(0) = T(L) = 0$, and the exact solution is

$$T = \frac{Q}{2k}(x^2 - xL) \tag{b}$$

obtained by integrating and substituting boundary conditions. Assume a three-term solution in the form of Eq. (*5.12*) and obtain an approximate solution using the Galerkin method.

The Galerkin method of numerical analysis will be illustrated. Assuming an approximate solution in the form of a second-order equation will give the exact solution since Eq. (b) above is second order. Assume

$$T_R = a_0 + a_1 x + a_2 x^2$$

Substituting boundary conditions gives

$$T(0) = 0 = a_0 \qquad \text{and} \qquad T(L) = 0 = a_0 + a_1 L + a_2 L^2$$

or $a_0 = 0$ and $a_2 = -a_1/L$, and the assumed solution is

$$T_R = a_1 \left(x - \frac{x^2}{L} \right) \tag{c}$$

Substitute into the governing equation as

$$\frac{d^2 T_R}{dx^2} - \frac{Q}{k} = R$$

to obtain

$$-2 \frac{a_1}{L} - \frac{Q}{k} = R \tag{d}$$

Since there is only one unknown a_1, there will be only one weighted residual equation. In the notation of Eq. (5.14),

$$\int_\Omega w_i(x) R(x : a_i) = \int_0^L w_1(x) R(x; a_1) dx = 0$$

or

$$\int_0^L \left(x - \frac{x^2}{L} \right) \left(-2 \frac{a_1}{L} - \frac{Q}{k} \right) dx = 0$$

Integrating, substituting limits, and solving for a_1 gives $a_1 = -QL/2k$. Substituting into Eq. (c) gives the exact solution, Eq. (b). The same result was obtained in Prob. 2.3 using the Rayleigh-Ritz method. As noted previously, when a variational principle exists for a given problem, the Rayleigh-Ritz solution and the Galerkin solution are the same.

5.11. Use the Galerkin method and the trial solution

$$v_R = a_0 + a_1 x + a_2 x^2 + a_4 x^4 \tag{a}$$

to obtain an approximate solution for the cable that was analyzed in Probs. 2.1 and 2.12.

The governing differential equation is

$$\frac{d^2 v}{dx^2} - \frac{kv}{T} + \frac{f}{T} = 0 \qquad v(0) = v(L) = 0 \tag{a}$$

The trial solution that satisfies the boundary conditions becomes

$$v_R = a_2(x^2 - xL) + a_4(x^4 - xL^3) \tag{b}$$

Substituting Eq. (b) into Eq. (a) gives the residual

$$2a_2 + 12a_4 x^2 - \left(\frac{k}{T} \right) [a_2(x^2 - xL) + a_4(x^4 - xL^3)] + \frac{f}{T} = R$$

There are two weighted equations corresponding to Eq. (5.14):

$$\int_0^L \left\{ 2a_2 + 12a_4 x^2 - \left(\frac{k}{T} \right) [a_2(x^2 - xL) + a_4(x^4 - xL^3)] + \frac{f}{T} \right\} (x^2 - xL) \, dx \tag{c}$$

$$\int_0^L \left\{ 2a_2 + 12a_4 x^2 - \left(\frac{k}{T} \right) [a_2(x^2 - xL) + a_4(x^4 - xL^3)] + \frac{f}{T} \right\} (x^4 - xL^3) \, dx \tag{d}$$

After integration Eqs. (c) and (d) become

$$a_2\left(-\frac{1}{3}+\frac{kL^2}{30T}\right)+a_4\left(-\frac{3L^2}{5}-\frac{5kL^4}{84T}\right)=\frac{f}{6T}$$

$$a_2\left(-\frac{6}{10}-\frac{5kL^2}{84T}\right)+a_4\left(-\frac{12L^2}{7}-\frac{kL^4}{9T}\right)=\frac{3f}{10T}$$

Substitute $L = 120$, $T = 600$, $f = 2$, and $k = \frac{1}{2}$ and solve for $a_2 = -7.1603(10^{-4})$ and $a_4 = -1.5816(10^{-9})$. The approximate solution is

$$v_R = -7.1603(10^{-4})(x^2 - xL) - 1.5816(10^{-9})(x^4 - xL^3)$$

Comparison with Prob. 2.12 gives

Node 2: $v_R(x = 24) = 1.7148$, $v(x = 24) = 1.8173$.

Node 3: $v_R(x = 48) = 2.5974$, $v(x = 48) = 2.5444$.

5.12. Derive a four-node finite element for two-dimensional steady-state heat conduction using the Galerkin method.

The element will be derived in detail to illustrate how shape functions are used as approximating functions similar to the polynomials of Eq. (*5.12*). The governing differential equation for steady-state heat conduction is

$$k_x \frac{\partial^2 T}{\partial x^2} + k_y \frac{\partial^2 T}{\partial y^2} - Q = 0 \qquad (a)$$

The interpolation function is the same as that used in Prob. 3.4. Let T_R be the approximate solution and

$$T_R = N_1(x, y)T_1 + N_2(x, y)T_2 + N_3(x, y)T_3 + N_4(x, y)T_4 = [N]\{T\} \qquad (b)$$

where the shape functions are defined in Eq. (*e*) of Prob. 3.1 and $\{T\}$ are the unknown temperatures at each finite element node and take the place of the unknown coefficients a_i in Eq. (*5.12*). Note that only the shape functions [N] are functions of x and y in Eq. (*b*) and that substituting Eq. (*b*) into Eq. (*a*) gives the residual as a function of x, y, and $\{T\}$. Formally, the result can be written as

$$k_x \frac{\partial^2 T_R}{\partial x^2} + k_y \frac{\partial^2 T_R}{\partial y^2} - Q = R(x, y: \{T\}) \qquad (c)$$

As an aid in visualizing the process of forming the residual, Eq. (*b*) is substituted into Eq. (a) and the results written in matrix format (see Prob. 3.4):

$$[\partial/\partial x \quad \partial/\partial y]\begin{bmatrix} k_x & 0 \\ 0 & k_y \end{bmatrix}\begin{bmatrix} \partial/\partial x \\ \partial/\partial y \end{bmatrix}[N_1 \quad N_2 \quad N_3 \quad N_4]\{T\} - [Q] = [R(x, y: \{T\})] \qquad (d)$$

The final result is a 1×1 matrix. The weighted residual of Eq. (*5.14*) is formulated as a matrix equation with w_i replaced by [N] and the integral taken over the volume, or $\int_V [N]^T [R(x, y: \{T\})]\, dV = 0$. The first term of Eq. (*d*) becomes

$$\int_V [N]^T\left[k_x \frac{\partial^2 N_1}{\partial x^2} + k_y \frac{\partial^2 N_1}{\partial y^2} \quad k_x \frac{\partial^2 N_2}{\partial x^2} + k_y \frac{\partial^2 N_2}{\partial y^2} \quad k_x \frac{\partial^2 N_3}{\partial x^2} + k_y \frac{\partial^2 N_3}{\partial y^2} \quad k_x \frac{\partial^2 N_4}{\partial x^2} + k_y \frac{\partial^2 N_4}{\partial y^2} \right]\{T\}\, dV = 0 \quad (e)$$

The final result is a 4×4 matrix postmultiplied by $\{T\}$. The second term of Eq. (*d*) is $\int_V [N]^T Q\, dV$. The first term of Eq. (*e*), after matrix multiplication, appears as follows and can be transformed using the divergence theorem (the Green-Gauss theorem) to give

$$\int_V \left(N_1 k_x \frac{\partial^2 N_1}{\partial x^2} + N_1 k_y \frac{\partial^2 N_1}{\partial y^2} \right) dV = -\int_V \left(\frac{\partial N_1}{\partial x} k_x \frac{\partial N_1}{\partial x} + \frac{\partial N_1}{\partial y} k_y \frac{\partial N_1}{\partial y} \right) dV$$

$$+ \int_S \left(N_1 k_x \frac{\partial N_1}{\partial x} n_x + N_1 k_y \frac{\partial N_1}{\partial x} n_y \right) dS \qquad (f)$$

The volume integral of Eq. (f) is the same as Eq. (c) of Prob. 3.5, and it becomes obvious that the final result for the local stiffness matrix will be the same as Eq. (e) of Prob. 3.5. The surface integrals of Eq. (f) correspond to the flux boundary conditions. As stated previously and illustrated here, the Galerkin method always gives information concerning natural boundary conditions. The heat source term is identical to Eq. (f) of Prob. 3.5.

In finite element literature a shorthand notation is often used that implies all of the above. Substituting Eq. (b) into Eq. (c) can be written as

$$k_x \frac{\partial^2 [N]}{\partial x^2} \{T\} + k_y \frac{\partial^2 [N]}{\partial y^2} \{T\} - Q = R(x, y: \{T\}) \qquad (g)$$

or, in a more abbreviated form, as

$$[k] \frac{\partial^2 [N]}{\partial x_i^2} \{T\} - Q = R(x, y: \{T\}) \qquad (h)$$

Using Eq. (h), Eq. (e) can be written as

$$\int_V \left([N]^T [k] \frac{\partial^2 [N]}{\partial x_i^2} \{T\} - [N]^T Q \right) dV = 0 \qquad (i)$$

The matrix [N] in Eqs. (h) and (i) cannot be the same as the matrix [N] defined in Eq. (d) and used in Eq. (e). The material constant matrix will remain 2×2 as defined in Eq. (d). Define the shape function matrix and an operator matrix as

$$[N] = \begin{bmatrix} N_1 & N_2 & N_3 & N_4 \\ N_1 & N_2 & N_3 & N_4 \end{bmatrix} \qquad [L] = \begin{bmatrix} \partial^2/\partial x^2 & 0 \\ 0 & \partial^2/\partial y^2 \end{bmatrix} \qquad (j)$$

The use of Eqs. (j) in Eq. (h) will give the equivalent of Eq. (d). It follows from Eq. (f) that the final result is

$$\int_V \left(\frac{\partial [N]^T}{\partial x_i} [k] \frac{\partial [N]}{\partial x_i} \{T\} + [N]^T Q \right) dV = \int_S [N]^T [k] \frac{\partial [N]}{\partial x_i} \{T\} n_i \, dS \qquad (k)$$

The analyst must interpret the specific form of [N], x_i, [k], and {T} that is to be used since they must correspond to the element and coordinate system being used. In subsequent developments the abbreviated form will be used for deriving local finite elements.

5.13. Use the Galerkin method to derive a four-node local finite element for mass transport with convection and chemical reactions. Compare the results with the model derived in Chaps. 2 and 3. This result will establish that the formulation of Chaps. 2 and 3 using the pseudovariational function was correct.

The governing differential equation is given as Eq. (3.2), and the development of the finite element follows from Prob. 5.12. Assume a trial solution as $T_R = [N]\{C\}$, where [N] is defined as in Eq. (d) of Prob. 5.12, and substitute into the governing differential equation:

$$u_x \frac{\partial [N]\{C\}}{\partial x} + u_y \frac{\partial [N]\{C\}}{\partial y} - D_x \frac{\partial^2 [N]\{T\}}{\partial x^2} - D_y \frac{\partial^2 [N]\{T\}}{\partial y^2} + K_r [N]\{C\} - m = R(x, y: \{C\})$$

Define the matrices

$$[u] = \begin{bmatrix} u_x & 0 \\ 0 & u_y \end{bmatrix} \qquad [D] = \begin{bmatrix} D_x & 0 \\ 0 & D_y \end{bmatrix} \qquad [K_r] = \begin{bmatrix} K_r/2 & 0 \\ 0 & K_r/2 \end{bmatrix}$$

Following the process used in the previous problem, multiply by the matrix of shape functions (the weighting functions), integrate over the volume, and set the result equal to zero:

$$\int_V \left([N]^T[u]\frac{\partial[N]}{\partial x_i}\{C\} - [N]^T[D]\frac{\partial^2[N]}{\partial x_i^2}\{C\} + [N]^T[K_r][N]\{C\} - [N]^T[m] \right) dV = 0 \tag{a}$$

Note that [N] of Eq. (a) is defined in Eq. (j) of Prob. 5.12 and that [K_r] above is defined as a 2×2 matrix using $K_r/2$ since [N] is defined as a 2×4 matrix. Again, use the Green-Gauss theorem to reduce the order of the second term in Eq. (a). The final result is

$$\int_V \left([N]^T[u]\frac{\partial[N]}{\partial x_i}\{C\} + \frac{\partial[N]^T}{\partial x_i}[D]\frac{\partial[N]}{\partial x_i}\{C\} + [N]^T[K_r][N]\{C\} - [N]^T[m] \right) dV$$

$$= \int_S \left([N]^T[D]\frac{\partial[N]}{\partial x_i}\{C\} \right) dS \tag{b}$$

Substitute shape functions or their derivatives into Eq. (b) and perform the indicated matrix multiplications. The integration is for a plane area, and the final result will be the same as that for Prob. 3.32. The diffusion-type flux boundary condition is obtained as a result of the Galerkin method.

5.14. A theory of coupled mass diffusion was discussed by Aifantis (1979) and can be applied to materials with multiple diffusivities. The time-dependent equations are (using subscript notation)

$$\frac{\partial C_1}{\partial t} + \frac{\partial j_{1i}}{\partial x_i} = m_1 = -k_1 C_1 + k_2 C_2 \tag{a}$$

$$\frac{\partial C_2}{\partial t} + \frac{\partial j_{2i}}{\partial x_i} = m_2 = k_1 C_1 - k_2 C_2 \tag{b}$$

where C_1 and C_2 are the concentrations, j_{1i} and j_{2i} are the corresponding flux terms, and $m_1 + m_2 = 0$. A general statement for the flux terms can be written as

$$j_{1i} = -D_{11}\frac{\partial C_1}{\partial x_i} + D_{12}\frac{\partial C_2}{\partial x_i} \tag{c}$$

$$j_{2i} = +D_{21}\frac{\partial C_1}{\partial x_i} - D_{22}\frac{\partial C_2}{\partial x_i} \tag{d}$$

with $D_{12} = D_{21}$. Use the Galerkin method to derive the corresponding finite element statement of this theory.

Substitute Eqs. (c) and (d) into Eqs. (a) and (b), respectively, to obtain the governing equations. There are two coupled second-order equations:

$$\frac{\partial C_1}{\partial t} - D_{11}\frac{\partial^2 C_1}{\partial x_i^2} + D_{12}\frac{\partial^2 C_2}{\partial x_i^2} + k_1 C_1 - k_2 C_2 = 0 \tag{e}$$

$$\frac{\partial C_2}{\partial t} - D_{22}\frac{\partial^2 C_2}{\partial x_i^2} + D_{12}\frac{\partial^2 C_1}{\partial x_i^2} - k_1 C_1 + k_2 C_2 = 0 \tag{f}$$

Assume approximate (trial) solutions as $C_{1R} = [N_1]\{C_1\}$ and $C_{2R} = [N_2]\{C_2\}$, where $[N_1]$ and $[N_2]$ are any matrices defining shape functions. They may be identical assumptions, but for the purpose of derivation they will be identified separately. The matrices $\{C_1\}$ and $\{C_2\}$ are the corresponding unknown nodal point values. Substitute into Eqs. (e) and (f):

$$[N_1]\frac{\partial\{C_1\}}{\partial t} - [D_{11}]\frac{\partial^2[N_1]}{\partial x_i^2}\{C_1\} + [D_{12}]\frac{\partial_2[N_2]}{\partial x_i^2}\{C_2\} + [k_1][N_1]\{C_1\} - [k_2][N_2]\{C_2\} = R_1 \qquad (g)$$

$$[N_2]\frac{\partial\{C_2\}}{\partial t} - [D_{22}]\frac{\partial^2[N_2]}{\partial x_i^2}\{C_2\} + [D_{12}]\frac{\partial_2[N_1]}{\partial x_i^2}\{C_1\} - [k_1][N_1]\{C_1\} + [k_2][N_2]\{C_2\} = R_2 \qquad (h)$$

Equations (g) and (h) are the residuals and are minimized by multiplying by the weight function. Equation (g) is premultiplied by $[N_1]$, and Eq. (h) is premultiplied by $[N_2]$. Integrating over the volume and setting the result equal to zero gives

$$\int_V \left([N_1]^T[N_1]\frac{\partial\{C_1\}}{\partial t} - [N_1]^T[D_{11}]\frac{\partial^2[N_1]}{\partial x_i^2}\{C_1\} + [N_1]^T[D_{12}]\frac{\partial_2[N_2]}{\partial x_i^2}\{C_2\}\right.$$

$$\left. - [N_1]^T[k_1][N_1]\{C_1\} + [N_1]^T[k_2][N_2]\{C_2\}\right) dV = 0$$

$$\int_V \left([N_2]^T[N_2]\frac{\partial\{C_2\}}{\partial t} - [N_2]^T[D_{22}]\frac{\partial^2[N_2]}{\partial x_i^2}\{C_2\} + [N_2]^T[D_{12}]\frac{\partial_2[N_1]}{\partial x_i^2}\{C_1\}\right.$$

$$\left. - [N_2]^T[k_1][N_1]\{C_1\} + [N_2]^T[k_2][N_2]\{C_2\}\right) dV = 0$$

The Green-Gauss theorem is used to reduce the second-order derivatives to first order and incorporate the flux boundary conditions. The final result is

$$\int_V \left([N_1]^T[N_1]\frac{\partial\{C_1\}}{\partial t} + \frac{\partial[N_1]^T}{\partial x_i}[D_{11}]\frac{\partial[N_1]}{\partial x_i}\{C_1\} - \frac{\partial[N_1]^T}{\partial x_i}[D_{12}]\frac{\partial[N_2]}{\partial x_i}\{C_2\} + [N_1]^T[k_1][N_1]\{C_1\}\right.$$

$$\left. - [N_1]^T[k_2][N_2]\{C_2\}\right) dV = \int_{S_1}[N_1]^T[D_{11}]\frac{\partial[N_1]}{\partial x_i}\{C_1\}n_i\, dS_1 - \int_{S_2}[N_1]^T[D_{12}]\frac{\partial[N_2]}{\partial x_i}\{C_2\}n_i\, dS_2 \quad (i)$$

$$\int_V \left([N_2]^T[N_1]\frac{\partial\{C_2\}}{\partial t} + \frac{\partial[N_2]^T}{\partial x_i}[D_{22}]\frac{\partial[N_2]}{\partial x_i}\{C_2\} - \frac{\partial[N_2]^T}{\partial x_i}[D_{12}]\frac{\partial[N_1]}{\partial x_i}\{C_1\} - [N_2]^T[k_1][N_1]\{C_1\}\right.$$

$$\left. + [N_2]^T[k_2][N_2]\{C_2\}\right) dV = \int_{S_2}[N_2]^T[D_{22}]\frac{\partial[N_2]}{\partial x_i}\{C_2\}n_i\, dS_2 - \int_{S_1}[N_2]^T[D_{12}]\frac{\partial[N_1]}{\partial x_i}\{C_1\}n_i\, dS_1 \quad (j)$$

The subscripts on the shape functions indicate the procedure for modeling C_1 and C_2 using different shape functions. For instance, N_1 could correspond to an eight-node element, while N_2 could represent the same element with only four nodes. This separation of shape functions is not so obvious when the variational formulation is used. However, for this coupled problem both unknowns should be modeled with the same shape function.

  The time derivatives remain in Eqs. (i) and (j); they will be discussed in detail in Chap. 7. Note that the time derivatives operate on the nodal point variables since the shape function is independent of time.

5.15. Assume steady-state conditions and a two-dimensional formulation in x and y for the coupled diffusion equations of Prob. 5.14. Discuss the formulation of the local stiffness matrix using four-node rectangular finite elements, assuming $N_1 = N_2 = N$.

  Delete the time derivative and write Eqs. (i) and (j) of Prob. 5.14 in a compact form that defines each term in the stiffness matrix

$$\begin{bmatrix} [K_{D11} + K_{k1}] & [-K_{D12} - K_{k2}] \\ [-K_{D21} - K_{k1}] & [K_{D22} + K_{k2}] \end{bmatrix} \begin{Bmatrix} \{C_1\} \\ \{C_2\} \end{Bmatrix} = \begin{Bmatrix} \{j_{11}\} \\ \{j_{22}\} \end{Bmatrix} - \begin{Bmatrix} \{j_{12}\} \\ \{j_{21}\} \end{Bmatrix} \qquad (a)$$

Each term in Eq. (a) has a counterpart in Chap. 3. For instance,

$$[K_{D11}] = \int_0^a \int_0^b [B]^T [D_{11}][B] \, dx \, dy \qquad (b)$$

where $[B] = [L][N]$ as in Probs. 3.4 and 3.5. The stiffness terms $[K_{D22}]$, $[K_{D12}]$, and $[K_{D21}]$ have similar definitions with a different diffusivity matrix, and the subscripts correspond to the diffusivity matrix. Note that $[K_{D21}] = [K_{D12}]^T$. Also,

$$[K_{k1}] = \int_0^a \int_0^b [N]^T [k_1][N] \, dx \, dy \qquad (c)$$

with a similar definition for $[K_{k2}]$. The local stiffness matrix given by Eq. (c) is identical to the result given in Prob. 3.34, with the chemical reaction coefficients K_r replaced by the mass interaction coefficients k_1. Note that the matrices $[k_1]$ and $[k_2]$ should be defined as 2×2 matrices, as illustrated in Prob. 5.13, since $[N]$ is defined as a 2×4 matrix.

The local stiffness matrix, formed from Eq. (a), is not symmetric. The boundary condition terms on the right-hand side of Eq. (a) correspond to applied boundary fluxes. Also, if the coupling defined by the stiffness terms $[K_{D12}]$ and $[K_{D21}]$ is neglected, it will be necessary to omit $\{j_{12}\}$ and $\{j_{21}\}$ since $[D_{12}]$ would be assumed as zero.

5.16. Reduce the equations of Prob. 5.15 to one dimension in x. Assume $D_{12} = 0$ and obtain an analytical solution following Aifantis (1979) for the resulting equations. Obtain a three-element and a five-element solution using the four-node rectangular local elements discussed in Prob. 5.15. Arrange the elements in a strip as shown in Fig. 5-1(a). For the analytical solution assume boundary conditions as $C_1(0) = C_2(0) = 0$, $C_1(L) = C_{1L}$, and $C_2(L) = C_{2L}$. Compare the solutions assuming $D_{11} = D_{22} = 0.005$, $k_1 = 0.04$, $k_2 = 0.03$, $C_{1L} = C_{2L} = 100$, $L = 1$, and $D_{12} = 0$.

The governing equations in one dimension are

$$D_{11} \frac{d^2 C_1}{dx^2} - k_1 C_1 + k_2 C_2 = 0 \qquad D_{22} \frac{d^2 C_2}{dx^2} + k_1 C_1 - k_2 C_2 = 0 \qquad (a)$$

with solution

$$C_1 = \gamma \sinh(\alpha x) + \frac{\beta x}{\alpha^2} \qquad C_2 = \frac{Ax - D_{11} C_1}{D_{22}} \qquad (b)$$

where $\qquad A = \dfrac{D_{11} C_{1L} + D_{22} C_{2L}}{L} \qquad \alpha^2 = \dfrac{D_{22} k_1 + D_{11} k_2}{D_{11} D_{22}} \qquad \beta = \dfrac{k_2 A}{D_{11} D_{22}} \qquad \gamma = \dfrac{C_{1L} - \beta L/\alpha^2}{\sinh(\alpha L)}$

The matrix equation of Prob. 5.15 lends itself to a local node numbering system that groups the C_1 unknowns first and then the C_2 unknowns. It follows that the local finite element will be as shown in Fig. 5-1(b). It is standard practice in finite element computer programming to arrange the unknowns (degrees of freedom) in the global system such that all unknowns at a node are grouped together. It is desirable to rearrange the rows and columns in the local matrix to conform to the global system before assembling the global matrix. This rearrangement can be accomplished using a transformation matrix of the type discussed in Prob. 3.24. The transformation, using the notation of Fig. 5-1(*b*), is written as $\{C\}_{new} = [T]\{C\}_{old}$:

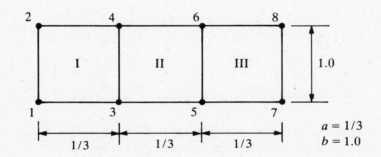

$$
\begin{Bmatrix} C_{1i} \\ C_{2i} \\ C_{1j} \\ C_{2j} \\ C_{1k} \\ C_{2k} \\ C_{1l} \\ C_{2l} \end{Bmatrix} = \begin{bmatrix} 1 & 0 & 0 & 0 & 0 & 0 & 0 & 0 \\ 0 & 0 & 0 & 0 & 1 & 0 & 0 & 0 \\ 0 & 1 & 0 & 0 & 0 & 0 & 0 & 0 \\ 0 & 0 & 0 & 0 & 0 & 1 & 0 & 0 \\ 0 & 0 & 1 & 0 & 0 & 0 & 0 & 0 \\ 0 & 0 & 0 & 0 & 0 & 0 & 1 & 0 \\ 0 & 0 & 0 & 1 & 0 & 0 & 0 & 0 \\ 0 & 0 & 0 & 0 & 0 & 0 & 0 & 1 \end{bmatrix} \begin{Bmatrix} C_{1i} \\ C_{1j} \\ C_{1k} \\ C_{1l} \\ C_{2i} \\ C_{2j} \\ C_{2k} \\ C_{2l} \end{Bmatrix}
$$

(*c*)

where $[K]_{new} = [T][K]_{old}[T]^{T}$.

The three-element solution is obtained using three identical finite elements with $a = \frac{1}{3}$ and $b = 1.0$. The local stiffness matrix is computed as follows before transformation.

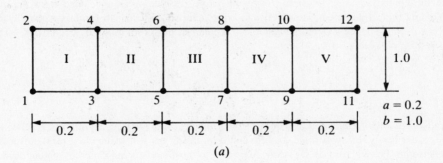

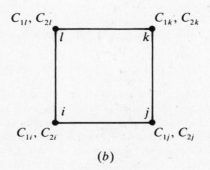

(*b*)

Fig. 5-1

$$
10^{-3}
\begin{bmatrix}
6.4815 & -4.2593 & -2.1297 & 3.2408 & -1.1111 & -0.5555 & -0.2777 & -0.5555 \\
-4.2593 & 6.4815 & 3.2408 & -2.1297 & -0.5555 & -1.1111 & -0.5555 & -0.2777 \\
-2.1297 & 3.2408 & 6.4815 & -4.2593 & -0.2777 & -0.5555 & -1.1111 & -0.5555 \\
3.2408 & -2.1297 & -4.2593 & 6.4815 & -0.5555 & -0.2777 & -0.5555 & -1.1111 \\
-1.4815 & -0.7407 & -0.3704 & -0.7407 & 6.1111 & -4.4445 & -2.2222 & 3.0556 \\
-0.7407 & -1.4815 & -0.7407 & -0.3704 & -4.4445 & 6.1111 & 3.0556 & -2.2222 \\
-0.3704 & -0.7407 & -1.4815 & -0.7407 & -2.2222 & 3.0556 & 6.1111 & -4.4445 \\
-0.7407 & -0.3704 & -0.7407 & -1.4815 & 3.0556 & -2.2222 & -4.4445 & 6.1111
\end{bmatrix}
$$

The three local stiffness matrices are assembled, and the results are tabulated in Table 5.1. The five-element solution is obtained using five identical local elements with $a = 0.2$ and $b = 1.0$. The three solutions are compared in Table 5.1.

Table 5.1   Coupled Diffusion Problem

x	Three-element solution		Five-element solution		Exact solution	
	C_1	C_2	C_1	C_2	C_1	C_2
0.0	0.0	0.0	0.0	0.0	0.0	0.0
0.2	—	—	17.664	22.335	17.699	22.301
0.333	29.475	37.191	—	—	29.654	37.013
0.4	—	—	35.651	44.349	35.734	44.276
0.6	—	—	54.480	65.520	54.593	65.407
0.667	60.848	72.486	—	—	61.221	72.111
0.8	—	—	75.195	84.805	75.318	84.682
1.0	100.0	100.0	100.0	100.0	100.0	100.0

5.17. Deduce the equations of steady-state thermoelasticity from Prob. 5.8. Use the Galerkin method to formulate a two-dimensional four-node finite element for this theory.

This formulation illustrates the coupling between the displacement solution and the temperature solution. The differential equation governing the temperature distribution is uncoupled from the differential equations governing the displacement. However, it is convenient and efficient to formulate one element for both effects. Refer to Prob. 5.8; the governing equations are obtained by substituting Eq. (c) into Eq. (a) and combining Eqs. (d) and (e) with $\partial/\partial t = 0$:

$$C_{ijkl}\epsilon_{kl,j} - \beta\theta_{,i} = -f_i, \tag{a}$$

$$k_{ij}\theta_{,ij} = -r \qquad k_{ij} = 0 \qquad \text{for } i \neq j \tag{b}$$

It follows that the formulation of a finite element for Eq. (b) is the same as in Prob. 5.12, with Q replaced by $-r$. The reader has probably guessed that the first term in Eq. (a) will reduce to a local stiffness matrix identical to that given in Prob. 3.10. It is, however, important to illustrate how the Galerkin method leads to the same result. It is quite elementary to translate the continuum mechanics statement for a physical problem directly into a general finite element statement of the problem. For instance, Eqs. (a) and (b) become

$$[K_{uu}]\{u\} - [K_{u\theta}]\{\theta\} = \{f\} + \{\bar{t}\} \tag{c}$$

$$[K_{\theta\theta}]\{\theta\} = \{r\} + \{\bar{Q}\} \tag{d}$$

The evaluation of each matrix is dependent upon the structure of the governing differential equations. The components $[K_{uu}]$ and $[K_{u\theta}]$ will be derived in detail using the Galerkin method.

Equation (a) of Prob. 5.8 is called an equilibrium equation in the theory of elasticity; in two dimensions there are two equations. The subscript tensor equation can be expanded to become Eqs. (3.4) and (3.5), the equilibrium equations of plane elasticity. Equation (c) of Prob. 5.8 would correspond to the set (3.6)–(3.8) or (3.10), (3.11), and (3.8), either plane stress or plane strain, with additional terms to relate stress and temperature. In general the stress-strain-temperature relations can be written

$$\sigma_{xx} = C_{11}\epsilon_{xx} + C_{12}\epsilon_{yy} - \beta\theta \tag{e}$$

$$\sigma_{yy} = C_{12}\epsilon_{xx} + C_{22}\epsilon_{yy} - \beta\theta \tag{f}$$

$$\sigma_{xy} = C_{33}\epsilon_{xy} \tag{g}$$

Strain-displacement relations given by Eq. (b) of Prob. 5.8 are the mathematical definition. In the finite element formulation the engineering definition given by Eq. (3.3) should be used.

Substituting Eqs. (e)–(g) into Eqs. (3.4) and (3.5) gives

$$C_{11}\frac{\partial\epsilon_{xx}}{\partial x} + C_{12}\frac{\partial\epsilon_{yy}}{\partial x} + C_{33}\frac{\partial\epsilon_{xy}}{\partial y} - \beta\frac{\partial\theta}{\partial x} + f_x = 0 \tag{h}$$

$$C_{12}\frac{\partial\epsilon_{xx}}{\partial y} + C_{22}\frac{\partial\epsilon_{yy}}{\partial y} + C_{33}\frac{\partial\epsilon_{xy}}{\partial x} - \beta\frac{\partial\theta}{\partial y} + f_y = 0 \tag{i}$$

The matrix equivalent of Eqs. (3.4) and (3.5) can be written as

$$\begin{bmatrix} \partial/\partial x & 0 & \partial/\partial x \\ 0 & \partial/\partial y & \partial/\partial y \end{bmatrix} \begin{Bmatrix} \sigma_{xx} \\ \sigma_{yy} \\ \sigma_{xy} \end{Bmatrix} = - \begin{Bmatrix} f_x \\ f_y \end{Bmatrix} \tag{j}$$

The matrix equivalent of Eqs. (e)–(g) becomes

$$\begin{Bmatrix} \sigma_{xx} \\ \sigma_{yy} \\ \sigma_{xy} \end{Bmatrix} = \begin{bmatrix} C_{11} & C_{12} & 0 \\ C_{12} & C_{22} & 0 \\ 0 & 0 & C_{33} \end{bmatrix} \begin{Bmatrix} \epsilon_{xx} \\ \epsilon_{yy} \\ \epsilon_{xy} \end{Bmatrix} - \begin{Bmatrix} \beta \\ \beta \\ 0 \end{Bmatrix} [\theta] \tag{k}$$

The strains are defined in Eq. (h) of Prob. 3.10 and are substituted into Eq. (k) and that result substituted into Eq. (j). The resulting differential equations appear in Prob. 5.27. Nodal displacements are defined by Eqs. (e)–(g) of Prob. 3.10. Strains are defined by Eq. (i) of Prob. 3.10, which can be written as

$$\{\epsilon\} = [L_u][N_u]\{u\}$$

where $[N_u]$ is defined by Eq. (b) of Prob. 3.11 for a four-node element. The displacements are assumed following Eq. (g) of Prob. 3.10, and the temperature is assumed as in Eq. (b) of Prob. 5.12. Both trial solutions are assumed in terms of shape functions for a four-node rectangular element, however, at this stage the matrices of the shape functions appear differently and are written

$$\begin{Bmatrix} u_e \\ v_e \end{Bmatrix}_R = [N_u]\{u\} \qquad \text{and} \qquad \theta_R = [N_\theta]\{\theta\} \tag{l}$$

The weighted residuals of Eqs. (a) and (b) are written using the shorthand notation introduced in Prob. 5.12. (See Prob. 5.29 for additional details.)

$$\int_V \left([N_u]^T[C]\frac{\partial^2[N_u]}{\partial x_i^2}\{u\} - [N_u]^T[\beta]\frac{\partial[N_\theta]}{\partial x_i}\{\theta\} + [N_u]^T[f] \right) dV = 0 \tag{m}$$

$$\int_V \left([N_\theta]^T[k]\frac{\partial^2[N_\theta]}{\partial x_i^2}\{\theta\} + [N_\theta]^T[r] \right) dV = 0 \tag{n}$$

Application of the Green-Gauss theorem to the terms containing derivatives gives the final result. The first term in Eq. (m) is modified to change the function $[N_u]$ to a first derivative, and the second derivative reduces to a first derivative as shown in Eq. (o). In addition, a traction boundary condition on displacement is introduced. The second term in Eq. (m) is modified using the Green-Gauss theorem [see the second term in Eq. (o)] and completes the traction boundary condition to add a temperature term in the form of the heat flux. The constitutive equation, Eq. (k), indicates that traction (stress) boundary conditions are a function of both displacement and temperature [see Eq. (c) of Prob. 5.8].

$$\int_V \left(\frac{\partial [N_u]^T}{\partial x_i} [C] \frac{\partial [N_u]}{\partial x_i} \{u\} - \frac{\partial [N_u]^T}{\partial x_i} [\beta][N_\theta]\{\theta\} \right) dV = \int_V [N_u]^T [f] \, dV$$

$$+ \int_S \left([N_u]^T [C] \frac{\partial [N_u]}{\partial x_i} \{u\} - [N_u]^T [\beta] \frac{\partial [N_\theta]}{\partial x_i} \{\theta\} \right) n_i \, dS \quad (o)$$

$$\int_V \left(\frac{\partial [N_\theta]^T}{\partial x_i} [k] \frac{\partial [N_\theta]}{\partial x_i} \{\theta\} \right) dV = \int_V [N_\theta]^T [r] \, dV + \int_S [N_\theta]^T [k] \frac{\partial [N_\theta]}{\partial x_i} \{\theta\} n_i \, dS \quad (p)$$

Equations (o) and (p) correspond directly to Eqs. (c) and (d). The stiffness matrix $[K_{uu}]$ was derived in Prob. 3.11; $[K_{\theta\theta}]$ corresponds to Prob. 5.12; and $\{f\}$, $\{\bar{t}\}$, $\{r\}$, and $\{\bar{Q}\}$ are the body force, surface traction boundary condition, heat source, and heat flux boundary condition matrices, respectively. The stiffness matrix $[K_{u\theta}]$ is sometimes called the coupling matrix and will be discussed in Probs. 5.18 and 5.30.

5.18. Formulate the two-dimensional local finite element for steady-state thermoelasticity defined in Prob. 5.17. Verify the accuracy of the element using a solution for the one-dimensional problem defined in Fig. 5-2. Use a strip of four elements as shown in the Fig. 5-2 and compute the displacement and temperature at each node. For displacement assume $u(0) = u(L) = 0$. For temperature assume $\theta(0) = \theta_0 = 100$ and $\theta(L) = 0$. Material constants can be assumed as Young's modulus $E = 1$ and Poisson's ratio $\nu = 0$; $\beta_{11} = 1$, $k_{11} = 1$, and $\beta_{22} = k_{22}$ can be assumed 0 in comparing the finite element and exact solutions. Refer to Prob. 3.13; the elastic constants become $C_{11} = C_{22} = C_{33} = 1$ and $C_{12} = 0$.

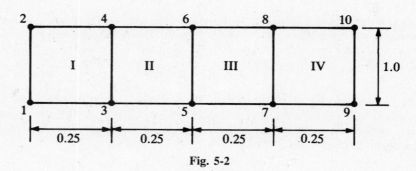

Fig. 5-2

The governing one-dimensional differential equations are

$$E \frac{d^2 u}{dx^2} - \beta \frac{d\theta}{dx} = 0 \qquad k \frac{d^2 \theta}{dx^2} = 0 \qquad (a)$$

with solution

$$\theta = \theta_0 \frac{L - x}{L} \qquad u = \beta \theta_0 \frac{xL - x^2}{2EL} \qquad (b)$$

The local finite element is formulated following Eqs. (o) and (p) of Prob. 5.17, and the coupling matrix is given in Prob. 5.30. All flux-type boundary conditions are zero. A transformation matrix is used to renumber the local element, numbering it in sequence corresponding to Eqs. (c) and (d) of Prob. 5.17. It is desirable to renumber the element such that all degrees of freedom at a node are consecutive. The transformation matrix is

$$
\begin{Bmatrix}
u_1 \\ v_1 \\ \theta_1 \\ u_2 \\ v_2 \\ \theta_2 \\ u_3 \\ v_3 \\ \theta_3 \\ u_4 \\ v_4 \\ \theta_4
\end{Bmatrix}
=
\begin{bmatrix}
1 & 0 & 0 & 0 & 0 & 0 & 0 & 0 & 0 & 0 & 0 & 0 \\
0 & 1 & 0 & 0 & 0 & 0 & 0 & 0 & 0 & 0 & 0 & 0 \\
0 & 0 & 0 & 0 & 0 & 0 & 0 & 0 & 1 & 0 & 0 & 0 \\
0 & 0 & 1 & 0 & 0 & 0 & 0 & 0 & 0 & 0 & 0 & 0 \\
0 & 0 & 0 & 1 & 0 & 0 & 0 & 0 & 0 & 0 & 0 & 0 \\
0 & 0 & 0 & 0 & 0 & 0 & 0 & 0 & 0 & 1 & 0 & 0 \\
0 & 0 & 0 & 0 & 1 & 0 & 0 & 0 & 0 & 0 & 0 & 0 \\
0 & 0 & 0 & 0 & 0 & 1 & 0 & 0 & 0 & 0 & 0 & 0 \\
0 & 0 & 0 & 0 & 0 & 0 & 0 & 0 & 0 & 0 & 1 & 0 \\
0 & 0 & 0 & 0 & 0 & 0 & 1 & 0 & 0 & 0 & 0 & 0 \\
0 & 0 & 0 & 0 & 0 & 0 & 0 & 1 & 0 & 0 & 0 & 0 \\
0 & 0 & 0 & 0 & 0 & 0 & 0 & 0 & 0 & 0 & 0 & 1
\end{bmatrix}
\begin{Bmatrix}
u_1 \\ v_1 \\ u_2 \\ v_2 \\ u_3 \\ v_3 \\ u_4 \\ v_4 \\ \theta_1 \\ \theta_2 \\ \theta_3 \\ \theta_4
\end{Bmatrix}
$$

where $[K]_{\text{new}} = [T][K]_{\text{old}}[T]^T$. The final results are given in Table 5.2, and for this elementary problem the finite element analysis agrees with the exact solution. The same problem can be solved using a strip of elements along the y axis in order to verify the finite element formulation in the y direction.

Table 5.2   Thermoelasticity Problem

	Finite element		Exact	
x	u	θ	u	θ
0.0	0.0	100	0.0	100
0.25	9.375	75	9.375	75
0.50	12.500	50	12.500	50
0.75	9.375	25	9.375	25
1.00	0.0	0	0.0	0

5.19. The theory that governs coupled electrical and deformation effects in materials is called piezoelectricity and is important for the study of microelectronic materials. The governing equations, with subscript tensor notation, are

$$\sigma_{ij,j} = 0 \qquad \sigma_{ij} = \sigma_{ji} \qquad \text{for } i \neq j \tag{a}$$

$$S_{ij} = \tfrac{1}{2}(u_{i,j} + u_{j,i}) \tag{b}$$

$$\sigma_{ij} = C_{ijkl}S_{kl} - e_{kij}E_k \tag{c}$$

$$D_{i,i} = 0 \tag{d}$$

$$E_i = -\phi_{,i} \tag{e}$$

$$D_i = e_{ikl}S_{kl} + \epsilon_{ik}E_k \tag{f}$$

where C_{ijkl}, e_{ikl}, and ϵ_{ik} are the elasticity, piezoelectric, and permittivity material constants, respectively. The strain is defined using S_{ij} to avoid confusion with the permittivity. The stress, electric field, electric displacement, and potential are defined by σ_{ij}, E_i, D_i, and ϕ, respectively, and the body force of Prob. 5.7 has been omitted. The boundary conditions are

$$u_i = \bar{u}_i \text{ on } \partial u \qquad \text{and} \qquad t_i = \sigma_{ij}n_j = \bar{t}_i \text{ on } \partial t \tag{g}$$

$$\phi = \bar{\phi} \text{ on } \partial \phi \qquad \text{and} \qquad d = D_i n_i = \bar{d} \text{ on } \partial d \tag{h}$$

Discuss the formulation of the corresponding finite element statement of the problem using the Galerkin method.

The finite element formulation will serve to illustrate the solution of a problem when the displacement equations of elasticity (discussed in Chap. 3) are coupled with a theory that can be described using the Laplace equation. In this theory the displacements and electric potential are fully coupled, whereas the steady-state thermoelasticity described in Prob. 5.17 represents an uncoupled theory. Piezoelectricity can be studied using finite elements in either two or three dimensions. However, when the problem is reduced to two dimensions, some important effects are uncoupled. The following discussion will pertain to the eight-node linear element shown in Fig. 5-3. Shape functions and local element matrices will not be developed since the amount of algebra is prohibitive. Three-dimensional elements will be discussed in Chapter 7.

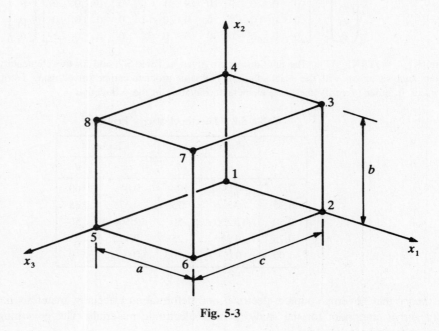

Fig. 5-3

Follow the procedure used in Prob. 5.17. Substitute Eqs. (c) and (e) into Eq. (a), keeping in mind that strain is formally defined in terms of displacement by Eq. (b). Also, substitute Eq. (e) into Eq. (f) and the result into Eq. (d) to obtain the governing equations:

$$C_{ijkl}S_{kl},_j + e_{kij}\phi,_{ij} = 0 \qquad (i)$$

$$e_{ikl}S_{kl},_i - \epsilon_{ik}\phi,_{ik} = 0 \qquad (j)$$

Strain is defined in terms of derivatives of displacement, and displacement and electric potential can be defined in terms of shape functions and nodal point values of displacement $\{u\}$ and potential $\{\phi\}$. Replace Eqs. (i) and (j) with equivalent matrix equations:

$$[C]\frac{\partial^2[N_u]}{\partial x_i^2}\{u\} + [e]\frac{\partial^2[N_\phi]}{\partial x_i^2}\{\phi\} = R_u$$

$$(k)$$

$$[e]\frac{\partial^2[N_u]}{\partial x_i^2}\{u\} - [\epsilon]\frac{\partial^2[N_\phi]}{\partial x_i^2}\{\phi\} = R_\phi$$

The weighted residuals are formed as

$$\int_V \left([N_u]^T [C] \frac{\partial^2 [N_u]}{\partial x_i^2} \{u\} + [N_u]^T [e] \frac{\partial^2 [N_\phi]}{\partial x_i^2} \{\phi\} \right) dV = 0$$

$$\int_V \left([N_\phi]^T [e] \frac{\partial^2 [N_u]}{\partial x_i^2} \{u\} - [N_\phi]^T [\epsilon] \frac{\partial^2 [N_\phi]}{\partial x_i^2} \{\phi\} \right) dV = 0$$

(l)

The Green-Gauss theorem is applied to all terms containing second-order derivatives:

$$\int_V \left(\frac{\partial [N_u]^T}{\partial x_i} [C] \frac{\partial [N_u]}{\partial x_i} \{u\} + \frac{\partial [N_u]^T}{\partial x_i} [e] \frac{\partial [N_\phi]}{\partial x_i} \{\phi\} \right) dV$$
$$= \int_S \left([N_u]^T [C] \frac{\partial [N_u]}{\partial x_i} \{u\} + [N_u]^T [e] \frac{\partial [N_\phi]}{\partial x_i} \{\phi\} \right) dS \quad (m)$$

$$\int_V \left(\frac{\partial [N_\phi]^T}{\partial x_i} [e] \frac{\partial [N_u]}{\partial x_i} \{u\} - \frac{\partial [N_\phi]^T}{\partial x_i} [\epsilon] \frac{\partial [N_\phi]}{\partial x_i} \{\phi\} \right) dV$$
$$= \int_S \left([N_\phi]^T [e] \frac{\partial [N_u]}{\partial x_i} \{u\} - [N_\phi]^T [\epsilon] \frac{\partial [N_\phi]}{\partial x_i} \{\phi\} \right) dS \quad (n)$$

The equivalent finite element equations are written in the abbreviated form

$$[K_{uu}]\{u\} + [K_{u\phi}]\{\phi\} = \{\bar{t}\}$$
$$[K_{u\phi}]^T \{u\} - [K_{\phi\phi}]\{\phi\} = \{\bar{d}\}$$

(o)

where $\quad [K_{uu}] = \int_V [B_u]^T [C][B_u]\, dV \quad [K_{\phi\phi}] = \int_V [B_\phi]^T [\epsilon][B_\phi]\, dV \quad [K_{u\phi}] = \int_V [B_u]^T [e][B_\phi]\, dV$

and $\{\bar{t}\}$ and $\{\bar{d}\}$ are applied boundary conditions. The B matrices are defined, as in previous problems, as an operator matrix times a shape function matrix. The element of Fig. 5-3 has eight shape functions, four degrees of freedom (unknowns) per node, three displacements, and the electric potential. Therefore, define $\{u\}$ and $\{\phi\}$ as

$$\{u\} = [u_1 \quad v_1 \quad w_1 \quad u_2 \quad v_2 \quad w_2 \quad \cdots \quad u_8 \quad v_8 \quad w_8]^T \quad \text{(24 terms)}$$
$$\{\phi\} = [\phi_1 \quad \phi_2 \quad \phi_3 \quad \phi_4 \quad \phi_5 \quad \phi_6 \quad \phi_7 \quad \phi_8]^T$$

(p)

Equation (b) defines the $[B_u]$ matrix, but the engineering definition of strain should be used. Recall from previous problems that $[B_u] = [L_u][N_u]$.

$$\begin{Bmatrix} \epsilon_{xx} \\ \epsilon_{yy} \\ \epsilon_{zz} \\ \epsilon_{yz} \\ \epsilon_{xz} \\ \epsilon_{xy} \end{Bmatrix} = \begin{bmatrix} \partial/\partial x & 0 & 0 \\ 0 & \partial/\partial y & 0 \\ 0 & 0 & \partial/\partial z \\ 0 & \partial/\partial z & \partial/\partial y \\ \partial/\partial z & 0 & \partial/\partial x \\ \partial/\partial y & \partial/\partial x & 0 \end{bmatrix} \begin{bmatrix} N_1 & 0 & 0 & N_2 & 0 & 0 & \cdots & N_8 & 0 & 0 \\ 0 & N_1 & 0 & 0 & N_2 & 0 & \cdots & 0 & N_8 & 0 \\ 0 & 0 & N_1 & 0 & 0 & N_2 & \cdots & 0 & 0 & N_8 \end{bmatrix}$$
(24 columns)

(q)

The right-hand side of Eq. (q) is $[L_u][N_u]$ for the eight-node brick element. Equation (e) for the electric field vector is used to define the $[B_\phi]$ matrix:

$$\begin{Bmatrix} E_x \\ E_y \\ E_z \end{Bmatrix} = \begin{bmatrix} \partial/\partial x \\ \partial/\partial y \\ \partial/\partial z \end{bmatrix} [N_1 \quad N_2 \quad N_3 \quad N_4 \quad N_5 \quad N_6 \quad N_7 \quad N_8]$$

(r)

The right-hand side of Eq. (r) is $[L_\phi][N_\phi]$. Note that the negative sign in Eq. (e) was previously included in the formulation. The material matrices are tabulated by Nye (1957) and can be specialized to correspond to any crystal class.

$$[C] = \begin{bmatrix} C_{11} & C_{12} & C_{13} & C_{14} & C_{15} & C_{16} \\ & C_{22} & C_{23} & C_{24} & C_{25} & C_{26} \\ & & C_{33} & C_{34} & C_{35} & C_{36} \\ & & & C_{44} & C_{45} & C_{46} \\ & \text{Symmetric} & & & C_{55} & C_{56} \\ & & & & & C_{66} \end{bmatrix} \quad [e] = \begin{bmatrix} e_{11} & e_{21} & e_{31} \\ e_{12} & e_{22} & e_{32} \\ e_{13} & e_{23} & e_{33} \\ e_{14} & e_{24} & e_{34} \\ e_{15} & e_{25} & e_{35} \\ e_{16} & e_{26} & e_{36} \end{bmatrix}$$

$$[\epsilon] = \begin{bmatrix} \epsilon_{11} & 0 & 0 \\ 0 & \epsilon_{22} & 0 \\ 0 & 0 & \epsilon_{33} \end{bmatrix}$$

Supplementary Problems

5.20. Use the Galerkin method to solve the cable problem presented in Prob. 2.1 and compare the results with the exact solution given in Prob. 2.12. Assume a trial solution $v_R = a_0 + a_1 x + a_2 x^2$.

5.21. Repeat Prob. 5.20 assuming a trial solution $v_R = a_1 \sin(\pi x/L)$, where L is the length of the cable.

5.22. Given the differential equation

$$u\frac{dC}{dx} - D\frac{d^2C}{dx} - m = 0 \qquad \text{with } C(0) = C(L) = 0$$

assume a solution $C_R = a_0 + a_1 x + a_2 x^2 + a_3 x^3$ and obtain an approximate solution using the Galerkin method. Compare the results with the exact solution using the parameters given in Prob. 2.18, $L = 1$ m, $D = 1.0(10^{-8})$, $u = 1.5(10^{-8})$, and assume $m = 1.0(10^{-4})$.

5.23. Repeat Prob. 5.22 assuming

$$C_R = a_1 x + a_2(1 - e^x)$$

5.24. Show that Eqs. (d) and (h) of Prob. 5.12 are equivalent. Note that as discussed in the problem [N] is defined differently in each equation. Explain the significance of redefining [N] while deriving a finite element model using the Galerkin method.

5.25. A one-dimensional mathematical model for two-phase diffusion in composite materials was suggested by Gurtin and Yatomi (1979). It defines a free-phase concentration that is assumed to diffuse through a rigid matrix material. Mass balance for the free phase C_f is given by

$$\frac{\partial C_f}{\partial t} + \frac{\partial j_f}{\partial x} = -m_f \qquad (a)$$

and for the rigid matrix, C_r is

$$\frac{\partial C_r}{\partial t} = m_f \qquad (b)$$

where j_f is the flux of the free phase and m_f is the rate at which free-phase material is being trapped. The constitutive equations are assumed as Fick's law for the free phase:

$$j_f = -D\frac{\partial C_f}{\partial x} \qquad (c)$$

and

$$m_f = \beta C_f - \alpha C_r \qquad (d)$$

for the rate term. Use the Galerkin method to formulate the corresponding finite element model for this theory of diffusion.

5.26. Derive a one-dimensional steady-state linear local finite element for the coupled diffusion theory given in Prob. 5.14.

5.27. Show that Eqs. (j) and (k) of Prob. 5.17 combine with Eq. (h) of Prob. 3.10 to give the governing differential equations

$$\left(C_{11} \frac{\partial^2}{\partial x^2} + C_{33} \frac{\partial^2}{\partial x\, \partial y} \right) u_e + \left(C_{12} \frac{\partial^2}{\partial x\, \partial y} + C_{33} \frac{\partial^2}{\partial x^2} \right) v_e - \beta \frac{\partial \theta}{\partial x} + f_x = 0$$

$$\left(C_{12} \frac{\partial^2}{\partial x\, \partial y} + C_{33} \frac{\partial^2}{\partial y^2} \right) u_e + \left(C_{22} \frac{\partial^2}{\partial y^2} + C_{33} \frac{\partial^2}{\partial x\, \partial y} \right) v_e - \beta \frac{\partial \theta}{\partial y} + f_y = 0$$

5.28. Formulate the matrix equation corresponding to the weighted residual of Eq. (a) of Prob. 5.17. Use shape functions for four-node elements as defined in that problem.

5.29. Formulate the matrix equation that corresponds to the first term of Eq. (m) of Prob. 5.17. Compare and discuss this matrix equation with the corresponding term given in the solution of Prob. 5.28.

5.30. Derive the coupling stiffness matrix defined as $[\mathrm{K}_{u\theta}]$ in Prob. 5.17 for a four-node rectangular element.

5.31. Derive the shape functions for the eight-node brick element of Fig. 5-3.

5.32. Construct the operators A and C that correspond to Eqs. (5.2) and (5.3) for the theory of piezoelectricity given in Prob. 5.19.

5.33. Derive the general variational principle for the theory of piezoelectricity given in Prob. 5.19.

5.34. Assume that Eq. (b), Eq. (e), the first Eq. (g), and the first Eq. (h) of Prob. 5.19 are identically satisfied and use the results of Prob. 5.33 to derive an extended variational principle for piezoelectricity that is suitable for finite element analysis.

Answers to Supplementary Problems

5.20. Satisfy the boundary conditions to obtain $v_R = a_2(x^2 - xL)$. Formulate the weighted residual

$$\int_0^L \left[2a_2 - \frac{k}{T} a_2(x^2 - xL) + \frac{f}{T} \right](x^2 - xL)\, dx$$

and $a_2(-\frac{1}{3} - kL^2/30T) = f/6T$. Compare with Table 2.2, $v(x = 24) = 1.7455$, $v(x = 48) = 2.6183$.

5.21. Substitute the trial solution into Eq. (e) of Prob. 2.1 to obtain the residual. Formulate the weighted residual

$$\int_0^L \sin\left(\frac{\pi x}{L}\right) R(x, a_1)\, dx = 0$$

$$\int_0^L \sin\left(\frac{\pi x}{L}\right)\left[-\left(\frac{\pi^2}{L^2}\right)a_1 \sin\left(\frac{\pi x}{L}\right) - \left(\frac{k}{T}\right)a_1 \sin\left(\frac{\pi x}{L}\right) + \frac{f}{T} \right] dx = 0$$

Integrate and solve for a_1:

$$a_1 = \frac{4L^2 f}{\pi(\pi^2 T + kL^2)}$$

Compare with Table 2.2, $a_1 = 2.7945$, $v(x = 24) = 1.6425$, $v(x = 48) = 2.6577$.

5.22. The exact solution is

$$C = \frac{m}{u}\left[x - \frac{L(1 - e^{ux/D})}{1 - e^{uL/D}}\right]$$

The trial solution that satisfies the boundary conditions is

$$C_R = a_2(x^2 - xL) + a_3(x^3 - xL^2)$$

The weighted residual equations after integration are

$$\frac{a_2 D}{3} + a_3\left(\frac{uL^2}{60} + \frac{DL}{2}\right) + \frac{m}{6} = 0$$

$$a_2\left(\frac{D}{2} - \frac{uL}{60}\right) + \frac{4DL}{5} + \frac{m}{4} = 0$$

and $a_2 = -1204.8$, $a_3 = -2409.6$. $C_R = (0.25L) = 790.65$, $C(0.25L) = 795.46$; $C_R(0.50L) = 1204.8$, $C(0.50L) = 1194.53$; $C_R(0.75L) = 1016.5$, $C(0.75L) = 1016.84$.

5.23. The trial solution that satisfies the boundary conditions is

$$C_R - a_1\left[x - \frac{L(1 - e^x)}{1 - e^L}\right]$$

The weighted residual equation after integration is

$$\frac{Da_1 L^2}{(1 - e^L)^2}\left(e^L - \frac{e^{2L}}{2} - \frac{1}{2}\right) + \frac{mL^2}{1 - e^L}(L - e^L + 1) - \frac{mL^2}{2} = 0$$

Substitute parameters and $a_1 = 1(10^4)$; $C_R(0.25L) = 847.04$, $C_R(0.50L) = 1224.95$, $C_R(0.75L) = 999.32$.

5.24. Perform the indicated multiplications and the equivalence is shown. Also, in actual application this redefinition of [N] is of no consequence since the finite element is derived using the first-order derivatives of the shape function. However, the redefinition of [N] is necessary for a proper derivation.

5.25. Combine the governing equations to give

$$\frac{\partial C_f}{\partial t} - D\frac{\partial^2 C_f}{\partial x^2} + \beta C_f - \alpha C_r = 0 \qquad \frac{\partial C_r}{\partial t} - \beta C_f + \alpha C_r = 0$$

Assume $C_f = [N]\{C_f\}$ and $C_r = [N]\{C_r\}$ and construct the weighted equations:

$$[N]^T[N]\frac{\partial\{C_f\}}{\partial t} + \frac{\partial[N]^T}{\partial x}[D]\frac{\partial[N]}{\partial x}\{C_f\} + [N]^T[\beta][N]\{C_f\} - [N]^T[\alpha][N]\{C_r\} = [N]j_r n$$

$$[N]^T[N]\frac{\partial\{C_r\}}{\partial t} - [N]^T[\beta][N]\{C_f\} + [N]^T[\alpha][N]\{C_r\} = 0$$

5.26. The governing equations are

$$-D_{11}\frac{d^2C_1}{dx^2} + D_{12}\frac{d^2C_2}{dx^2} + k_1C_1 - k_2C_2 = 0$$

$$-D_{22}\frac{d^2C_1}{dx^2} + D_{12}\frac{d^2C_2}{dx^2} - k_1C_1 + k_2C_2 = 0$$

Follow Eq. (e) of Prob. 2.9 and Eq. (d) of Prob. 2.11. Assume a two-node linear element with node numbering, i, j:

$$A\begin{bmatrix} \dfrac{D_{11}}{L}+\dfrac{k_1L}{3} & -\dfrac{D_{11}}{L}+\dfrac{k_1L}{6} & -\dfrac{D_{12}}{L}-\dfrac{k_2L}{3} & \dfrac{D_{12}}{L}-\dfrac{k_2L}{6} \\[2mm] -\dfrac{D_{11}}{L}+\dfrac{k_1L}{6} & \dfrac{D_{11}}{L}+\dfrac{k_1L}{3} & \dfrac{D_{12}}{L}-\dfrac{k_2L}{6} & -\dfrac{D_{12}}{L}-\dfrac{k_2L}{3} \\[2mm] -\dfrac{D_{12}}{L}-\dfrac{k_1L}{3} & \dfrac{D_{12}}{L}-\dfrac{k_1L}{6} & \dfrac{D_{22}}{L}+\dfrac{k_2L}{3} & -\dfrac{D_{22}}{L}+\dfrac{k_2L}{6} \\[2mm] \dfrac{D_{12}}{L}-\dfrac{k_1L}{6} & -\dfrac{D_{12}}{L}-\dfrac{k_1L}{3} & -\dfrac{D_{22}}{L}+\dfrac{k_2L}{6} & \dfrac{D_{22}}{L}+\dfrac{k_2L}{3} \end{bmatrix}\begin{Bmatrix} C_{1i} \\ C_{1j} \\ C_{2i} \\ C_{2j} \end{Bmatrix}$$

5.28.   Take the integral over the volume of

$$\begin{bmatrix} N_1 & 0 \\ 0 & N_1 \\ N_2 & 0 \\ 0 & N_2 \\ N_3 & 0 \\ 0 & N_3 \\ N_4 & 0 \\ 0 & N_4 \end{bmatrix}\begin{bmatrix} \partial/\partial x & 0 & \partial/\partial x \\ 0 & \partial/\partial y & \partial/\partial y \end{bmatrix}\begin{bmatrix} C_{11} & C_{12} & 0 \\ C_{12} & C_{22} & 0 \\ 0 & 0 & C_{33} \end{bmatrix}\begin{bmatrix} \partial/\partial x & 0 \\ 0 & \partial/\partial y \\ \partial/\partial y & \partial/\partial x \end{bmatrix}$$

$$\times \begin{bmatrix} N_1 & 0 & N_2 & 0 & N_3 & 0 & N_4 & 0 \\ 0 & N_1 & 0 & N_2 & 0 & N_3 & 0 & N_4 \end{bmatrix}\begin{Bmatrix} u_1 \\ v_1 \\ u_2 \\ v_2 \\ u_3 \\ v_3 \\ u_4 \\ v_4 \end{Bmatrix}$$

$$-\begin{bmatrix} N_1 & 0 \\ 0 & N_1 \\ N_2 & 0 \\ 0 & N_2 \\ N_3 & 0 \\ 0 & N_3 \\ N_4 & 0 \\ 0 & N_4 \end{bmatrix}\begin{bmatrix} \beta_x & 0 \\ 0 & \beta_y \end{bmatrix}\begin{bmatrix} \partial/\partial x \\ \partial/\partial y \end{bmatrix}[N_1 \quad N_2 \quad N_3 \quad N_4]\begin{Bmatrix} \theta_1 \\ \theta_2 \\ \theta_3 \\ \theta_4 \end{Bmatrix} + \begin{bmatrix} N_1 & 0 \\ 0 & N_1 \\ N_2 & 0 \\ 0 & N_2 \\ N_3 & 0 \\ 0 & N_3 \\ N_4 & 0 \\ 0 & N_4 \end{bmatrix}f = 0$$

Note that the shape function $[N_u]$ used in Eq. (m) of Prob. 5.17 must be redefined as a 3×8 matrix in that equation. See the discussion in Probs. 5.12, 5.25, and 5.29.

5.29.   The shape function matrix and the operator matrix must be redefined. The proof of the equivalence of the two matrix equations is obtained by performing the matrix multiplications and comparing the results.

$$
\begin{bmatrix}
N_1 & 0 & N_1 \\
0 & N_1 & N_1 \\
N_2 & 0 & N_2 \\
0 & N_2 & N_2 \\
N_3 & 0 & N_3 \\
0 & N_3 & N_3 \\
N_4 & 0 & N_4 \\
0 & N_4 & N_4
\end{bmatrix}
\begin{bmatrix}
C_{11} & C_{12} & 0 \\
C_{12} & C_{22} & 0 \\
0 & 0 & C_{33}
\end{bmatrix}
\begin{bmatrix}
\partial^2/\partial x^2 & 0 & 0 \\
0 & \partial^2/\partial y^2 & 0 \\
0 & 0 & \partial^2/\partial x^2
\end{bmatrix}
$$

$$
\times
\begin{bmatrix}
N_1 & 0 & N_2 & 0 & N_3 & 0 & N_4 & 0 \\
0 & N_1 & 0 & N_2 & 0 & N_3 & 0 & N_4 \\
N_1 & N_1 & N_2 & N_2 & N_3 & N_3 & N_4 & N_4
\end{bmatrix}
\begin{Bmatrix}
u_1 \\ v_1 \\ u_2 \\ v_2 \\ u_3 \\ v_3 \\ u_4 \\ v_4
\end{Bmatrix}
$$

After matrix multiplication and application of the Green-Gauss theorem, the matrices can be written in the format of Eq. (*o*) of Prob. 5.17.

5.30. The stiffness matrix is defined as the second term in Eq. (*o*) of Prob. 5.17, and the integral to be evaluated is

$$
-\int_V \frac{\partial [N_u]^T}{\partial x_i} [\beta][N_\theta]\, dV
$$

The first term is Eq. (*c*) of Prob. 3.11, and the third term is Eq. (*e*) of Prob. 3.4. The material matrix must connect the 8×3 matrix with the 2×4 matrix. Formally, the matrices appear as the integral of

$$
\begin{bmatrix}
\partial N_{u1}/\partial x & 0 & \partial N_{u1}/\partial y \\
0 & \partial N_{u1}/\partial y & \partial N_{u1}/\partial x \\
\partial N_{u2}/\partial x & 0 & \partial N_{u2}/\partial y \\
0 & \partial N_{u2}/\partial y & \partial N_{u2}/\partial x \\
\partial N_{u3}/\partial x & 0 & \partial N_{u3}/\partial y \\
0 & \partial N_{u3}/\partial y & \partial N_{u3}/\partial x \\
\partial N_{u4}/\partial x & 0 & \partial N_{u4}/\partial y \\
0 & \partial N_{u4}/\partial y & \partial N_{u4}/\partial x
\end{bmatrix}
\begin{bmatrix}
\beta_x & 0 \\
0 & \beta_y \\
0 & 0
\end{bmatrix}
\begin{bmatrix}
N_{\theta 1} & N_{\theta 2} & N_{\theta 3} & N_{\theta 4} \\
N_{\theta 1} & N_{\theta 2} & N_{\theta 3} & N_{\theta 4}
\end{bmatrix}
$$

The shape functions are subscripted u and θ to indicate the difference in the weight functions in the formulation. In this application they are the same and are given in Chap. 3, Probs. 3.1 and 3.5. Note that the first matrix could have been written as an 8×2 and the $[\beta]$ matrix written as a 2×2 and the same result would have been obtained. However, the formulation above makes the first matrix correspond to the first matrix in Eq. (*o*) of Prob. 5.17, and that is convenient for computer implementation. Substituting shape functions and integrating $0 \to a$ and $0 \to b$ gives the final stiffness matrix:

$$
\begin{bmatrix}
-\beta_x b/6 & -\beta_x b/6 & -\beta_x b/12 & -\beta_x b/12 \\
-\beta_y a/6 & -\beta_y a/12 & -\beta_y a/12 & -\beta_y a/6 \\
\beta_x b/6 & \beta_x b/6 & \beta_x b/12 & \beta_x b/12 \\
-\beta_y a/12 & -\beta_y a/6 & -\beta_y a/6 & -\beta_y a/12 \\
\beta_x b/12 & \beta_x b/12 & \beta_x b/6 & \beta_x b/6 \\
\beta_y a/12 & \beta_y a/6 & \beta_y a/6 & \beta_y a/12 \\
-\beta_x b/12 & -\beta_x b/12 & -\beta_x b/6 & -\beta_x b/6 \\
\beta_y a/6 & \beta_y a/12 & \beta_y a/12 & \beta_y a/6
\end{bmatrix}
$$

5.31. Derive the three-dimensional shape functions using three one-dimensional shape functions at each node. Let $N_{1\text{-}2}$ be the one-dimensional shape function between nodes 1 and 2. Then, $N_{1\text{-}2} = (a - x_1)/a$. Similarly, $N_{1\text{-}4} = (b - x_2)/b$ and $N_{1\text{-}5} = (c - x_3)/c$. Then, $N_1 = N_{1\text{-}2}N_{1\text{-}4}N_{1\text{-}5}$, or

$$N_1 = \frac{(a - x_1)(b - x_2)(c - x_3)}{abc} \qquad N_2 = \frac{x_1(b - x_2)(c - x_3)}{abc}$$

$$N_3 = \frac{x_1 x_2 (c - x_3)}{abc} \qquad N_4 = \frac{(a - x_1)x_2(c - x_3)}{abc}$$

$$N_5 = \frac{(a - x_1)(b - x_2)x_3}{abc} \qquad N_6 = \frac{x_1(b - x_2)x_3}{abc}$$

$$N_7 = \frac{x_1 x_2 x_3}{abc} \qquad N_8 = \frac{(a - x_1)x_2 x_3}{abc}$$

5.32.
$$\{u\} = \{u_m, S_{ij}, \sigma_{ij}, D_i, E_i, \phi; u_i, t_i, \phi, d\}$$

$$\{f; g\} = \{0, 0, 0, 0, 0, 0; \bar{t}_i, -\bar{u}_i, \bar{d}, -\bar{\phi}\}$$

$$\begin{bmatrix} 0 & 0 & -L & 0 & 0 & 0 \\ 0 & C_{ijkl}\delta_{ik}\delta_{jl} & -1 & 0 & -e_{kij}\delta_{ik} & 0 \\ L & -1 & 0 & 0 & 0 & 0 \\ 0 & 0 & 0 & 0 & 1 & \partial/\partial x_i \\ 0 & -e_{ijk}\delta_{ik} & 0 & 1 & -\epsilon_{ik}\delta_{ik} & 0 \\ 0 & 0 & 0 & -\partial/\partial x_i & 0 & 0 \end{bmatrix} \qquad \begin{matrix} 0 & 1 & 0 & 0 \\ -1 & 0 & 0 & 0 \\ 0 & 0 & 0 & 1 \\ 0 & 0 & -1 & 0 \end{matrix}$$

$$L = \tfrac{1}{2}(\delta_{ik}\partial/\partial x_j + \delta_{jk}\partial x_i)$$

5.33.
$$J_1(u) = \{-[u_m, L\sigma_{ij}] + [S_{ij}, C_{ijkl}S_{kl}] - [S_{ij}, \sigma_{ij}] - [S_{ij}, e_{kij}E_k]$$

$$+ [\sigma_{ij}, Lu_m] - [\sigma_{ij}, S_{ij}] + [D_i, E_i] + [D_i, \partial\phi/\partial x_i] - [E_i, e_{ijk}S_{ij}]$$

$$+ [E_i, D_i] - [E_i, \epsilon_{ik}E_k] - [\phi, \partial D_i/\partial x_i]\}_\Omega$$

$$+ [u_i, t_i]_{\partial t} - [t_i, u_i]_{\partial u} + [\phi, d]_{\partial d} - [d, \phi]_{\partial \phi}$$

$$- 2[u_i, \bar{t}_i]_{\partial t} + 2[t_i, \bar{u}_i]_{\partial u} - 2[\phi, \bar{d}]_{\partial d} + 2[d, \bar{\phi}]_{\partial \phi}$$

5.34.
$$J_2(u) = \{[S_{ij}, C_{ijkl}S_{kl}] - [E_i, \epsilon_{ik}E_k] - 2[S_{ij}, e_{kij}E_k]\}_\Omega - 2[u_i, \bar{t}_i]_{\partial t} - 2[\phi, \bar{d}]_{\partial d}$$

Chapter 6

Isoparametric Finite Elements

6.1. INTRODUCTION

The previous chapters have been concerned with the derivation of local finite elements for various physical problems, and applications have centered about the rectangular finite element. The geometry of the element and the relation between interpolation functions and shape functions will be emphasized in this chapter. Several example problems will require computing the elements of the stiffness matrix. They will enable analysts writing computer code to check their computations at various stages of code development.

The concept of an isoparametric finite element will be discussed in detail. The derivations and applications will center about the four-node quadrilateral element, but other element configurations will also be introduced. The isoparametric quadrilateral element is not required to conform to a cartesian coordinate system. The power of the finite element method as a method of numerical analysis can be emphasized since it will now be possible to model physical problems with more complicated boundary conditions. The previous chapters have been somewhat academic from an applications viewpoint, but this chapter will demonstrate that the finite element method can be applied to complex problems in science and engineering.

Numerical integration plays a significant role in the formulation of an isoparametric finite element. On the one hand, the numerical integration causes the development to appear more complicated, but on the other hand, the entire process of developing a local element is streamlined since the area integrals need not be evaluated analytically.

Axisymmetric field problems were introduced in Chap. 3 and will be studied in greater detail in this chapter. Numerical integration and isoparametric finite elements are logical choices for solving this class of important problems. Again, the four-node quadrilateral element will be used to illustrate these applications of the finite element method.

6.2. NUMERICAL INTEGRATION

Numerical integration is used extensively in finite element analysis. Elementary integration formulas, such as the trapezoidal rule, often assume equally spaced data and can become somewhat limited in applicability and accuracy when used in finite element analysis. The *Gauss quadrature* has become the accepted numerical integration scheme in the majority of finite element applications. The term *quadrature* means numerical integration. In general, an integral is evaluated approximately as

$$I = \int_a^b f(x)\, dx = \sum_{k=1}^{n} w_k f(x_k) \tag{6.1}$$

where the x_k are the *sampling points* (*Gauss points*) and the w_k are called the *weights*. Numerous Gauss quadrature formulas can be used in conjunction with Eq. (*6.1*), however, the *Gauss-Legendre* quadrature is the most popular for finite element analysis. The development of the weights and sampling points is based upon the Legendre polynomial, and that derivation is given in most books on numerical analysis (see, for instance, Scheid, 1988). Gauss-Legendre numerical integration requires that the integration limits be from -1 to $+1$, or

$$I = \int_{-1}^{+1} f(r)\, dr = \sum_{k=1}^{n} w_k f(r_k) \tag{6.2}$$

The integrals in most finite element applications are of the form of Eq. (*6.2*). Numerical integration using Gauss-Legendre quadratures is easily extended to two or three dimensions.

Area integration for linear triangular finite elements was discussed in Chap. 3. Additional integration formulas for linear and higher-order triangular elements will be given in this chapter. (See Prob. 6.22.)

6.3. INTERPOLATION FORMULAS AND SHAPE FUNCTION FORMULAS

Interpolation formulas and shape functions for higher-order finite elements are often derived by inspection. In what follows, two families of shape functions that can be derived using a formula approach will be discussed.

Lagrange Polynomial Formula

The Lagrange polynomial formula is an interpolation formula that is useful for generating shape functions and is defined as

$$L_k = \prod_{\substack{m=0 \\ k \neq m}}^{n} \frac{x - x_m}{x_k - x_m} = \frac{(x - x_0)(x - x_1) \cdots (x - x_{k-1})(x - x_{k+1}) \cdots (x - x_n)}{(x_k - x_0)(x_k - x_1) \cdots (x_k - x_{k-1})(x_k - x_{k+1}) \cdots (x_k - x_n)} \tag{6.3}$$

and represents the product of all terms. When $x = x_k$, the product becomes unity. However, when $x = x_m$ with $m \neq k$, the product becomes zero. It follows that L_k of Eq. (6.3) has properties similar to a shape function N_k. The Lagrange polynomial can be used to construct an interpolation formula or shape function corresponding to any line element, and the line element is easily extended to higher dimensions. (See, for instance, Probs. 6.3 and 6.23.)

Triangular Shape Function Formulas

Shape functions for triangular finite elements of any order can be derived using the area coordinates introduced in Chap. 3. The three-node triangular element is shown in Fig. 3-12 with area coordinates defined relative to the sides of the triangular element. Area coordinates are independent of the number of nodes used to define the triangular element, however, the number of nodes and the placement of the nodes for higher-order elements must satisfy certain requirements (see Prob. 6.18). The shape function can be derived in terms of area coordinates using the formula

$$N_k = \prod_{m=1}^{n} \frac{F_m(L_1, L_2, L_3)}{F_{m|L_1, L_2, L_3}} \tag{6.4}$$

where n is the order of the triangle and is equal to 1 less than the number of nodes along a side. The function F_m is obtained from the equations of n lines that pass through all the nodes except the one of interest. The denominator is the value of F_m when evaluated at the coordinates of node k.

6.4. GENERALIZED COORDINATES

Area coordinates discussed in the preceding section were used in Chap. 3 to facilitate the area integration required for deriving a three-node triangular finite element. Area coordinates are a type of generalized coordinate that can be referred to as *normalized coordinates*. The four-node rectangular elements of Chap. 3 were derived using a local coordinate system that was identical to the global system, and all area integrations were elementary and were carried out in closed form. However, applications were limited to geometries that could be described using connected rectangles. In this chapter the rectangular element will be replaced with a four-node quadrilateral element that is not required to conform with the global coordinate system. Applications using a four-sided element can now be extended to nonrectangular areas. Area integration, even for a four-node element, would become extremely tedious if attempted analytically using the global coordinate system. A generalized coordinate system that has been used extensively in finite element analysis simplifies the area integration. The local element, described in global coordinates, is integrated numerically in a normalized element space that is always defined by boundaries lying at ± 1 from the origin of the element space. Area integrations are accomplished numerically using the normalized element having boundaries of -1 and $+1$, and it follows that the Gauss-Legendre quadrature of Eq. (6.2) is an obvious choice for numerical integration. The element shape functions must be derived in the normalized space. Also, the actual element geomerty is defined in the global coordinate system using an interpolation function that transforms the normalized geometry of the element into the global space. The coordinate system described above and the elements used are often referred to as *serendipity coordinates* and *serendipity elements*, respectively.

6.5.   ISOPARAMETRIC ELEMENTS

Isoparametric Quadrilateral Elements

Isoparametric is a name that implies certain properties for an element that is integrated in a normalized space such as that described in Sec. 6.4. An interpolation function is used to approximate the physical parameter along the boundaries and interior of the element. When the same node locations are used for both approximations, the element is said to be isoparametric. It follows that the orders of the two approximations are not required to be the same. When fewer nodes are used to define the geometry than are used to define the shape function, the element is termed *subparametric*. Also, when there are more geometry nodes than shape function nodes, the element is called *superparametric*. Subparametric and superparametric elements will not be discussed in detail.

The development of a local stiffness matrix for any meaningful problem requires that the derivatives of the shape functions be evaluated. In the isoparametric formulation the local element geometry is transformed into the generalized space using an interpolation formula. The derivatives must also be evaluated in the generalized space, and that requires a formal mathematical coordinate transformation written in functional form that can later be evaluated numerically. The reader should be careful not to confuse mathematical operations pertaining to coordinate transformation with those pertaining to numerical integration, such as evaluating the derivatives of the shape functions. This confusion arises because the polynomial form of the shape function and the coordinate interpolation function are identical for an isoparametric finite element, and the same notation is often used in both situations.

The element in the undistorted space, using ξ, η coordinates, is often referred to as the *parent element*. The distorted element, the local finite element, is defined in ξ, η coordinates using a coordinate transformation or interpolation function that has the same form as the shape function (see Prob. 6.5). Derivatives of physical parameters are computed in the x, y system, but the element has been defined in the ξ, η system. It is necessary to use the chain rule for partial derivatives to define derivatives in the ξ, η system and then relate these functions back to the x, y system. This transformation is accomplished using the *jacobian matrix* (see Prob. 6.6). After these transformations are complete, the geometry is completely defined in the ξ, η system, and the local stiffness matrix is evaluated using numerical integration for shape functions and their derivatives that are defined in the ξ, η system.

Isoparametric Triangular Elements

The discussion of quadrilateral elements is equally valid for triangular elements as far as the fundamental concepts of coordinate transformations are concerned. However, there are some significant differences if area coordinates are to be used to define the shape functions for triangular elements. In a two-dimensional space there are three area coordinates for defining a triangle, but there are only two coordinates, ξ and η, for the parent element described above. In this situation the construction of a jacobian matrix for transformation would give a rectangular matrix with no inverse. This difficulty can be circumvented by recognizing that the area coordinates are not independent and that one coordinate can be written in terms of the remaining two. In addition, the integration limits in the undistorted space, ξ and η, must be changed to correspond to triangular limits. Fortunately, numerical integration formulas have been derived that are quite accurate for evaluating the integrals. Triangular isoparametric elements are discussed in Prob. 6.21.

6.6. AXISYMMETRIC FORMULATIONS

Isoparametric finite elements for axisymmetric problems are derived in the same manner as all other two-dimensional formulations. A primary difference is that the volume integration must be replaced with $dV = 2\pi r\, dr\, dz$.

The axisymmetric finite element formulation for physical problems governed by equations of the heat-conduction type was outlined in Prob. 3.25 using the variational function to derive the basic finite element equation. In this chapter the reader will find that the isoparametric finite element and numerical integration can simplify the analysis of this type of axisymmetric problem.

The equations governing axisymmetric elasticity problems were discussed briefly in Chap. 3. The strain-displacement equations to be modeled are obtained from Eqs. (3.17) assuming $v = \partial/\partial\theta = 0$:

$$\epsilon_{rr} = \frac{\partial u}{\partial r} \qquad \epsilon_{\theta\theta} = \frac{u}{r} \qquad \epsilon_{zz} = \frac{\partial w}{\partial z} \qquad \epsilon_{rz} = \frac{\partial w}{\partial r} + \frac{\partial u}{\partial z} \tag{6.5}$$

The equations of equilibrium are independent of the tangential coordinate θ and are

$$\frac{\partial \sigma_{rr}}{\partial r} + \frac{\partial \sigma_{rz}}{\partial z} + \frac{\sigma_{rr} - \sigma_{\theta\theta}}{r} = 0 \tag{6.6}$$

$$\frac{\partial \sigma_{rz}}{\partial r} + \frac{\partial \sigma_{zz}}{\partial z} + \frac{\sigma_{rz}}{r} = 0 \tag{6.7}$$

The matrix of material constants is given in Prob. 3.26. The formulation for an isoparametric axisymmetric element in r, z coordinates is similar to the formulation of the corresponding element in cartesian coordinates with the exception of the strain term $\epsilon_{\theta\theta}$ that has the r coordinate in the denominator. The formulation of the isoparametric element for axisymmetric elasticity is outlined in Probs. 6.12–6.14.

The Laplace operator in cylindrical coordinates (r, θ, z) is written as

$$\nabla^2 = \frac{\partial^2}{\partial r^2} + \frac{1}{r}\frac{\partial}{\partial r} + \frac{1}{r^2}\frac{\partial^2}{\partial^2\theta} + \frac{\partial^2}{\partial z^2} \tag{6.8}$$

The dependence upon the θ coordinate is deleted in the axisymmetric formulation. A finite element was derived for this type of problem in Chap. 3, Prob. 3.25, using a variational formulation. Identical results can be obtained using the Galerkin formulation. Axisymmetric problems involving heat conduction, diffusion, electrostatics, and so on, are discussed in Prob. 6.17.

Solved Problems

6.1. Use Gauss-Legendre numerical integration to integrate the body force term for the one-dimensional linear finite element derived in Prob. 2.9.

The matrix equation to be evaluated is written below for reference.

$$A \int_0^L \begin{Bmatrix} (x - L)/L \\ x/L \end{Bmatrix} f\, dx \tag{a}$$

Exact results can be obtained using a two-point integration formula or $n = 2$ in Eq. (6.2). The sampling points and weights for $n = 2$ through $n = 5$ are given in Table 6.1; see Scheid (1988).

Table 6.1   Sampling Points and Weights for Gauss-Legendre Integration

n	x_k	W_k
2	$\pm 0.577\,350\,269$	$1.000\,000\,000$
3	$0.774\,596\,669$	$0.555\,555\,555$
	$0.000\,000\,000$	$0.888\,888\,888$
	$-0.774\,596\,669$	$0.555\,555\,555$
4	$\pm 0.861\,136\,312$	$0.347\,854\,845$
	$\pm 0.339\,981\,043$	$0.652\,145\,155$
5	$\pm 0.538\,469\,310$	$0.478\,628\,670$
	$0.000\,000\,000$	$0.568\,888\,889$
	$\pm 0.906\,179\,845$	$0.236\,726\,885$

The argument of the integral must be changed to represent integration limits -1 to $+1$, or $x = (rL + L)/2$ with $dx = L\,dr/2$. The integral to be evaluated is

$$AfL \int_{-1}^{+1} \left\{ \begin{matrix} (1 - r)/4 \\ (r + 1)/4 \end{matrix} \right\} dr \qquad (b)$$

Substituting the sampling point and weight values for $n = 2$ into the first term gives (with five-digit accuracy)

$$\left(\frac{AfL}{4} \right) \{ [1 - (0.577735)](1) + [1 - (-0.577735)](1) \} = \frac{AfL}{2}$$

The integration is exact. Similarly, the second term gives the exact answer, $AfL/2$.

6.2.   Use Gauss-Legendre quadratures with $n = 2$ to numerically integrate

$$\int_0^a \int_0^b xy\,dx\,dy = \frac{a^2 b^2}{4}$$

The integration arguments are modified using $x = (r + 1)a/2$, $dx = a\,dr/2$, and $y = (s + 1)b/2$, $dy = b\,ds/2$. The equation to be integrated is

$$\int_{-1}^{+1} \int_{-1}^{+1} \frac{a}{2}(r + 1) \frac{b}{2}(s + 1) \frac{ab}{4}\,dr\,ds$$

or, in the form of Eq. (6.2),

$$\frac{a^2 b^2}{16} \sum_{i=1}^{2} \sum_{j=1}^{2} (r_i + 1)w_i(s_j + 1)w_j \qquad (a)$$

Evaluate Eq. (a) with $i = 1$ while j equals 1 and 2, then with $i = 2$ while j equals 1 and 2:

$$\frac{a^2 b^2}{16} [\{(-0.577735 + 1)w_1[(-0.577735 + 1)w_1 + (0.577735 + 1)w_2]\}$$

$$+ \{(0.577735 + 1)w_2[(-0.577735 + 1)w_1 + (0.577735 + 1)w_2]\}] \quad (b)$$

Evaluating Eq. (b) gives the exact result $a^2 b^2/4$.

6.3. Use the Lagrangian interpolation formula for deriving one-dimensional three-node shape functions for the element illustrated in Fig. 6-1. Specialize the shape function for an element of length L with a node at its center.

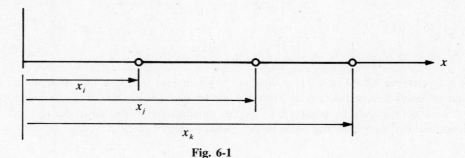

Fig. 6-1

Substitute into Eq. (6.3):

$$L_i = N_i = \frac{(x - x_j)(x - x_k)}{(x_i - x_j)(x_i - x_k)} \qquad L_j = N_j = \frac{(x - x_i)(x - x_k)}{(x_j - x_i)(x_j - x_k)} \qquad L_k = N_k = \frac{(x - x_i)(x - x_j)}{(x_k - x_i)(x_k - x_j)}$$

Let $x_i = 0$, $x_j = L/2$, and $x_k = L$ and let corresponding node numbers be 1, 2, and 3:

$$N_1 = \frac{(x - L/2)(x - L)}{(-L/2)(-L)} \qquad N_2 = \frac{x(x - L)}{(L/2)(L/2 - L)} \qquad N_3 = \frac{x(x - L/2)}{L(L - L/2)}$$

6.4. Derive the interpolation functions for a four-node isoparametric quadrilateral element.

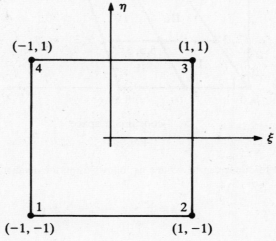

Fig. 6-2

Use the Lagrange polynomial formula and Fig. 6-2. The interpolation function for node 1 is the product of shape functions along the lines $\xi = -1$ and $\eta = -1$. Note that the interpolation function is identical to the shape function and that the notation for the shape function is used:

$$N_1 = L_{1\xi}L_{1\eta} = \frac{\xi - 1}{-1 - 1} \frac{\eta - 1}{-1 - 1} = \frac{(1 - \xi)(1 - \eta)}{4}$$

The remaining functions are derived in a similar manner.

$$N_2 = L_{2\xi}L_{2\eta} = \frac{\xi+1}{1+1}\frac{\eta-1}{-1-1} = \frac{(1+\xi)(1-\eta)}{4}$$

$$N_3 = L_{3\xi}L_{3\eta} = \frac{\xi+1}{1+1}\frac{\eta+1}{1+1} = \frac{(1+\xi)(1-\eta)}{4}$$

$$N_4 = L_{4\xi}L_{4\eta} = \frac{\xi-1}{-1-1}\frac{\eta+1}{1+1} = \frac{(1-\xi)(1+\eta)}{4}$$

6.5. A four-element model of a plane area is shown in Fig. 6-3. Use the interpolation functions for a four-node quadrilateral derived in Prob. 6.4 and for element III show that coordinate location ($x = 7.0$, $y = 6.0$) corresponds to point $(1, 1)$ in the generalized space. Also, for $\xi = 0.5$ and $\eta = -0.5$, determine the corresponding point in the global system.

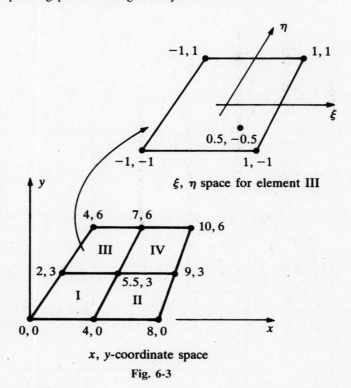

ξ, η space for element III

x, y-coordinate space

Fig. 6-3

The interpolation functions can be written as follows using the notation of Prob. 6.4:

$$x = \sum_{i=1}^{4} N_i x_i \qquad y = \sum_{i=1}^{4} N_i y_i \tag{a}$$

or

$$x = \tfrac{1}{4}(1-\xi)(1-\eta)x_1 + \tfrac{1}{4}(1+\xi)(1-\eta)x_2 + \tfrac{1}{4}(1+\xi)(1+\eta)x_3 + \tfrac{1}{4}(1-\xi)(1+\eta)x_4 \tag{b}$$

$$y = \tfrac{1}{4}(1-\xi)(1-\eta)y_1 + \tfrac{1}{4}(1+\xi)(1-\eta)y_2 + \tfrac{1}{4}(1+\xi)(1+\eta)y_3 + \tfrac{1}{4}(1-\xi)(1+\eta)y_4 \tag{c}$$

Substitute the global coordinate values for element III:

$$x = \tfrac{1}{4}(1-\xi)(1-\eta)(2) + \tfrac{1}{4}(1+\xi)(1-\eta)(5.5) + \tfrac{1}{4}(1+\xi)(1+\eta)(7) + \tfrac{1}{4}(1-\xi)(1+\eta)(4) \tag{d}$$

$$y = \tfrac{1}{4}(1-\xi)(1-\eta)(3) + \tfrac{1}{4}(1+\xi)(1-\eta)(3) + \tfrac{1}{4}(1+\xi)(1+\eta)(6) + \tfrac{1}{4}(1-\xi)(1+\eta)(6) \tag{e}$$

Substitute $\xi = 1$ and $\eta = 1$ into Eqs. (d) and (e) and compute the corresponding x and y. Note that all terms in Eq. (d) are zero except the third term and that it corresponds to node 3. Similarly, all terms in Eq. (e) are zero except the third that corresponds to node 3. It can be seen that the interpolation function is similar to the shape function. For $\xi = 1$ and $\eta = 1$, $N_3 = 1$ with $N_1 = N_2 = N_3 = 0$.

Substituting $\xi = 0.5$ and $\eta = -0.5$ into Eqs. (d) and (e) gives $x = 5.0313$ and $y = 3.75$. It is easily verified that these solutions represent a linear approximation for coordinate locations along the element boundaries and on the interior of the element.

6.6. An isoparametric parent element is shown in Fig. 6-4(a), and a corresponding isoparametric distorted element is shown in Fig. 6-4(b). Discuss the transformation that relates partial derivatives in the original x, y coordinates to the generalized ξ, η coordinates.

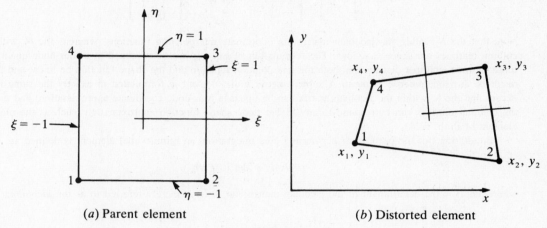

(a) Parent element   (b) Distorted element

Fig. 6-4

The shape functions, as ilustrated in Prob. 6.4, are functions of ξ and η, or

$$N_i = N_i(\xi, \eta) \qquad \text{(the } N_i \text{ are shape functions)} \tag{a}$$

The x, y coordinates are defined in terms of ξ, η coordinates by Eq. (a) of Prob. 6.5, or

$$x = x(N_i) = x(\xi, \eta) \qquad y = y(N_i) = y(\xi, \eta) \qquad \text{(the } N_i \text{ are the interpolation functions)} \tag{b}$$

The derivatives of the shape functions can be written as follows, using the chain rule:

$$\frac{\partial N_i}{\partial \xi} = \frac{\partial N_i}{\partial x} \frac{\partial x}{\partial \xi} + \frac{\partial N_i}{\partial y} \frac{\partial y}{\partial \xi}$$

$$\frac{\partial N_i}{\partial \eta} = \frac{\partial N_i}{\partial x} \frac{\partial x}{\partial \eta} + \frac{\partial N_i}{\partial y} \frac{\partial y}{\partial \eta} \tag{c}$$

Equation (c) is written in matrix form as

$$\begin{bmatrix} \partial N_i / \partial \xi \\ \partial N_i / \partial \eta \end{bmatrix} = \begin{bmatrix} \partial x / \partial \xi & \partial y / \partial \xi \\ \partial x / \partial \eta & \partial y / \partial \eta \end{bmatrix} \begin{bmatrix} \partial N_i / \partial x \\ \partial N_i / \partial y \end{bmatrix} \tag{d}$$

The first matrix on the right-hand side is defined as J and is referred to as the *jacobian matrix*:

$$J = \begin{bmatrix} \dfrac{\partial x}{\partial \xi} & \dfrac{\partial y}{\partial \xi} \\[2mm] \dfrac{\partial x}{\partial \eta} & \dfrac{\partial y}{\partial \eta} \end{bmatrix} \tag{e}$$

Multiplying Eq. (d) by J^{-1} gives the form of the equation that can be used to compute derivatives of the shape functions in the x, y system:

$$\begin{bmatrix} \dfrac{\partial N_i}{\partial x} \\[3mm] \dfrac{\partial N_i}{\partial y} \end{bmatrix} = J^{-1} \begin{bmatrix} \dfrac{\partial N_i}{\partial \xi} \\[3mm] \dfrac{\partial N_i}{\partial \eta} \end{bmatrix} = \begin{bmatrix} \dfrac{\partial x}{\partial \xi} & \dfrac{\partial y}{\partial \xi} \\[3mm] \dfrac{\partial x}{\partial \eta} & \dfrac{\partial y}{\partial \eta} \end{bmatrix}^{-1} \begin{bmatrix} \dfrac{\partial N_i}{\partial \xi} \\[3mm] \dfrac{\partial N_i}{\partial \eta} \end{bmatrix} \qquad (f)$$

Substituting Eqs. (a) of Prob. 6.5 into (f) gives the final form

$$\begin{bmatrix} \dfrac{\partial N_i}{\partial x} \\[3mm] \dfrac{\partial N_i}{\partial y} \end{bmatrix} = \begin{bmatrix} \sum \dfrac{\partial N_i}{\partial \xi} x_i & \sum \dfrac{\partial N_i}{\partial \xi} y_i \\[3mm] \sum \dfrac{\partial N_i}{\partial \eta} x_i & \sum \dfrac{\partial N_i}{\partial \eta} y_i \end{bmatrix}^{-1} \begin{bmatrix} \dfrac{\partial N_i}{\partial \xi} \\[3mm] \dfrac{\partial N_i}{\partial \eta} \end{bmatrix} \qquad (g)$$

Note that the N_i within the jacobian matrix are coordinate interpolation functions, whereas the N_i within the column matrices are shape functions. The formulation of Eq. (g) has been related to a four-node quadrilateral finite element, and that dictates that both the interpolation function and the shape function be linear and that the result be an isoparametric element. A subparametric finite element is formulated in exactly the same manner except that the N_i within the jacobian matrix can be linear (a four-node coordinate approximation) and the N_i in the column matrices can correspond to any higher node shape function approximation (such as the nine-node element of Prob. 6.24).

In addition, for the purpose of performing area integration, an infinitesimal element is defined as

$$dx\, dy = |\det \mathrm{J}|\, d\xi\, d\eta \qquad (h)$$

where $|\det \mathrm{J}|$ is the determinant of the jacobian matrix and is often merely referred to as the *jacobian*.

6.7. Discuss the evaluation of the jacobian matrix for a four-node isoparametric finite element.

The interpolation functions are defined in Prob. 6.4, and the jacobian matrix is defined by Eq. (g) of Prob. 6.6. The jacobian matrix can be written as a product of two matrices:

$$J = \begin{bmatrix} \sum (\partial N_i / \partial \xi) \\[2mm] \sum (\partial N_i / \partial \eta) \end{bmatrix} [x_i \quad y_i]$$

or

$$J = \frac{1}{4} \begin{bmatrix} -(1-\eta) & (1-\eta) & (1+\eta) & -(1+\eta) \\ -(1-\xi) & -(1+\xi) & (1+\xi) & (1-\xi) \end{bmatrix} \begin{bmatrix} x_1 & y_1 \\ x_2 & y_2 \\ x_3 & y_3 \\ x_4 & y_4 \end{bmatrix} \qquad (a)$$

This definition of J can be extended to any interpolation function. The matrix multiplication would be a $[2 \times n][n \times 2]$, where n is the number of nodes to be used for geometry transformation. Equation (a) can be used to evaluate J at any ξ, η location for an element defined in the x, y system. In finite element analysis Eq. (a) is used in conjunction with the numerical integration and the evaluation of the local stiffness matrix.

6.8. The fundamental theory for the development of a local stiffness matrix for heat transfer is given by Probs. 3.4 and 3.5 as

$$[\mathrm{K}] = \int_A [\mathrm{B}]^T [\mathrm{k}][\mathrm{B}] t\, dx\, dy \qquad (a)$$

where $[\mathrm{B}] = [\mathrm{L}][\mathrm{N}]$ [Eq. (e) of Prob. 3.4]. The $[\mathrm{B}]$ matrix corresponds to the left-hand side of Eq. (g) of Prob. 6.6. Discuss the formulation of the $[\mathrm{B}]$ matrix for a four-node quadrilateral isoparametric element.

Expand Eq. (g) of Prob. 6.6 for $i = 4$:

$$\begin{bmatrix} \partial N_1/\partial x & \partial N_2/\partial x & \partial N_3/\partial x & \partial N_4/\partial x \\ \partial N_1/\partial y & \partial N_2/\partial y & \partial N_3/\partial y & \partial N_4/\partial y \end{bmatrix} = \begin{bmatrix} J_{11} & J_{12} \\ J_{21} & J_{22} \end{bmatrix}^{-1} \begin{bmatrix} \partial N_1/\partial \xi & \partial N_2/\partial \xi & \partial N_3/\partial \xi & \partial N_4/\partial \xi \\ \partial N_1/\partial \eta & \partial N_2/\partial \eta & \partial N_3/\partial \eta & \partial N_4/\partial \eta \end{bmatrix} \quad (b)$$

The components of the inverse of J are computed using Eq. (a) of Prob. 6.7. The derivatives of the shape functions on the right-hand side of Eq. (b) above are a matrix that is identical to the matrix of Eq. (a) of Prob. 6.7. Note that these two matrices are always identical for a *four-node isoparametric element with one degree of freedom per node*, but that Eq. (b) above represents shape functions whereas Eq. (a) of Prob. 6.7 represents interpolation functions for geometry transformation. Equation (b) is the formal representation of the [B] matrix.

6.9. A quadrilateral element is shown in Fig. 6-5. Evaluate the stiffness matrix for heat transfer using the definition given by Eq. (a) of Prob. 6.8. Assume the thermal conductivity as $k_x = k_y = k = 1$ Btu/ (hr·in·°F). Assume a unit thickness for the element.

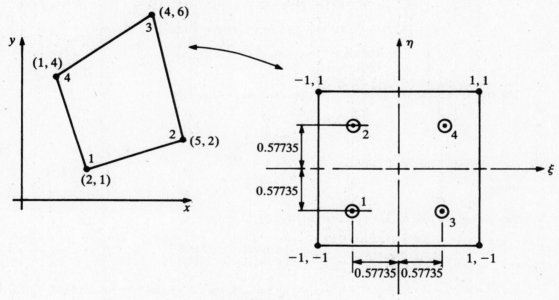

Fig. 6-5

A 2×2 gaussian quadrature will be used, and the parent element is shown in Fig. 6-5. The numerical integration procedure is similar to that illustrated in Eq. (a) of Prob. 6.2. The [B] matrix, in the form given by Eq. (b) of Prob. 6.8, must be evaluated for each *Gauss point* of Fig. 6-5. Each time the [B] matrix is computed, the contribution to the stiffness matrix is computed as $[B]^T[k][B]$, and the final stiffness matrix is the sum of the four contributions. In what follows, the computations will be shown in detail, and the reader should keep in mind that they are being done using a computer code with a nested DO loop where I = 1 TO 2 as J = 1 TO 2, where I and J correspond to the 2×2 gaussian integration. In order to evaluate [B] at each Gauss point, the jacobian of the transformation for that Gauss point must be computed. Equation (a) of Prob. 6.7 is used to compute the jacobian in matrix format. The matrix that defines the element in the x, y coordinate system corresponds to the second matrix of that equation and is evaluated using Fig. 6-5:

$$\begin{bmatrix} 2 & 1 \\ 5 & 2 \\ 4 & 6 \\ 1 & 4 \end{bmatrix} \quad (a)$$

Gauss point 1:

The location of the Gauss point is obtained from Table 6.1 and is shown in Fig. 6-5. Let $\xi_{i=1} = -0.57735$ and $\eta_{j=1} = -0.57735$. Use Eq. (a) of Prob. 6.7 to evaluate the jacobian matrix:

$$J_{11} = \tfrac{1}{4}[-(1-\eta)x_1 + (1-\eta)x_2 + (1+\eta)x_3 - (1+\eta)x_4]$$

$$J_{11} = \tfrac{1}{4}[-(1+0.57735)(2) + (1+0.57735)(5) + (1-0.57735)(4) - (1-0.57735)(1)]$$

$$= \tfrac{1}{4}[-3.15470 + 7.88675 + 1.69060 - 0.42265] = 1.50$$

Similarly,

$$J_{12} = \tfrac{1}{4}[-(1+0.57735)(1) + (1+0.57735)(2) + (1-0.57735)(6) - (1-0.57735)(4)]$$

$$= \tfrac{1}{4}[-1.57735 + 3.15470 + 2.53590 - 1.69060] = 0.60566$$

$$J_{21} = \tfrac{1}{4}[-(1+0.57735)(2) - (1-0.57735)(5) + (1-0.57735)(4) + (1+0.57735)(1)]$$

$$= \tfrac{1}{4}[-3.15470 - 2.11325 + 1.69060 + 1.57735] = -0.5$$

$$J_{22} = \tfrac{1}{4}[-(1+0.57735)(1) - (1-0.57735)(2) + (1-0.57735)(6) + (1+0.57735)(4)]$$

$$= \tfrac{1}{4}[-1.57735 - 0.84530 + 2.53590 + 6.3094] = 1.60566$$

$$J = \begin{bmatrix} 1.5 & 0.60566 \\ -0.5 & 1.60566 \end{bmatrix}$$

The inverse of J can be computed in an elementary way as (see Prob. 1.14)

$$J^{-1} = \frac{\begin{bmatrix} J_{22} & -J_{12} \\ -J_{21} & J_{11} \end{bmatrix}}{|\det J|} \tag{b}$$

where

$$|\det J| = J_{11}J_{22} - J_{21}J_{21} \tag{c}$$

Substituting into Eqs. (c) and (b) gives $|\det J| = 2.71133$ and

$$J^{-1} = \begin{bmatrix} 0.59221 & -0.22338 \\ 0.18441 & 0.55324 \end{bmatrix} \tag{d}$$

Let $dA = dx\, dy = |\det J|\, d\xi\, d\eta = |\det J| w_{i=1} w_{j=1} = 2.71133$, and the weights have been included in this calculation. Note that the weight functions for 2×2 integration are equal to 1.0.

The [B] matrix of Eq. (b) of Prob. 6.8 is evaluated using the derivatives of the shape functions in the ξ, η system. These numbers are the same as those used above to compute J (see Prob. 6.8).

$$[B]_1 = \begin{bmatrix} 0.59221 & -0.22338 \\ 0.18441 & 0.55324 \end{bmatrix} \begin{bmatrix} -1.57735 & 1.57735 & 0.42265 & -0.42265 \\ -1.57735 & -0.42265 & 0.42265 & 1.57735 \end{bmatrix} \tag{e}$$

$$[B]_1 = \begin{bmatrix} -0.14544 & 0.25713 & 0.03897 & -0.15066 \\ -0.29088 & 0.01426 & 0.07794 & 0.19868 \end{bmatrix}$$

The contribution to the stiffness matrix for Gauss point 1 can be written as

$$[K]_1 = [B]_1^T [k][B]_1 \, dA$$

where dA is defined above.

$$[K]_1 = \begin{bmatrix} 0.28676 & -0.11265 & -0.07684 & -0.09728 \\ & 0.17982 & 0.03018 & -0.09735 \\ & \text{Symmetric} & 0.02059 & 0.02606 \\ & & & 0.16857 \end{bmatrix} \tag{f}$$

The computation for the remaining Gauss points will be shown but with less detail.

Gauss point 2:

$$\xi_{i=1} = -0.57735 \qquad \text{and} \qquad \eta_{j=2} = 0.57735$$

$$J = \begin{bmatrix} 1.5 & 0.89434 \\ -0.5 & 1.60566 \end{bmatrix} \qquad J^{-1} = \begin{bmatrix} 0.56227 & -0.31318 \\ 0.17509 & 0.52527 \end{bmatrix}$$

$$|\det J| = 2.85567$$

$$[B]_2 = \begin{bmatrix} 0.06409 & 0.09250 & 0.18863 & -0.34522 \\ -0.22564 & -0.03700 & 0.12455 & 0.13809 \end{bmatrix}$$

$$[K]_2 = \begin{bmatrix} 0.15711 & 0.04077 & -0.04573 & -0.15216 \\ & 0.02834 & 0.03667 & -0.10578 \\ & \text{Symmetric} & 0.14591 & -0.13685 \\ & & & 0.39479 \end{bmatrix}$$

(g)

Gauss point 3:

$$\xi_{i=2} = 0.57735 \qquad \text{and} \qquad \eta_{j=1} = -0.57735$$

$$J = \begin{bmatrix} 1.5 & 0.60566 \\ -0.5 & 1.89434 \end{bmatrix} \qquad J^{-1} = \begin{bmatrix} 0.60246 & -0.19262 \\ 0.15902 & 0.47705 \end{bmatrix}$$

$$|\det J| = 3.14434$$

$$[B]_3 = \begin{bmatrix} -0.21722 & 0.31353 & -0.01230 & -0.08401 \\ -0.11311 & -0.12541 & 0.20492 & 0.03360 \end{bmatrix}$$

$$[K]_3 = \begin{bmatrix} 0.18859 & -0.16954 & -0.06448 & 0.45428 \\ & 0.35855 & -0.09293 & -0.09607 \\ & \text{Symmetric} & 0.13251 & 0.02490 \\ & & & 0.02574 \end{bmatrix}$$

(h)

Gauss point 4:

$$\xi_{i=2} = 0.57735 \qquad \text{and} \qquad \eta_{j=2} = 0.57735$$

$$J = \begin{bmatrix} 1.5 & 0.89434 \\ -0.5 & 1.89434 \end{bmatrix} \qquad J^{-1} = \begin{bmatrix} 0.57602 & -0.27945 \\ 0.15204 & 0.45611 \end{bmatrix}$$

$$|\det J| = 3.28868$$

$$[B]_4 = \begin{bmatrix} -0.03213 & 0.16810 & 0.11991 & -0.25588 \\ -0.06426 & -0.16380 & 0.23982 & -0.01176 \end{bmatrix}$$

$$[K]_4 = \begin{bmatrix} 0.01697 & 0.01685 & -0.06334 & 0.02952 \\ & 0.18116 & -0.06289 & -0.13512 \\ & \text{Symmetric} & 0.23642 & -0.11018 \\ & & & 0.21577 \end{bmatrix}$$

(i)

The final stiffness matrix is the sum of Eqs. (f)–(i):

$$[K] = \begin{bmatrix} 0.64945 & -0.22456 & -0.25040 & -0.17449 \\ & 0.74787 & -0.08897 & -0.43433 \\ & \text{Symmetric} & 0.53543 & -0.19606 \\ & & & 0.80488 \end{bmatrix}$$

(j)

6.10. Assume a four-node isoparametric finite element and compare the formulation of the [B] matrix for a plane elasticity problem with the heat transfer formulation of Prob. 6.8.

Review Probs. 3.10 and 3.11. The [B] matrix for plane elasticity is formulated as [B] = [L][N], where [L] is defined by Eq. (h) of Prob. 3.10 and [N] is defined by Eq. (b) of Prob. 3.11. The [B] matrix for a plane elasticity problem corresponds to Eq. (c) of Prob. 3.11 and is written as

$$[B] = \begin{bmatrix} \partial N_1/\partial x & 0 & \partial N_2/\partial x & 0 & \partial N_3/\partial x & 0 & \partial N_4/\partial x & 0 \\ 0 & \partial N_1/\partial y & 0 & \partial N_2/\partial y & 0 & \partial N_3/\partial y & 0 & \partial N_4/\partial y \\ \partial N_1/\partial y & \partial N_1/\partial x & \partial N_2/\partial y & \partial N_2/\partial x & \partial N_3/\partial y & \partial N_3/\partial x & \partial N_4/\partial y & \partial N_4/\partial x \end{bmatrix} \qquad (a)$$

The elements of the [B] matrix are computed using Eq. (b) of Prob. 6.8. After matrix multiplication of that equation, the proper terms must be assigned to the proper locations in the 3×8 matrix above.

6.11. Use the element of Fig. 6-5 and derive the stiffness matrix for the plane stress elasticity problem assuming $E = 1.0$, $\nu = 0.25$ and unit thickness $t = 1$. Use the linear isoparametric formulation.

   A 2×2 gaussian quadrature will be used, and it follows that the computations for the jacobian matrix and the derivatives of the shape functions are identical to those of Prob. 6.9. The B matrix for the first Gauss integration point will be evaluated to illustrate the procedure. Let $\xi_{i=1} = -0.57735$ and $\eta_{j=1} = -0.57735$ to correspond with Gauss point 1 of Fig. 6-5, and the first contribution to the [B] matrix is formulated as follows (note the similarity to the terms in $[B]_1$ of Prob. 6.9):

$$[B]_1 = \begin{bmatrix} -0.14544 & 0 & 0.25713 & 0 & 0.03897 & 0 & -0.15066 & 0 \\ 0 & -0.29088 & 0 & 0.01426 & 0 & 0.07794 & 0 & 0.19868 \\ -0.29088 & -0.14544 & 0.01426 & 0.25713 & 0.07794 & 0.03897 & 0.19868 & -0.15066 \end{bmatrix}$$

The elasticity matrix of material constants [C] is evaluated using Eqs. (3.6)–(3.8):

$$[C] = \begin{bmatrix} C_{11} & C_{12} & 0 \\ C_{12} & C_{22} & 0 \\ 0 & 0 & C_{33} \end{bmatrix} = \begin{bmatrix} E/(1-\nu^2) & \nu E/(1-\nu^2) & 0 \\ \nu E/(1-\nu^2) & E/(1-\nu^2) & 0 \\ 0 & 0 & E/2(1+\nu) \end{bmatrix} = \begin{bmatrix} 1.0667 & 0.2667 & 0 \\ 0.2667 & 1.0667 & 0 \\ 0 & 0 & 0.4 \end{bmatrix}$$

The first contribution to the stiffness matrix is $[K]_1 = [B]_1^T[C][B]_1 t\, dA$, where $dA = 2.71133$ is defined in Prob. 6.9. This contribution is given by the symmetric matrix

$$[K]_1 = \begin{bmatrix} 0.15294 & 0.07647 & -0.11266 & -0.08262 & -0.04098 & -0.02049 & 0.00069 & 0.02664 \\ & 0.26765 & -0.05633 & -0.05256 & -0.02049 & -0.07172 & 0.00035 & -0.14337 \\ & & 0.19144 & 0.00663 & 0.03019 & 0.01509 & -0.10897 & 0.03461 \\ & & & 0.07229 & 0.02214 & 0.01408 & 0.05385 & -0.03382 \\ & & & & 0.01098 & 0.00549 & -0.00019 & -0.00714 \\ & & & & & 0.01922 & -0.00009 & 0.03842 \\ & & & & & & 0.10846 & -0.05411 \\ & & & & & & & 0.13877 \end{bmatrix}$$

The computations are repeated for the symmetric matrix $[K]_2$ using Gauss point 2 with $\xi_{i=1} = -0.57735$ and $\eta_{j=2} = 0.57735$:

$$[K]_2 = \begin{bmatrix} 0.07066 & -0.02753 & 0.02753 & -0.02565 & 0.00472 & -0.04254 & -0.10298 & 0.09572 \\ & 0.15977 & -0.01860 & 0.03220 & -0.02329 & -0.07179 & 0.06943 & -0.12018 \\ & & 0.02763 & -0.00652 & 0.04789 & 0.00080 & -0.10311 & 0.02432 \\ & & & 0.01394 & 0.00784 & 0.00589 & 0.02432 & -0.05204 \\ & & & & 0.12611 & 0.04473 & -0.17872 & -0.02928 \\ & & & & & 0.08789 & -0.00299 & -0.02120 \\ & & & & & & 0.38481 & -0.09076 \\ & & & & & & & 0.19422 \end{bmatrix}$$

Similarly, for Gauss point 3, $\xi_{i=2} = 0.57735$ and $\eta_{j=1} = -0.57735$, and that portion of the stiffness matrix is the symmetric matrix

$$[K]_3 = \begin{bmatrix} 0.17435 & 0.05150 & -0.21058 & -0.02176 & -0.02019 & -0.03557 & 0.05642 & 0.00583 \\ & 0.10226 & 0.00453 & -0.03808 & -0.05482 & -0.07438 & -0.00121 & 0.01020 \\ & & 0.34948 & -0.08242 & -0.04526 & 0.05581 & -0.09364 & 0.02209 \\ & & & 0.17639 & 0.08210 & -0.09105 & 0.02209 & -0.04726 \\ & & & & 0.05332 & -0.00528 & 0.01213 & -0.02120 \\ & & & & & 0.14103 & -0.01496 & 0.02440 \\ & & & & & & 0.02509 & -0.00592 \\ & & & & & & & 0.01266 \end{bmatrix}$$

And, for Gauss point 4, $\xi_{i=2} = 0.57735$ and $\eta_{j=1} = 0.57735$, and that portion of the (symmetric) stiffness matrix is

$$[K]_4 = \begin{bmatrix} 0.00905 & 0.00453 & -0.00510 & -0.00959 & -0.03379 & -0.01689 & 0.02983 & 0.02196 \\ & 0.01584 & -0.00255 & 0.02982 & -0.01689 & -0.05913 & 0.01492 & 0.01347 \\ & & 0.13442 & -0.06037 & 0.01903 & 0.00952 & -0.14836 & 0.05340 \\ & & & 0.13128 & 0.03581 & -0.11128 & 0.03416 & -0.04983 \\ & & & & 0.12609 & 0.06305 & -0.11134 & -0.08196 \\ & & & & & 0.22066 & -0.05567 & -0.05025 \\ & & & & & & 0.22986 & 0.00660 \\ & & & & & & & 0.08661 \end{bmatrix}$$

The final (symmetric) stiffness matrix is the sum of the four component matrices, or

$$[K] = \begin{bmatrix} 0.40701 & 0.10497 & -0.30074 & -0.13962 & -0.09023 & -0.11550 & -0.01603 & 0.15014 \\ & 0.54552 & -0.07296 & -0.02862 & -0.11550 & -0.27701 & 0.08348 & -0.23988 \\ & & 0.70296 & -0.14268 & 0.05185 & 0.08122 & -0.45407 & 0.13441 \\ & & & 0.39392 & 0.14789 & -0.18235 & 0.13441 & -0.18295 \\ & & & & 0.31650 & 0.10798 & -0.27812 & -0.14037 \\ & & & & & 0.46880 & -0.07371 & -0.00944 \\ & & & & & & 0.74822 & -0.14418 \\ & & & & & & & 0.43227 \end{bmatrix}$$

6.12. Discuss the derivation of the [B] matrix for axisymmetric elasticity.

The governing strain-displacement relationships are given by Eqs. (*3.17*) with $v = \partial/\partial\theta = 0$. The matrix of material constants [C] is given by Eq. (*d*) of Prob. 3.26. The [B] matrix is derived, as usual, as a shape function matrix postmultiplied by an operator matrix. The form of the operator matrix is dictated, in the case of axisymmetric elasticity, by the order of the stresses in the stress matrix or the order of the strains in the strain matrix. In this case, use the same strain matrix given by Eq. (*a*) of Prob. 3.26. Then, as in Chap. 3, $\{\epsilon\} = [L][N]\{u\}$, where $\{u\}$, in this application, is a matrix of eight unknown displacements corresponding to a four-node quadrilateral element:

$$\begin{Bmatrix} \epsilon_{rr} \\ \epsilon_{\theta\theta} \\ \epsilon_{zz} \\ \epsilon_{rz} \end{Bmatrix} = \begin{bmatrix} \partial/\partial r & 0 \\ 1/r & 0 \\ 0 & \partial/\partial z \\ \partial/\partial z & \partial/\partial r \end{bmatrix} \begin{bmatrix} N_1 & 0 & N_2 & 0 & N_3 & 0 & N_4 & 0 \\ 0 & N_1 & 0 & N_2 & 0 & N_3 & 0 & N_4 \end{bmatrix} \{u\} \qquad (a)$$

$$[B] = [L][N] = \begin{bmatrix} \partial N_1/\partial r & 0 & \partial N_2/\partial r & 0 & \partial N_3/\partial r & 0 & \partial N_4/\partial r & 0 \\ N_1/r & 0 & N_2/r & 0 & N_3/r & 0 & N_4/r & 0 \\ 0 & \partial N_1/\partial z & 0 & \partial N_2/\partial z & 0 & \partial N_3/\partial z & 0 & \partial N_4/\partial z \\ \partial N_1/\partial z & \partial N_1/\partial r & \partial N_2/\partial z & \partial N_2/\partial r & \partial N_3/\partial z & \partial N_3/\partial r & \partial N_4/\partial z & \partial N_4/\partial r \end{bmatrix} \qquad (b)$$

The terms containing partial derivatives are obtained from Eq. (*b*) of Prob. 6.8 and substituted into Eq. (*b*) above, with r and z replacing x and y, respectively. The terms containing the shape function divided by r are computed directly for each node (shape function). For instance, let the x coordinate correspond to the radial coordinate r; Eq. (*b*) of Prob. 6.5 is used to compute the r in Eq. (*b*) above. The ξ and η of Prob. 6.5 correspond to the coordinates of the integration point in the ξ, η system. The shape functions are evaluated by substituting the coordinates of the integration point (Gauss point) into the corresponding shape function equation (see Prob. 6.4 for a four-node quadrilateral).

6.13. (*a*) Assume a five-element model as shown in Fig. 6-6 to solve the long cylinder problem described in Probs. 6.29 and 6.30. The cylinder has an inside radius of 1 in and outside radius of 2 in. Assume an axisymmetric internal pressure loading of 1000 psi. Show the results for the $[B]_1$ matrix and the local stiffness matrix for element I. Compare results for radial displacement with the exact solution. Assume $E = 1.0$ psi and $\nu = 0.3$.

(*b*) Compare the results for the 5-element model with those for a 10-element model (see Fig. 6-6).

(*a*) The five-element model of Fig. 6-6 is constructed of square elements 0.2×0.2 in. In the derivation for the axisymmetric finite element the differential area used was $2\pi r\,dr\,dz$, and that requires that the internal pressure be distributed around the inside circumference of the cylinder. The pressure loading is converted to nodal point loading as $(1000\ \text{psi})(2\pi)(1\ \text{in})(0.2\ \text{in}) = 1265.64$ lb, and one-half of that is applied to nodes 1 and 2.

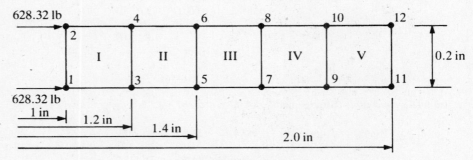

Five-element model

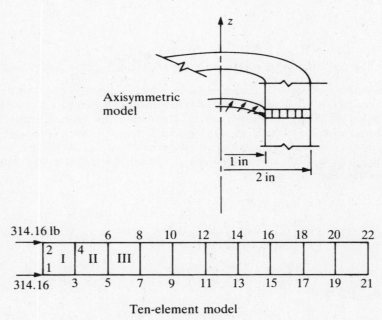

Ten-element model

Fig. 6-6

The $[B]_1$ matrix for element I is a 4×8 matrix corresponding to Eq. (*b*) of Prob. 6.12 and is computed for Gauss point 1 using $\xi_{i=1} = -0.57735$ and $\eta_{j=1} = -0.57735$:

$$[B]_1 = \begin{bmatrix} -3.9434 & 0 & 3.9434 & 0 & 1.0566 & 0 & -1.0566 & 0 \\ 0.5968 & 0 & 0.1599 & 0 & 0.0428 & 0 & 0.1599 & 0 \\ 0 & -3.9434 & 0 & -1.0566 & 0 & 1.0566 & 0 & 3.9434 \\ -3.9434 & -3.9434 & -1.0566 & 3.9434 & 1.0566 & 1.0566 & 3.9434 & -1.0566 \end{bmatrix}$$

Note that the complete stiffness matrix is made up of four parts as in Prob. 6.11. There is a separate B matrix corresponding to each integration point, and the stiffness matrix for element I is the 8×8 symmetric matrix

$$[K]_I = \begin{bmatrix} 3.74129 & 1.49024 & -2.64114 & 0.32221 & -1.98513 & -1.61107 & 0.60193 & -0.20136 \\ & 3.84644 & -0.34235 & 0.66457 & -1.71177 & -1.99370 & 0.20138 & -2.51730 \\ & & 4.30205 & -1.83260 & 0.76148 & 0.46318 & -1.98514 & 1.71177 \\ & & & 4.12838 & -0.46318 & -2.79924 & 1.61107 & -1.99370 \\ & & & & 4.30205 & -1.83260 & -2.64114 & 0.34235 \\ & & & & & 4.12838 & -0.32221 & 0.66457 \\ & & & & & & 3.74129 & -1.49024 \\ & & & & & & & 3.84644 \end{bmatrix}$$

The exact solution is computed using the results of Prob. 6.30. Results for displacement are given in Table 6.2. Note that all displacements in the z direction are zero for this problem and were entered into the computer solution as zero displacement boundary conditions.

(b) The 10-element solution is computed using square elements 0.1×0.1, and the nodal point loading is computed as 314.16 lb at nodes 1 and 2. The results are given in Table 6.2. The results for displacement tabulated in Table 6.2 are for illustration purposes, and the large values are a result of assuming the material constant $E = 1$ psi.

Table 6.2   Radial Displacements for a Thick-Walled Cylinder (in)

r	Five elements	Nodes	Ten elements	Nodes	Exact
1.0	1894.35	1, 2	1903.52	1, 2	1906.67
1.1			1763.64	3, 4	1766.42
1.2	1642.63	3, 4	1649.94	5, 6	1652.44
1.3			1556.37	7, 8	1558.60
1.4	1472.46	5, 6	1478.64	9, 10	1480.76
1.5			1413.57	11, 12	1415.55
1.6	1353.35	7, 8	1358.79	13, 14	1360.67
1.7			1312.50	15, 16	1314.27
1.8	1268.72	9, 10	1273.26	17, 18	1274.96
1.9			1239.98	19, 20	1241.61
2.0	1207.18	11, 12	1211.76	21, 22	1213.33

6.14. Compute the stresses in the radial direction σ_{rr} and the tangential direction $\sigma_{\theta\theta}$ for the thick-walled cylinder described in Prob. 6.13.

Strains are computed as $\{\epsilon\} = [B]\{u\}$ using the [B] matrix of Prob. 6.12, and stresses are computed as $\{\sigma\} = [C]\{\epsilon\}$ as discussed in Prob. 3.10, except that

$$\{\sigma\} = [\sigma_{rr} \quad \sigma_{\theta\theta} \quad \sigma_{zz} \quad \sigma_{rz}]^T \tag{a}$$

and corresponds to the ordering of the strains in Eq. (a) of Prob. 6.12. Note that in the finite element formulation all four stresses are computed. The matrix of material constants is given in Prob. 3.26. Stresses are usually computed at the Gauss points when isoparametric finite elements are used but can be computed at any location within the element (ξ, η must have values corresponding to the x, y location within the element). The strains at Gauss point 1 of element I for the five-element model are obtained by multiplying the $[B]_1$ matrix of Prob. 6.13 by the corresponding displacement vector $\{u\}_I$. Note that the displacement vector corresponds to the node numbering of element I:

$$\{u\}_I = [u_1 \quad w_1 \quad u_3 \quad w_3 \quad u_4 \quad w_4 \quad u_2 \quad w_2]^T$$

$$= [1894.35 \quad 0 \quad 1642.63 \quad 0 \quad 1642.63 \quad 0 \quad 1894.35 \quad 0]^T \qquad (b)$$

$$[\epsilon_{rr} \quad \epsilon_{\theta\theta} \quad \epsilon_{zz} \quad \epsilon_{rz}]_1^T = [B]_1\{u\}_I = [-1258.60 \quad 1766.42 \quad 0 \quad 0]^T \qquad (c)$$

The results for strains and displacements are too large to be practical, but in this example the material constant E was taken as 0.1 psi. A reasonable value for an actual material would be of the order 10^7 psi. The stresses (in psi) are computed as

$$\begin{Bmatrix} \sigma_{rr} \\ \sigma_{\theta\theta} \\ \sigma_{zz} \\ \sigma_{rz} \end{Bmatrix} = \begin{bmatrix} 1.34615 & 0.57693 & 0.57692 & 0 \\ 0.57692 & 1.34615 & 0.57692 & 0 \\ 0.57692 & 0.57692 & 1.34615 & 0 \\ 0 & 0 & 0 & 0.38462 \end{bmatrix} \begin{Bmatrix} -1258.60 \\ 1766.42 \\ 0 \\ 0 \end{Bmatrix} = \begin{Bmatrix} -675.16 \\ 1651.75 \\ 292.97 \\ 0 \end{Bmatrix}$$

The location of Gauss point 1 within element I is computed using Eqs. (b) and (c) of Prob. 6.5 with x and y replaced by r and z, respectively. Also, referring to Fig. 6-5, $\xi = -0.57735$ and $\eta = -0.57735$.

$$r = \tfrac{1}{4}[(1.57735)(1.57735)(1.0) + (0.42265)(1.57735)(1.2) + (0.42265)(0.42265)(1.2)$$

$$+ (1.57735)(0.42265)(1.0)] = 1.04226 \text{ in}$$

$$z = \tfrac{1}{4}[(1.57735)(1.57735)(0.0) + (0.42265)(1.57735)(0.2) + (0.42265)(0.42265)(0.2)$$

$$+ (1.57735)(0.42265)(0.0)] = 0.04226 \text{ in}$$

The same results would be found for Gauss point 2 of element I. The strains and stresses can be computed for any coordinate location within or on the boundary of the element. However, the B matrix utilized for constructing the stiffness matrix cannot be used here. A new B matrix must be developed using Eq. (b) of Prob. 6.12 with ξ, η coordinates corresponding to the proper r, z coordinates. For instance, the strains at node 3, element I, correspond to $\xi = +1$ and $\eta = +1$.

The remaining stresses are given in Tables 6.3 and 6.4 and are computed at the Gauss point for each element. The exact axial stress is computed using Eq. (3.12) as $\sigma_{zz} = \nu(\sigma_{rr} + \sigma_{\theta\theta})$ and is constant, $\sigma_{zz} = 200$.

Table 6.3   Stresses for a Thick-Walled Cylinder (psi) Five-Element Model

Element	Gauss point, r (in)	Finite element			Exact	
		σ_{rr}	$\sigma_{\theta\theta}$	σ_{zz}	σ_{rr}	$\sigma_{\theta\theta}$
1	1.04226	−675.13	1651.86	293.02	−894.07	1560.73
1	1.15774	−849.19	1245.71	118.95	−661.43	1328.19
2	1.24226	−399.23	1250.15	255.28	−530.66	1197.33
2	1.35774	−504.43	1004.68	150.07	−389.95	1056.62
3	1.44226	−222.76	1007.26	235.35	−307.66	974.32
3	1.55774	−291.15	847.70	166.96	−216.14	882.81
4	1.64226	−103.18	849.33	223.85	−161.40	827.71
4	1.75774	−150.11	739.83	176.92	−98.22	764.88
5	1.84226	−18.44	740.94	216.75	−59.53	726.19
5	1.95774	−52.03	662.56	183.16	−14.55	681.21

Table 6.4   Stresses for a Thick-Walled Cylinder (psi) Ten-Element Model

Element	Gauss point, r (in)	Finite element			Exact	
		σ_{rr}	$\sigma_{\theta\theta}$	σ_{zz}	σ_{rr}	$\sigma_{\theta\theta}$
1	1.02113	-824.22	1663.44	251.76	-945.38	1612.05
1	1.07887	-924.07	1430.47	151.92	-812.19	1478.85
2	1.12113	-635.44	1432.78	239.20	-727.44	1394.11
2	1.17887	-711.41	1255.52	163.23	-626.08	1292.75
3	1.22113	-489.33	1257.28	230.38	-560.83	1227.49
3	1.27887	-548.47	1119.29	171.24	-481.91	1148.57
4	1.32113	-373.94	1120.65	224.01	-430.59	1097.25
4	1.37887	-420.88	1011.13	177.08	-367.95	1034.62
5	1.42113	-281.23	1012.21	219.29	-326.86	993.53
5	1.47887	-319.11	923.84	181.42	-276.31	942.98
6	1.52113	-205.63	924.71	215.72	-242.91	909.58
6	1.57887	-236.63	852.37	184.72	-201.53	868.20
7	1.62113	-143.16	853.08	212.98	-174.01	840.68
7	1.67887	-168.86	793.13	187.28	-139.70	806.38
8	1.72113	-90.97	793.72	210.83	-116.77	783.44
8	1.77887	-112.50	743.47	189.29	-88.02	754.69
9	1.82113	-46.91	743.96	209.12	-68.70	735.36
9	1.87887	-65.13	701.43	190.89	-44.37	711.03
10	1.92113	-9.36	701.85	207.75	-27.93	694.60
10	1.97887	-24.93	665.54	192.18	-7.10	673.82

6.15. The displacements computed for the axisymmetric cylinder of Prob. 6.13 are quite accurate, even for the five-element model. However, the stresses computed in Prob. 6.14 do not share that same degree of accuracy and in fact appear to oscillate about the exact solution. Discuss the computation of stresses for the thick-walled cylinder problem.

The lack of accuracy for the computed stresses does not indicate that the finite element method does not give satisfactory results. The fault lies with the choice of element. In some applications *mesh refinement* can improve the results. That is, more elements can be used to attempt to improve the answer. In the limit the finite element result should approach the exact solution. In this application of the finite element method, more elements will improve the result but will not eliminate the oscillatory behavior of the radial stress σ_{rr}. The computation for the radial stress is $\sigma_{rr} = C_{11} \, du/dr + C_{12} u/r$, where the first term is negative and the second term is positive. The term du/dr is constant for a linear element at all locations within an element. The displacement u decreases as r increases, and the result u/r tends to decrease. The final result for σ_{rr} tends to increase as r increases within a single element. However, the reader should refer to Prob. 6.31 and compare the results of stress computations at the Gauss points with stress computations at the center of the element.

The finite element analysis can be improved significantly by using a higher-order element in the radial direction since that will allow the term du/dr to vary within the element. Compare the four-node element solutions with the eight-node element solution given in Prob. 6.32. Note that only two elements are used for that analysis. Linear elements can give good results for problems such as temperature distribution, where the nodal point results are of interest. In any application where the derivative is the significant result, the analyst should be aware of the limitations of linear element formulations.

6.16. Assume the eight-node quadrilateral isoparametric element shown in Fig. 6-7 is loaded with a uniform pressure loading, $p_x = 1.0$, acting in the x direction along the side defined by nodes 1, 8, and 4. Compute the distribution of the pressure loading to each node. Use a three-point gaussian quadrature.

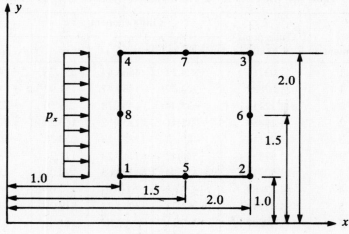

Fig. 6-7

The formulation is similar to Prob. 3.12. The surface loading is represented as

$$\int_{S} [\mathbf{N}]^{T}\{\mathbf{T}\}\, dS \qquad (a)$$

where $\{\mathbf{T}\}$ is the surface traction matrix and $[\mathbf{N}]$ is the shape function matrix and is similar to Eq. (b) of Prob. 3.12 except that there are 16 rows rather than 8. The uniform pressure loading along side 1-8-4 corresponds to the isoparametric coordinate $\xi = -1$, and all shape functions will compute as zero except at nodes 1, 8, and 4. Formally, Eq. (a) can be written as

$$\int_{-1}^{+1} \begin{bmatrix} N_1 & 0 & 0 & 0 & 0 & 0 & N_4 & 0 & 0 & 0 & 0 & 0 & 0 & 0 & N_8 & 0 \\ 0 & N_1 & 0 & 0 & 0 & 0 & 0 & N_4 & 0 & 0 & 0 & 0 & 0 & 0 & 0 & N_8 \end{bmatrix}^{T} \begin{Bmatrix} p_x = 1.0 \\ p_y = 0 \end{Bmatrix} d\eta \qquad (b)$$

The term $\int_{-1}^{+1} N_1 p_x\, d\eta$ is evaluated numerically with $\xi = -1$ and η and the weight functions taken from Table 6.1. The shape functions are given in Prob. 6.25.

$$N_1 \ (\text{with } \xi = -1) = \frac{(1+1)(1-\eta)(1-\eta-1)}{4} = \frac{\eta^2 - \eta}{2}$$

$$\int_{-1}^{+1} N_1 p_x\, d\eta = \{[(-0.774597)^2 + 0.774597)](0.555555) + (0 - 0)(0.888888)$$

$$+ [(0.774597)^2 - 0.774597)](0.555555)\} p_x/2 = 0.333333 p_x \qquad (c)$$

Similarly, the integral involving N_4 becomes $0.333333 p_x$, and the integral representing the contribution to the center node 8 becomes $0.666667 p_x$. The total length of the side of the isoparametric element is 2, and the results are interpreted to mean that each corner node is assigned one-sixth of the uniform load and the center node is assigned two-thirds of the uniform load. Note that the result given by Eq. (c) could have been obtained by direct integration.

While the preceding computations illustrate distributing the force to each node, the analyst should take advantage of the isoparametric formulation to compute the node loading. Consider the plane element of Fig. 6-4(b). In the isoparametric formulation the term dS should be expressed in terms of the ξ, η system using an interpolation function. Assume a unit thickness ($t = 1$) or carry t through the derivation and write $dS = t\, dL$, where dL defines a curve corresponding to the boundary of the plane area. In the cartesian coordinate system,

$$dL = [(dx)^2 + (dy)^2]^{1/2} \qquad (d)$$

Refer to Eq. (b) of Prob. 6.6 and relate dx and dy to the ξ, η system:

$$dx = \frac{\partial x}{\partial \xi} d\xi + \frac{\partial x}{\partial \eta} d\eta \qquad dy = \frac{\partial y}{\partial \xi} d\xi + \frac{\partial y}{\partial \eta} d\eta \qquad (e)$$

The boundary of the element will always correspond to $\xi = \pm 1$ or $\eta = \pm 1$. Let $\xi = -1$, corresponding to the element of Fig. 6-4, and it follows that $\partial/\partial\xi = 0$. Use Eq. (a) of Prob. 6.5 and rewrite Eq. (e):

$$dx = \frac{\partial x}{\partial \eta} d\eta = \sum \frac{\partial N_i}{\partial \eta} x_i \, d\eta \qquad (i = 1 \text{ to } 8) \qquad (f)$$

$$dy = \frac{\partial y}{\partial \eta} d\eta = \sum \frac{\partial N_i}{\partial \eta} y_i \, d\eta \qquad (i = 1 \text{ to } 8) \qquad (g)$$

where the N_i are interpolation functions corresponding to the eight-node isoparametric element. Assume that a three-point integration is to be used along the curve $\xi = -1$ for the element of Fig. 6-4(a). For the first integration point let $\xi = -1$ and $\eta = -0.774597$. Equations (f) and (g) are evaluated, then substituted into Eq. (d) to obtain dL. Equation (a) is evaluated with $\xi = -1$ and $\eta = -0.774597$. In the axisymmetric formulation dS is $2\pi r \, dL$ to simulate the pressure distributed around the entire circumference of the cylinder. The computations are repeated for the remaining Gauss points, and the final result is the sum of the three computations.

For illustration, use the element of Fig. 6-7 in x, y coordinates. Let $\xi = -1$ and $\eta = -0.774597$, then substitute into Eqs. (f) and (g) to obtain $dx = 0$ and $dy = 0.5$ and by Eq. (d), $dL = 0.5$. The first contribution to the pressure loading is computed as follows.

Node 1: $(N_1)(p_x)(dL)(w_1) = (0.687298)p_x(0.5)(0.555555) = 0.190916p_x$

Node 4: $(N_4)(p_x)(dL)(w_1) = (-0.087298)p_x(0.5)(0.555555) = -0.024249p_x$

Node 8: $(N_8)(p_x)(dL)(w_1) = (0.400)p_x(0.5)(0.555555) = 0.111111p_x$

Similarly, let $\xi = -1$ and $\eta = 0.0$, $dx = 0.0$ and $dy = 0.5$, and $dL = 0.5$. The second contribution to the pressure loading is (note that $w_2 = 0.888888$) as follows.

Node 1: $N_1 = 0.0$

Node 4: $N_4 = 0.0$

Node 8: $(1.0)p_x(0.5)(0.88888) = 0.44444p_x$

The third integration point corresponds to $\xi = -1$ and $\eta = +0.774597$ with $dx = 0.0$, $dy = 0.5$, $dL = 0.5$, and $w_3 = 0.55555$, and the contribution to the pressure loading is computed as follows.

Node 1: $-0.024249p_x$

Node 4: $0.190916p_x$

Node 8: $0.111111p_x$

The total loading is the sum of the three contributions and is computed as $0.166667p_x$ at nodes 1 and 4, and as $0.666666p_x$ at node 8.

6.17. Discuss the formulation of an isoparametric finite element for axisymmetric electrostatics. Refer to Probs. 2.19 and 6.37.

Axisymmetric elasticity was analyzed in previous problems before any discussion of the Laplace operator. The finite element formulation for physical problems involving the Laplace operator can be misleading. Refer to Eq. (6-8) and note that the governing equation has both shape functions and derivatives of shape functions as well as division by the r coordinate. However, the formulation given in Probs. 3.25 and 6.37 indicates that the formulation of the [B] matrix and subsequent formulation of the stiffness matrix for the axisymmetric problem is identical to the formulation in cartesian coordinates with the exception of the definition of dV. It is merely good fortune that the formulations are similar, and the reader should not be mislead into thinking that there is no difference between axisymmetric cylindrical coordinates and cartesian coordinates for problems involving the Laplace operator.

6.18. Discuss node placement and area coordinates for linear and higher-order triangular isoparametric elements.

Linear, quadratic, and cubic elements are shown in Fig. 6-8. The node numbering sequence is arbitrary, but very often the corner nodes are numbered 1, 2, and 3, with the intermediate nodes numbered in sequence starting with 4 along the side between nodes 1 and 2. The area coordinates then correspond to Fig. 3-12 for triangular elements of any order. For instance, area coordinate L_2 emanates from the side opposite node 2. The intermediate nodes are placed equidistant from the end nodes, and for a cubic element the side nodes are placed at the one-third points along the side.

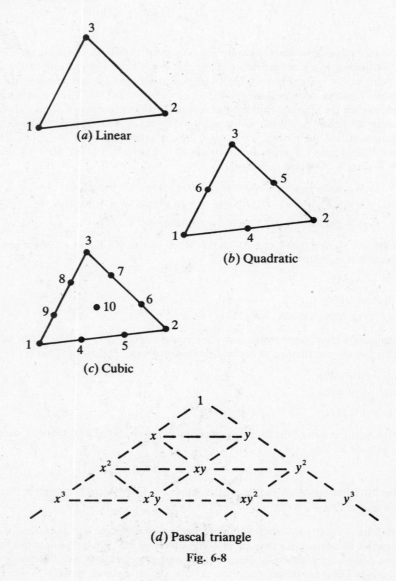

(a) Linear

(b) Quadratic

(c) Cubic

(d) Pascal triangle

Fig. 6-8

The derivation for the linear three-node triangular finite element discussed in Prob. 3.2 began by assuming an interpolation function that was linear in terms of the x, y coordinates, or

$$\phi = C_1 + C_2 x + C_3 y \tag{a}$$

It follows that a quadratic triangular finite element must contain all possible linear and quadratic coordinate functions, or

$$\phi = C_1 + C_2 x + C_3 y + C_4 x^2 + C_5 xy + C_6 y^2 \tag{b}$$

This complete polynomial representation for interpolation functions corresponds to the *Pascal triangle* shown in Fig. 6-8(d). The cubic element should contain all linear, quadratic, and cubic coordinate terms, and it can be seen from the Pascal triangle that there are 10 such terms. The cubic element, Fig. 6.8(c), must have 10 corresponding nodes. Therefore, the tenth node is located at the centroid of the triangular element. The derivation of shape functions using interpolation functions and x, y coordinates can become a tedious algebraic task. The use of area coordinates and Eq. (6.4) simplifies the derivation and formulation of a stiffness matrix.

6.19. Discuss higher-order isoparametric elements in terms of number of nodes versus the complete polynomial representation that satisfies the Pascal triangle requirement. In particular, compare the six-node quadratic triangular element, the eight-node serendipity element, and the nine-node Lagrangian element.

All these elements can be classified as quadratic elements because they have three nodes along each side. The triangular element was discussed in Prob. 6.18, and the corresponding interpolation function contains all possible quadratic terms [see Eq. (b) of Prob. 6.18] and none higher than quadratic.

A study of the shape functions for the eight-node serendipity element shows that all six constant, linear, and quadratic terms are represented with the addition of two cubic terms. It follows that an eight-node element must have eight terms in its corresponding interpolation function, and the cubic terms in this case are $\xi^2 \eta$ and $\xi \eta^2$. The nine-node Lagrangian element shape functions are given in Prob. 6.24, and a study of these shape functions indicates that there are two additional third-order terms and one fourth-order term that can be shown to be $\xi^2 \eta$, $\xi \eta^2$, and $\xi^2 \eta^2$.

It can be concluded that quadrilateral elements do not satisfy the Pascal triangle requirement that the shape function be represented by a complete polynomial. However, the idea is to satisfy the Pascal triangle (completeness) requirement in the best possible way. In higher-order elements the excessive interior nodes in the Lagrange family of elements can cause difficulty with convergence and should be avoided. For additional study see Burnett (1987) or Zienkiewicz and Taylor (1989).

6.20. Use area coordinates to determine the shape function N_4 for both the quadratic and cubic triangular finite elements of Fig. 6-8.

The quadratic element is shown in Fig. 6-9(a). The order of the triangle is $n = 3 - 1$, and it follows that two lines should be sufficient to pass through all nodes except node 4. These lines are shown in the figure as $L_1 = 0$ passing through nodes 2, 5, and 3 and $L_2 = 0$ passing through nodes 1, 6, and 3. The location of node 4 is defined using all three area coordinates as $L_1 = \frac{1}{2}$, $L_2 = \frac{1}{2}$, $L_3 = 0$. There are two terms in Eq. (6.4):

$$N_4 = \left[\frac{L_1 - 0}{\frac{1}{2}} \right] \left[\frac{L_2 - 0}{\frac{1}{2}} \right] = 4L_1 L_2$$

where the numerator in each term is an equation of a line and the denominator is the equation of the line evaluated using the coordinates of node 4.

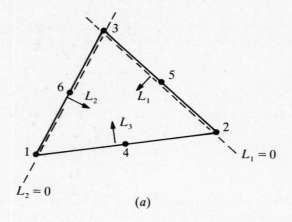

(a)

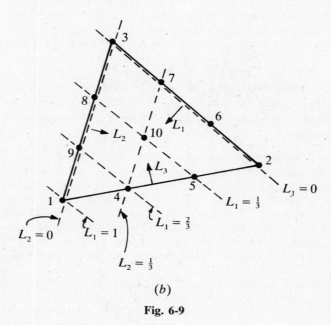

(b)

Fig. 6-9

The order of the cubic element is 3. The lines $L_1 = 0$, $L_1 = \frac{1}{3}$, and $L_2 = 0$, shown in Fig. 6.9(b), intersect all nodes except node 4. The coordinate location of node 4 is ($L_1 = \frac{2}{3}$, $L_2 = \frac{1}{3}$, $L_3 = 0$). Substituting into Eq. (6.4) gives

$$N_4 = \left[\frac{L_1 - 0}{\frac{2}{3}}\right]\left[\frac{L_2 - 0}{\frac{1}{3}}\right]\left[\frac{L_1 - \frac{1}{3}}{\frac{2}{3} - \frac{1}{3}}\right] = \frac{9}{2}L_1 L_2 (3L_1 - 1)$$

6.21. Area coordinates and area integration formulas for three-node triangular elements were introduced in Chap. 3, Probs. 3.16 and 3.17. The isoparametric formulation is advantageous for higher-order triangular elements. Like the quadrilateral element, the distorted element in x, y coordinates can be mapped into a parent element, and standard integration routines can be used to evaluate the stiffness integrals. Discuss the coordinate transformations for isoparametric triangular elements using area coordinates.

The method used by Zienkiewicz and Taylor (1989) may help to avoid some confusion in the derivation. There are three area coordinates for a two-dimensional element, but the original coordinate is 2-space (x and y). It follows that the parent element should be 2-space (ξ and η for the quadrilateral). Recall from Prob. 3.16 that there are only two independent area coordinates, or $L_1 + L_2 + L_3 = 1$. Identify

$$\xi = L_1 \qquad \eta = L_2 \qquad \text{then } L_3 = 1 - \xi - \eta \tag{a}$$

and then Eqs. (c)–(f) and (h) of Prob. 6.6 are valid. The shape functions are written in terms of L_1, L_2, and L_3, or

$$N_i = N_i(L_1, L_2, L_3) \tag{b}$$

Hold ξ, defined in Eq. (a), independent of L_1 in Eq. (b) and write the partial derivative

$$\frac{\partial N_i}{\partial \xi} = \frac{\partial N_i}{\partial L_1} \frac{\partial L_1}{\partial \xi} + \frac{\partial N_i}{\partial L_2} \frac{\partial L_2}{\partial \xi} + \frac{\partial N_i}{\partial L_3} \frac{\partial L_3}{\partial \xi} \tag{c}$$

The right-hand side of Eq. (c), in view of (a), can be evaluated as

$$\frac{\partial L_1}{\partial \xi} = 1 \qquad \frac{\partial L_2}{\partial \xi} = 0 \qquad \frac{\partial L_3}{\partial \xi} = -1$$

and Eq. (c) becomes

$$\frac{\partial N_i}{\partial \xi} = \frac{\partial N_i}{\partial L_1} - \frac{\partial N_i}{\partial L_3} \tag{d}$$

A similar analysis gives

$$\frac{\partial N_i}{\partial \eta} = \frac{\partial N_i}{\partial L_2} - \frac{\partial N_i}{\partial L_3} \tag{e}$$

The jacobian can be written the same as Eq. (g) of Prob. 6.6:

$$J = \begin{bmatrix} \sum \dfrac{\partial N_i}{\partial \xi} x_i & \sum \dfrac{\partial N_i}{\partial \xi} y_i \\ \sum \dfrac{\partial N_i}{\partial \eta} x_i & \sum \dfrac{\partial N_i}{\partial \eta} y_i \end{bmatrix} \tag{f}$$

Equation (f) is evaluated using the definitions given by Eqs. (d) and (e).

6.22. A triangular element is shown in Fig. 6-10(a). Use the isoparametric formulation to compute the terms in the stiffness matrix corresponding to convection as discussed in Probs. 3.18 and 3.39. Integration results for triangles are given in Fig. 6-10(b). Illustrate the use of both linear and quadratic integration formulas.

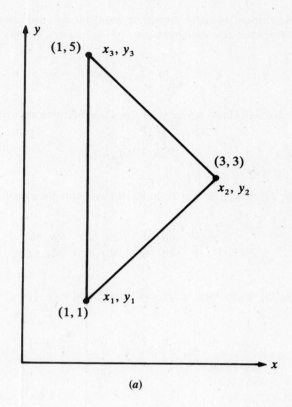

(a)

Order	Points	Area coordinates			Weights
		L_1	L_2	L_3	w_i
Linear	a	$\frac{1}{3}$	$\frac{1}{3}$	$\frac{1}{3}$	1
Quadratic	a	$\frac{1}{2}$	$\frac{1}{2}$	0	$\left.\begin{array}{c}\\\\\\\end{array}\right\}\frac{1}{3}$
	b	$\frac{1}{2}$	0	$\frac{1}{2}$	
	c	0	$\frac{1}{2}$	$\frac{1}{2}$	
Cubic	a	$\frac{1}{3}$	$\frac{1}{3}$	$\frac{1}{3}$	$-\frac{27}{48}$
	b	0.6	0.2	0.2	$\left.\begin{array}{c}\\\\\\\end{array}\right\}\frac{25}{48}$
	c	0.2	0.6	0.2	
	d	0.2	0.2	0.6	

(b)

Fig. 6-10

The function to be integrated is Eq. (b) of Prob. 3.18. Assume a unit value of t; for reference, the function is

$$\int_A \begin{bmatrix} N_1 & N_1 \\ N_2 & N_2 \\ N_3 & N_3 \end{bmatrix} \begin{bmatrix} u_x & 0 \\ 0 & u_y \end{bmatrix} \begin{bmatrix} \partial N_1/\partial x & \partial N_2/\partial x & \partial N_3/\partial x \\ \partial N_1/\partial y & \partial N_2/\partial y & \partial N_3/\partial y \end{bmatrix} dA \qquad (a)$$

Recall from Chap. 3 or Prob. 6.42 that there is an equality between the linear shape functions and area coordinates:

$$x = [N]\{x\} = [L_1 \quad L_2 \quad L_3][x_1 \quad x_2 \quad x_3]^T = L_1 + 3L_2 + L_3$$
$$y = L_1 + 3L_2 + 5L_3$$

There are partial derivatives in Eq. (a), and that will require evaluating the inverse of the jacobian given by Eq. (f) of Prob. 6.21. Refer to Eqs. (d) and (e) of Prob. 6.21; it follows that

$$\frac{\partial N_1}{\partial \xi} = \frac{\partial N_1}{\partial L_1} - \frac{\partial N_1}{\partial L_3} = 1 - 0$$

$$\frac{\partial N_2}{\partial \xi} = \frac{\partial N_2}{\partial L_1} - \frac{\partial N_2}{\partial L_3} = 0 - 0$$

$$\frac{\partial N_3}{\partial \xi} = \frac{\partial N_3}{\partial L_1} - \frac{\partial N_3}{\partial L_3} = 0 - 1$$

$$\frac{\partial N_1}{\partial \eta} = \frac{\partial N_1}{\partial L_2} - \frac{\partial N_1}{\partial L_3} = 0 - 0 \qquad (b)$$

$$\frac{\partial N_2}{\partial \eta} = \frac{\partial N_2}{\partial L_2} - \frac{\partial N_2}{\partial L_3} = 1 - 0$$

$$\frac{\partial N_3}{\partial \eta} = \frac{\partial N_3}{\partial L_2} - \frac{\partial N_3}{\partial L_3} = 0 - 1$$

$$J_{11} = (1)(1) + (0)(3) + (-1)(1) = 0 \qquad J_{12} = (1)(1) + (0)(3) + (-1)(5) = -4$$

$$J_{21} = (0)(1) + (1)(3) + (-1)(1) = 2 \qquad J_{22} = (0)(1) + (1)(3) + (-1)(5) = -2$$

and
$$J = \begin{bmatrix} 0 & -4 \\ 2 & -2 \end{bmatrix} \qquad J^{-1} = \begin{bmatrix} -\frac{1}{4} & \frac{1}{2} \\ -\frac{1}{4} & 0 \end{bmatrix} \qquad |\det J| = 8 = 2A \qquad (c)$$

The jacobian for the coordinate transformation is complete, and now the right-hand matrix of Eq. (a) must be evaluated. The computation is analogous to Eq. (g) of Prob. 6.6. Using Eqs. (b) and (c) above,

$$\begin{bmatrix} \partial N_1/\partial x & \partial N_2/\partial x & \partial N_3/\partial x \\ \partial N_1/\partial y & \partial N_2/\partial y & \partial N_3/\partial y \end{bmatrix} = J^{-1} \begin{bmatrix} 1 & 0 & -1 \\ 0 & 1 & -1 \end{bmatrix} = \begin{bmatrix} -\frac{1}{4} & \frac{1}{2} & \frac{3}{4} \\ -\frac{1}{4} & 0 & \frac{1}{4} \end{bmatrix} \qquad (d)$$

Combining Eq. (d) with Eq. (a) gives the final form of the equation that is to be integrated numerically:

$$\begin{bmatrix} (-L_1/4)(u_x + u_y) & L_1 u_x/2 & (L_1/4)(3u_x + u_y) \\ (-L_2/4)(u_x + u_y) & L_2 u_x/2 & (L_2/4)(3u_x + u_y) \\ (-L_3/4)(u_x + u_y) & L_3 u_x/2 & (L_3/4)(3u_x + u_y) \end{bmatrix} \qquad (e)$$

Numerical integration for the stiffness matrix is accomplished using the formulas of Fig. 6-10(b) corresponding to area coordinates. The limits of integration are written in terms of area coordinates, and the weighted integral approximation follows as

$$\int_0^1 \int_0^{1-L_2} F \, dL_1 \, dL_2 = \sum_{i=1}^n \frac{1}{2} w_i F_i(L_{1i} L_{2i} L_{3i}) \qquad (f)$$

where w_i are the weights and L_{1i}, L_{2i}, and L_{3i} are the sampling points. Keep in mind that the function of Eq. (a) is defined in the x, y system, $dA = dx\, dy$, and in the new system, $dA = |\det J|\, dL_1\, dL_2$. Consider the first term of Eq. (e) and a linear order integration formula. For each L_i, substitute $\frac{1}{3}$ and note that $w_i = 1$. Of course, only L_1 appears in the first term, or

$$K_{11} = -\frac{1}{2}(1)\frac{\frac{1}{3}}{4}(u_x + u_y)(8) = -\frac{u_x + u_y}{3}$$

Similarly,

$$K_{22} = \frac{1}{2}(1)\frac{1}{2}\frac{1}{3}u_x(8) = \frac{2u_x}{3}$$

The remaining terms can be computed in the same manner and should be verified by the reader. The correctness of the linear integration can be checked using the exact solution given in Prob. 3.39:

$$K_{11} = \frac{u_x b_1 + u_y c_1}{6} \qquad b_1 = y_2 = y_3 = -2 \qquad c_1 = x_3 - x_2 = -2$$

$$K_{11} = -\frac{u_x + u_y}{3}$$

The linear integration gives an exact result for this term. Similarly,

$$K_{22} = \frac{u_x b_2 + u_y c_2}{6} \qquad b_2 = y_3 - y_1 = 4 \qquad c_2 = x_1 - x_3 = 0$$

$$K_{22} = \frac{2u_x}{3}$$

Again, the exact result is found.

The quadratic integration formula will be illustrated for the K_{11} term. Referring to Fig. 6-10(b), L_1 takes the values $\frac{1}{2}$, $\frac{1}{2}$, and 0, and $w_i = \frac{1}{3}$.

$$K_{11} = \frac{1}{2}\left\{-\left[\frac{1}{4}\frac{1}{2}(u_x + u_y)\frac{1}{3}\right] - \left[\frac{1}{4}\frac{1}{2}(u_x + u_y)\frac{1}{3}\right] - \left[\frac{1}{4}(0)(u_x + u_y)\frac{1}{3}\right]\right\}$$

$$K_{11} = -\frac{u_x + u_y}{3}$$

(8)

Again, the exact result is obtained. The remaining terms are computed in a similar manner and will not be recorded. Note that linear integration was sufficient to give exact results for the linear triangular finite element. However, the reader should study Prob. 6.43, where shape functions (no derivatives appear in that problem) are integrated, and discover that linear integration is not sufficient to give exact results in that case.

Supplementary Problems

6.23. Use the Lagrangian interpolation formula to derive shape functions for the six-node element shown in Fig. 6-11.

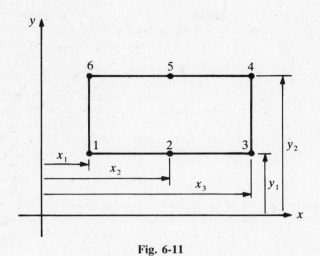

Fig. 6-11

6.24. Use the Lagrangian interpolation formula to derive shape functions for a nine-node element using ξ, η coordinates and the node numbering sequence shown in Fig. 6-12.

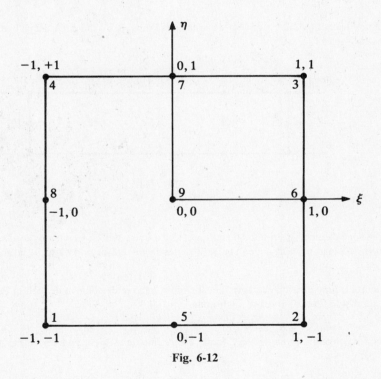

Fig. 6-12

6.25. Derive shape functions for an eight-node isoparametric element in ξ, η coordinates using the numbering sequence of Fig. 6-7.

6.26. A rectangular element has sides that are parallel to the x and y axes of a cartesian coordinate system. Derive the jacobian matrix and compute the value of the determinant of the jacobian.

6.27. Evaluate Eq. (h) of Prob. 6.7 for the rectangular element defined in Prob. 6.26.

6.28. Assume a four-node isoparametric formulation and derive the stiffness matrix for the plane elasticity case defined by Prob. 3.13. Assume $\nu = 0.3$ and give the result before and after substitution of displacement boundary conditions.

6.29. The analysis of a thick-walled cylinder with axisymmetric pressure loading is an important problem in the theory of elasticity. Because of the axisymmetric loading, it can be solved in one dimension. Conditions of *plane strain*, $\epsilon_{zz} = 0$, can be assumed if the ends of the cylinder are restrained. (*a*) Deduce the equation of equilibrium. (*b*) Derive the stress-strain equations using the plane strain assumption. (*c*) Substitute the results of part (*b*) into the equation obtained in part (*a*) and use Eqs. (*6.5*) to obtain one governing equation in terms of radial displacement u.

6.30. (*a*) Obtain the general solution for the differential equation derived in part (*c*) of Prob. 6.29. (*b*) Displacement boundary conditions are not specified for the general problem, however, stress boundary conditions in the form of compressive pressures $\sigma_{rr}(a) = -p_a$ and $\sigma_{rr}(b) = -p_b$ can be used to evaluate the constants of integration. Derive the displacement solution. (*c*) Complete the analysis by deriving a solution for the stresses. Where $r = a$ and $r = b$ are the inside and outside radius, respectively.

6.31. Refer to the thick-walled cylinder of Prob. 6.14 and compute the stresses at the center of each element using the 10-element model. Compare the result to the exact solution.

6.32. Solve the thick-walled cylinder problem described in Probs. 6.13 and 6.14 using two eight-node quadrilateral elements as shown in Fig. 6-13.

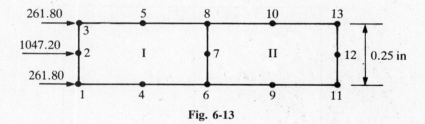

Fig. 6-13

6.33. For the eight-node element of Prob. 6.16 and Fig. 6-7 assume a uniformly varying load on side 1-8-4 and find the distribution of the load to each node. The load varies from zero at node 1 to p_x at node 4.

6.34. Assume a uniform load distributed along the side of the nine-node element defined in Prob. 6.24 (Fig. 6-12) and determine the distribution of force to each node.

6.35. Assume a surface loading along $\eta = \pm 1$ for an isoparametric element and derive the expression for dS.

6.36. Use the element defined in Fig. 6-5 with midside nodes defined at the center of the midsides and obtain the distribution for a uniform pressure loading p_x in the x direction along the side defined by nodes 1, 8, and 4. Use the isoparametric formulation and the node numbering and shape functions defined in Prob. 6.25.

6.37. Use the Galerkin method to derive the finite element equations for axisymmetric heat conduction.

6.38. An axisymmetric element is defined in Fig. 6-14. Compute the distribution of a unit charge density ($\rho_0 = 1$) to each node assuming both a four-node element and an eight-node element.

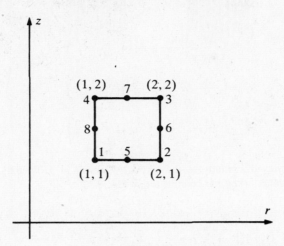

Fig. 6-14 Axisymmetric element.

6.39. Given an axisymmetric space defined by $r_{in} = 1.0$ and $r_{out} = 2.0$. The one-dimensional electrostatics problem is defined by Eq. (2.22) and can be written in polar coordinates by referring to Eq. (6.8) and neglecting the dependence on θ and z. (a) Assume constant charge density ρ_0 and boundary conditions $\phi(r_{in}) = \phi(r_{out}) = 0$ to obtain the analytical solution. (b) Divide the axisymmetric space into six equally spaced four-node axisymmetric finite elements and compare the results for the potential ϕ at each node with the exact solution. Assume $\epsilon = 1.0$ and $\rho_0 = 1.0$.

6.40. Repeat Prob. 6.39 using eight-node isoparametric elements and compare the results with the four-node element solution. (a) Assume a one-element model. (b) Use three equally spaced elements.

6.41. Solve Prob. 2.19 using two eight-node axisymmetric isoparametric finite finite elements, one for each material. Repeat the analysis using two equally spaced eight-node isoparametric elements for each material, a total of four elements.

6.42. Derive shape functions for the linear, quadratic, and cubic triangular finite elements of Fig. 6-8. The node numbering sequence is arbitrary, but in many applications the corner nodes are numbered 1, 2, and 3, and the side nodes are then numbered sequentially starting with the side between nodes 1 and 2.

6.43. The stiffness matrix for the chemical reaction term appearing in Eq. (3.2) is derived in Prob. 3.17 for a linear triangular finite element. Refer to Prob. 3.17 and repeat the analysis using a linear isoparametric triangular finite element. Compare linear and quadratic numerical integration.

Answers to Supplementary Problems

6.23. Combine two- and three-node interpolation formulas using Eq. (6.3). $N_1 = L_{1x}L_{1y}$, $N_3 = L_{3x}L_{3y}$, and so on.

$$N_1 = \left(\frac{(x - x_2)(x - x_3)}{(x_1 - x_2)(x_1 - x_3)} \right)\left(\frac{y - y_2}{y_1 - y_2} \right)$$

$$N_2 = \left(\frac{(x - x_1)(x - x_3)}{(x_2 - x_1)(x_2 - x_3)} \right)\left(\frac{y - y_2}{y_1 - y_2} \right)$$

$$N_3 = \left(\frac{(x - x_1)(x - x_2)}{(x_3 - x_1)(x_3 - x_2)} \right)\left(\frac{y - y_2}{y_1 - y_2} \right)$$

$$N_4 = \left(\frac{(x - x_1)(x - x_2)}{(x_3 - x_1)(x_3 - x_2)}\right)\left(\frac{y - y_1}{y_2 - y_1}\right)$$

$$N_5 = \left(\frac{(x - x_1)(x - x_3)}{(x_2 - x_1)(x_2 - x_3)}\right)\left(\frac{y - y_1}{y_2 - y_1}\right)$$

$$N_6 = \left(\frac{(x - x_2)(x - x_3)}{(x_1 - x_2)(x_1 - x_3)}\right)\left(\frac{y - y_1}{y_2 - y_1}\right)$$

6.24. The shape functions are constructed similarly to the six-node element of Prob. 6.23. Three-point Lagrange interpolation formulas are combined using the coordinates of Fig. 6-12.

$$N_1 = \xi\eta(\xi - 1)(\eta - 1)/4 \qquad N_2 = \xi\eta(\xi + 1)(\eta - 1)/4 \qquad N_3 = \xi\eta(\xi + 1)(\eta + 1)/4$$

$$N_4 = \xi\eta(\xi - 1)(\eta + 1)/4 \qquad N_5 = -\eta(\xi + 1)(\xi - 1)(\eta - 1)/2$$

$$N_6 = -\xi(\xi + 1)(\eta + 1)(\eta - 1)/2 \qquad N_7 = -\eta(\xi + 1)(\xi - 1)(\eta + 1)/2$$

$$N_8 = -\xi(\xi - 1)(\eta + 1)(\eta - 1)/2 \qquad N_9 = (\xi + 1)(\xi - 1)(\eta + 1)(\eta - 1)$$

6.25.

$$N_1 = (1 - \xi)(1 - \eta)(-\xi - \eta - 1)/4 \qquad N_2 = (1 + \xi)(1 - \eta)(\xi - \eta - 1)/4$$

$$N_3 = (1 + \xi)(1 + \eta)(\xi + \eta - 1)/4 \qquad N_4 = (1 - \xi)(1 + \eta)(-\xi + \eta - 1)/4$$

$$N_5 = (1 - \xi^2)(1 - \eta)/2 \qquad N_6 = (1 + \xi)(1 - \eta^2)/2$$

$$N_7 = (1 - \xi^2)(1 + \eta)/2 \qquad N_8 = (1 - \xi)(1 - \eta^2)/2$$

6.26. Substitute into Eq. (*a*) of Prob. 6.7:

$$J = \frac{1}{4}\begin{bmatrix} x_2 - x_1 & 0 \\ 0 & y_4 - y_1 \end{bmatrix} \qquad |\det J| = \frac{ab}{4}$$

6.27.

$$\int_{A_{xy}} dx\, dy = \int_{-1}^{+1}\int_{-1}^{+1} |\det J|\, d\xi\, d\eta = ab$$

6.28. The stiffness matrix before substitution of boundary conditions is the symmetric matrix

$$\begin{bmatrix}
4334555 & 1785714 & -1129426 & -137362 & -2167277 & -1785714 & -1037851 & 137362 \\
 & 6874238 & 137362 & 2283273 & -1785714 & -3437119 & -137362 & -5720391 \\
 & & 4334554 & -1785714 & -1037851 & -137362 & -2167277 & 1785714 \\
 & & & 6874237 & -137362 & -5720391 & 1785714 & -3437119 \\
 & & & & 4334554 & 1785714 & -1129426 & -137362 \\
 & & & & & 6874237 & 137362 & 2283273 \\
 & & & & & & 4334554 & -1785714 \\
 & & & & & & & 6874238
\end{bmatrix}$$

See Prob. 3.13 for the stiffness matrix after substitution of boundary conditions.

6.29. (*a*) Because of the axisymmetric nature of the problem $\sigma_{rz} = 0$ and because of the plane strain assumption $\partial/\partial z = 0$. The governing equation becomes

$$\frac{d\sigma_{rr}}{dr} + \frac{\sigma_{rr} - \sigma_{\theta\theta}}{r} = 0$$

(*b*) The stress-strain equations are obtained from Eqs. (*3.18*) with $\epsilon_{zz} = 0$:

$$\sigma_{rr} = \frac{E}{(1 + \nu)(1 - 2\nu)}[(1 - \nu)\epsilon_{rr} + \nu\epsilon_{\theta\theta}]$$

$$\sigma_{\theta\theta} = \frac{E}{(1 + \nu)(1 - 2\nu)}[(1 - \nu)\epsilon_{\theta\theta} + \nu\epsilon_{rr}]$$

The third of Eqs. (3.18) can be solved to give $\sigma_{zz} = \nu(\sigma_{rr} + \sigma_{\theta\theta})$.

(c)   Substitute the strains into the equation above and substitute the stresses into the equation of part (a). The result is

$$\frac{d^2u}{dr^2} + \frac{1}{r}\frac{du}{dr} - \frac{u}{r^2} = 0$$

6.30.   (a) $u = C_1 r + C_2/r$.

(b)   Substitute into part (b) of Prob. 6.29:

$$\sigma_{rr}(a) = -p_a = \frac{E}{(1 + \nu)(1 - 2\nu)}\left[C_1 - (1 - 2\nu)\frac{C_2}{a^2}\right]$$

$$\sigma_{rr}(b) = -p_b = \frac{E}{(1 + \nu)(1 - 2\nu)}\left[C_1 - (1 - 2\nu)\frac{C_2}{b^2}\right]$$

Solving these equations gives

$$C_1 = \frac{(1 + \nu)(1 - 2\nu)}{E}\frac{p_a a^2 - p_b b^2}{b^2 - a^2} \qquad C_2 = \frac{1 + \nu}{E}\frac{(a - b)a^2 b^2}{b^2 - a^2}$$

(c)   Substitute into the stress-displacement equations:

$$\sigma_{rr} = C_3 - \frac{C_4}{r^2} \qquad \sigma_{\theta\theta} = C_3 + \frac{C_4}{r^2}$$

$$C_3 = \frac{p_a a^2 - p_b b^2}{b^2 - a^2} \qquad \text{and} \qquad C_4 = \frac{(p_a - p_b)a^2 b^2}{b^2 - a^2}$$

6.31.   The results are given in Table 6.5. Compare these results with those in Prob. 6.14 and the discussion given in Prob. 6.15.

Table 6.5   Stresses for a Thick-Walled Cylinder (psi) Ten-Element Model

Element	r (in)	Finite element			Exact	
		σ_{rr}	$\sigma_{\theta\theta}$	σ_{zz}	σ_{rr}	$\sigma_{\theta\theta}$
1	1.05	−875.52	1543.75	200.47	−876.04	1542.71
2	1.15	−674.39	1341.93	200.26	−674.86	1341.52
3	1.25	−519.58	1186.69	200.13	−520.00	1186.67
4	1.35	−397.91	1064.72	200.04	−398.26	1064.93
5	1.45	−300.55	963.14	199.98	−300.83	967.50
6	1.55	−221.42	887.87	199.93	−221.64	888.31
7	1.65	−156.24	822.58	199.90	−156.41	823.08
8	1.75	−101.91	768.18	199.18	−102.04	768.71
9	1.85	−56.16	722.37	199.86	−56.25	722.91
10	1.95	−17.26	683.43	199.85	−17.31	683.98

These results for stress are surprisingly accurate when compared with the results given in Prob. 6.14. At the center of the element the radial stress oscillation is misleadingly accurate when compared with the accuracy of the total formulation.

6.32. The node numbering is shown in Fig. 6-13. The element dimension in the z direction is 0.25, and the total pressure loading on the inside surface of the cylinder is $(1000)(2\pi)(0.25)(r = 1.0) = 1570.80$. The load is distributed to nodes 1, 2, and 3 according to the results of Prob. 6.16. Nodes 1 and 3 have $1570.80/6 = 261.80$, and node 2 has $1570.80(\frac{2}{3}) = 1047.20$. The results are given in Tables 6.6 and 6.7.

Table 6.6 Radial Displacements for a Thick-Walled Cylinder (in) Two-Element Model, Eight-Node Elements

r (in)	Nodes	Finite element	Exact
1.00	1, 2, 3	1905.09	1906.67
1.25	4, 5	1602.57	1603.33
1.50	6, 7, 8	1414.60	1415.55
1.75	9, 10	1293.03	1293.81
2.00	11, 12, 13	1212.55	1213.33

Table 6.7 Stresses for a Thick-Walled Cylinder (psi) Two-Element Model, Eight-Node Elements

Element	r (in)	Finite element			Exact	
		σ_{rr}	$\sigma_{\theta\theta}$	σ_{zz}	σ_{rr}	$\sigma_{\theta\theta}$
1	1.0564	−800.58	1557.38	227.04	−861.43	1528.10
1	1.25	−580.92	1159.87	173.69	−520.00	1186.67
1	1.4436	−264.54	988.05	217.05	−306.47	973.13
2	1.5564	−202.40	890.19	206.34	−217.09	883.76
2	1.75	−117.70	761.51	193.14	−102.04	768.71
2	1.9437	−8.43	690.17	204.52	−19.59	686.27

6.33. Zero at node 1, two-thirds of the total distributed load at node 8, and one-third of the total distributed load at node 4.

6.34. The results are the same as for the eight-node element of Prob. 6.16.

6.35.
$$dL = \left[\left(\sum \frac{\partial N_i}{\partial \xi} x_i \, d\xi \right)^2 + \left(\sum \frac{\partial N_i}{\partial \xi} y_i \, d\xi \right)^2 \right]^{1/2} \qquad (i = 1 \text{ to the number of nodes})$$

6.36. Refer to Prob. 6.16. The length of the element side between nodes 1 and 4 is 3.16228, and the total uniform load is $3.16228p_x$. The isoparametric element computation gives $0.52704p_x$ distributed to nodes 1 and 4, and $2.10819p_x$ applied at node 8. The distribution is the same as in Prob. 6.16, one-sixth of the load at the corner nodes and two-thirds of the load at the midside node. This problem illustrates that the isoparametric formulation takes into account the shape of the boundary surface.

6.37. See Prob. 7.4.

6.38.

Eight-node element		Four-node element	
Node	ϕ	Node	ϕ
1	-0.87266	1	2.0944
2	-0.69813	2	2.6180
3	-0.69813	3	2.6180
4	-0.87266	4	2.0944
5	3.14159		
6	3.49066		
7	3.14159		
8	2.79253		

In both computations the total charge density acting on the element is 9.4248 and can also be computed as $2\pi r_{avg}\rho_0 = 9.4248$, where $r_{avg} = 1.5$.

6.39.
$$\phi(r) = \frac{\rho}{4\epsilon}(r_{in}^2 - r^2) - \frac{\rho}{4\epsilon}(r_{in}^2 - r_{out}^2)\frac{\ln(r_{in}^2/r)}{\ln(r_{in}^2/r_{out}^2)}$$

6.40. The results for three eight-node elements are given in Table 6.8. Note that there is a slight increase in accuracy. The one-element solution has a node at $r = 1.5$, and ϕ is computed as 0.125; this compares fairly well with the exact solution.

Table 6.8 Axisymmetric Electrostatics Problem (V)

Node location, r	ϕ, four-node finite element	ϕ, exact	ϕ, eight-node finite element
1.0	0.0	0.0	0.0
1.1667		0.07640	0.07648
1.3333		0.11668	0.11683
1.5	0.12622	0.12608	0.12620
1.6667	0.10829	0.10817	0.10828
1.8333		0.06552	0.06557
2.0	0.0	0.0	0.0

6.41. The results are given in Table 6.9. Compare with Table 2.19.

Table 6.9 Coaxial Cable Analysis (V)

r	ϕ, two elements	ϕ, four elements
5.00	500.0	500.0
6.25		1136.53
7.50	1364.61	1373.23
8.75		1283.52
10.00	910.76	914.10
13.75		596.59
17.50	357.80	355.92
21.25		162.26
25.00	0	0

6.42. Linear:
$$N_1 = L_1 \qquad N_2 = L_2 \qquad N_3 = L_3$$

Quadratic:
$$N_1 = L_1(2L_1 - 1) \qquad N_2 = L_2(2L_2 - 1) \qquad N_3 = L_3(2L_3 - 1)$$
$$N_4 = 4L_1L_2 \qquad\qquad N_5 = 4L_2L_3 \qquad\qquad N_6 = 4L_1L_3$$

Cubic:
$$N_1 = \tfrac{1}{2}L_1(3L_1 - 1)(3L_1 - 2) \qquad N_2 = \tfrac{1}{2}L_2(3L_2 - 1)(3L_2 - 2)$$
$$N_3 = \tfrac{1}{2}L_3(3L_3 - 1)(3L_3 - 2) \qquad N_4 = \tfrac{9}{2}L_1L_2(3L_1 - 1)$$
$$N_5 = \tfrac{9}{2}L_1L_2(3L_2 - 1) \qquad N_6 = \tfrac{9}{2}L_2L_3(3L_2 - 1)$$
$$N_7 = \tfrac{9}{2}L_2L_3(3L_3 - 1) \qquad N_8 = \tfrac{9}{2}L_1L_3(3L_3 - 1)$$
$$N_9 = \tfrac{9}{2}L_1L_3(3L_1 - 1) \qquad N_{10} = 27L_1L_2L_3$$

6.43. The stiffness matrix is of the form [Eq. (*b*) of Prob. 3.17]:

$$[K] = \int_A [N]^T[K_r][N]t \, dx \, dy$$

Since there are no derivatives, the inverse of the jacobian is not required; it can be shown, using coordinates x_1, y_1, x_2, y_2, and x_3, y_3, that $|\det J| = 2A$. Formulate the solution as in Eq. (*d*) of Prob. 3.17; it follows that the function to be integrated is

$$\begin{bmatrix} L_1 & L_1 \\ L_2 & L_2 \\ L_3 & L_3 \end{bmatrix} \begin{bmatrix} K_r/2 & 0 \\ 0 & K_r/2 \end{bmatrix} \begin{bmatrix} L_1 & L_2 & L_3 \\ L_1 & L_2 & L_3 \end{bmatrix} |\det J| = 2A(K_{rx}) \begin{bmatrix} L_1^2 & L_1L_2 & L_1L_3 \\ L_2L_1 & L_2^2 & L_2L_3 \\ L_3L_1 & L_3L_2 & L_3^2 \end{bmatrix}$$

Linear integration gives

$$K_{rx} \begin{bmatrix} 1 & 1 & 1 \\ 1 & 1 & 1 \\ 1 & 1 & 1 \end{bmatrix} \frac{A}{9}$$

and is not exact when compared with Eq. (*e*) of Prob. 3.17. However, quadratic integration gives the exact result.

Chapter 7

Selected Topics in Finite Element Analysis

7.1. INTRODUCTION

Fundamental theory and applied problems have been discussed in previous chapters. The intent of each chapter has been to outline and discuss a particular topic and to demonstrate that finite element theory is merely an extension of numerical methods that have been available to the analyst for many years. The computer brought about the possibility of solving large systems of equations and manipulating massive amounts of data. As a result, the existing numerical methods became useful to scientists and engineers. It has been illustrated in this book that finite element analysis is nothing more than a modern application of the Rayleigh-Ritz method and/or the Galerkin method of numerical analysis.

Each year many new applications and concepts appear in the scientific literature, and a user or researcher often finds a real challenge in trying to stay abreast of new developments. Most analysts work within an area of application corresponding to their area of scientific interest. Of course, research continues to be applied toward a more thorough understanding of the fundamental mathematics of the finite element method.

This chapter is an introduction to a variety of problems of practical interest to the user. The solution of initial-value problems, some that could not even be attempted a few decades ago, is of interest in many fields of study and research. Methods of analysis have been available for approximating time derivatives for many years, but the difficulty in modeling boundary conditions has been a major limitation in analysis. The finite element method of modeling coupled with time-dependent analysis has had an impact on the analysis of time-dependent problems. Finite difference approximations for first-order time derivatives are introduced in this chapter, and a variety of applications are illustrated.

The classical eigenvalue problem is discussed in this chapter, and several applications are illustrated. The differential equations analyzed here correspond to a variety of applied problems that occur in engineering. Methods of analysis for computing the eigenvalues for large systems of equations are not discussed but can be found in numerous textbooks.

Several sections are included in this chapter that extend finite element concepts to three dimensions and introduce the idea of higher-order elements. These topics can be found in more extensive treatments of finite element theory, and the reader who has mastered this book should have little difficulty studying the more exhaustive literature. Finally, this chapter ends with an introduction to plate finite elements. Plate finite elements have been a topic of analysis since inception of the finite element method and continue to be the subject of a major share of the current published literature. The element presented here is typical of plate finite elements and at the same time applicable to both thin-plate theory and moderately thick-plate theory.

7.2. INITIAL-VALUE PROBLEMS

Initial-value problems were mentioned briefly in Chap. 5, where the Galerkin method was used to formulate the finite element counterpart of the differential equation. Methods for the numerical analysis of a time-dependent differential equation will be discussed in this chapter. It is possible to use a finite element to approximate the time derivative. However, the standard finite difference approximation will be used here since a finite element approximation in time does not offer any particular advantage. Recall that in Chap. 6 the computational power of the finite element was demonstrated using the isoparametric formulation where complicated boundary conditions could be modeled. Initial-value problems usually begin with a time-equal-zero condition, and that does not justify the finite element formulation.

There is an abundance of literature available to the reader concerning numerical analysis of initial-value problems. In this chapter the emphasis will be on the classical *transient* problem, that is, an equation of the form

$$\alpha(x)\frac{\partial\phi(x,t)}{\partial t} - \frac{\partial}{\partial x}\left[\beta(x)\frac{\partial\phi(x,t)}{\partial x}\right] + \gamma(x)\phi(x,t) = \delta(x,t) \qquad (7.1)$$

where $\alpha(x)$, $\beta(x)$, $\gamma(x)$, and $\delta(x,t)$ are functions that pertain to a particular application. The fact that they may be functions of x offers no more obstacle than in previous chapters. They will be assumed constant within an element but can vary from element to element. Also, since the time solution is obtained in an incremental form, the material constants can be functions of time and allowed to vary in an incremental manner.

Differential equations containing the second derivative in time are often referred to as *dynamic* problems or *wave propagation* problems. That class of problems will not be discussed in detail, except that in the next section it will be shown that the problem can be formulated, with certain assumptions, as an eigenvalue problem. Solutions for that important class of problems will be illustrated.

Finite difference models for transient problems can be formulated in any manner that approximates a first derivative. Classically, these are called forward difference, backward difference, and central difference approximations. The θ method (theta method) is quite general and contains the three previous methods as special cases. The θ method will be derived and used in this chapter (see Prob. 7.3).

The Galerkin method is employed in this chapter to derive finite element models for a variety of initial-value problems. In particular, transient heat conduction is illustrated in Probs. 7.2, 7.19, and 7.20. A similar problem for axisymmetric transient diffusion through the walls of a hollow cylinder is demonstrated in Probs. 7.4, 7.22, and 7.23. A problem involving coupled two-phase diffusion is given by Probs. 5.25, 7.5, and 7.21. In each case the finite element formulation, analytical solution, and finite element solution are discussed and compared.

7.3.   EIGENVALUE PROBLEMS

Eigenvalue problems are sometimes called *characteristic-value* problems and occur in the analysis of homogeneous differential equations. In mathematical physics and engineering an important class of eigenvalue problems is the *vibration of continuous systems*. Critical buckling loads for beam columns, elastic plates, and shells constitute another class of significant eigenvalue problems. The finite element formulation of an eigenvalue problem follows the concepts that have been developed; in fact, no new concepts are needed. The solution for the eigenvalue problem, after formulation, offers a new challenge. Methods of analysis for large systems of equations will not be discussed, but the reader can find numerous textbooks that deal with this important topic. Also, excellent computer software is available for the analysis of eigenvalue problems.

A continuous system can be illustrated by referring to the cable of Eq. (*e*) of Prob. 2.1. Neglecting the foundation modulus k and external loading f, the dynamic problem is obtained by summing the forces in the y direction, Eq. (*a*), and including the inertia force. Recall Newton's law, $\Sigma F_y = \rho a_y$, where ρ is the mass or density, depending upon the formulation, and a_y is the acceleration in the y direction. The final differential equation has the form

$$T\frac{\partial^2 v}{\partial x^2} = \rho\frac{\partial^2 v}{\partial t^2} \qquad (7.2)$$

Equation (*7.2*) is a hyperbolic differential equation and governs the dynamic motion of the cable. The problem that tracks the motion of the cable as a function of x and t is often called a wave propagation problem and requires a solution in x and t. The eigenvalue problem occurs with the assumption that the free transverse motion of the cable is oscillatory and periodic in nature and can be described using a periodic function. The assumption is supported mathematically using the separation of variables solution technique that will lead to a periodic solution. Several examples are illustrated in the solved problems.

7.4. THREE-DIMENSIONAL FINITE ELEMENTS

Three-dimensional finite elements are an extension of the two-dimensional isoparametric elements of Chap. 6. The theory of piezoelectricity that was discussed in Prob. 5.19 required three-dimensional modeling, and the idea of an eight-node three-dimensional element was introduced. The Lagrange family of shape functions includes three-dimensional elements as an extension of two-dimensional elements, and the derivation is similar to Prob. 6.4. Numerical integration using the Gauss-Legendre quadrature to evaluate volume elements follows directly from the area integrations of Chap. 6. The reader should review the numerical integration procedure outlined in Prob. 6.9 where the integral is numerically evaluated as

$$I = \int_{-1}^{+1} \int_{-1}^{+1} f(\xi, \eta)\, d\xi\, d\eta \tag{7.3}$$

The three-dimensional counterpart of Eq. (7.3) is

$$I = \int_{-1}^{+1} \int_{-1}^{+1} \int_{-1}^{+1} f(\xi, \eta, \psi)\, d\xi\, d\eta\, d\psi \tag{7.4}$$

where ξ, η, ψ are the coordinates of the three-dimensional parent element. The evaluation of the volume integral requires a transformation between the parent element using interpolation functions and a three-dimensional jacobian matrix to give the form of Eq. (7.4). The numerical integration is carried out using three nested summations. The equivalent of Eq. (7.4) becomes

$$I \simeq \sum_{i=1}^{m} \sum_{j=1}^{n} \sum_{k=1}^{l} w_i w_j w_k f(\xi_i, \eta_j, \psi_k) \tag{7.5}$$

where, usually, $m = n = l$.

Derivation of shape functions for the rectangular brick element is discussed in Prob. 7.14, and several additional elements are presented in the supplementary problems.

The three-dimensional counterpart of the isoparametric triangular element is called a *tetrahedral* element. The tetrahedral element is four-sided and can be classified as linear (4 nodes), quadratic (10 nodes), or cubic (20 nodes). A special set of integration quadrature is used to numerically evaluate the volume integrals that are an extension of those used in Chap. 6 for two-dimensional triangular isoparametric elements. The reader can refer to an advanced text on finite element analysis such as Zienkiewicz and Taylor (1989) or Stasa (1985) for integration formulas and additional information on three-dimensional isoparametric finite elements.

7.5. HIGHER-ORDER FINITE ELEMENTS

Linear and quadratic isoparametric finite elements were discussed and used in Chap. 6. Eight- and nine-node isoparametric elements are usually sufficient to model any two-dimensional problem. Three-dimensional problems can be modeled quite nicely using the linear and quadratic elements of the previous section. The Lagrange family of elements can be extended to include cubic elements but will have the disadvantage that more internal nodes may be required than are desirable. For instance, even though derivation of the two-dimensional Lagrange cubic element would be simple, the element would have 12 exterior nodes and 4 interior nodes. The cubic element without interior nodes (see Probs. 7.15 and 7.33) would probably be a more efficient choice.

The order of integration using gaussian quadratures should be dependent upon the order of the finite element. According to Zienkiewicz and Taylor (1989) a single integration point is satisfactory for linear quadrilaterals and triangles. For parabolic quadrilaterals 2×2 Gauss point integration is adequate, and for parabolic bricks $2 \times 2 \times 2$ Gauss point integration will suffice. The author has used 2×2 and 3×3 Gauss point integration for parabolic finite elements and has always obtained identical results.

7.6. ELEMENT CONTINUITY

Continuity in finite element analysis refers to the continuity of the solution along element boundaries. The very nature of the finite element method implies a piecewise solution of the problem. Application of the finite element method to various physical problems in previous chapters has been accomplished without studying or even suggesting that there may be problems with interelement continuity. The majority of the elements that have been studied were linear. Visualize a one-dimensional space divided into two elements. The node that connects the two elements acts as a connecting boundary point for the two elements. Obviously, the function being modeled is single-valued at the connecting node. Continuity questions arise concerning the derivatives of the function, i.e., continuity of the derivatives of the shape functions being used to model the physical problem.

Mathematical functions are often assigned to a class and are said to be of *class C^n* within a certain domain. A function that is of class C^0 is continuous. If the first derivative of the function is continuous, it is classified as class C^1, and if the second derivative is also continuous, it is of class C^2. It follows that a function specified within some domain is of class C^n if the function and its first n derivatives are continuous.

7.7. PLATE FINITE ELEMENTS

The analysis of problems that occur in the bending of flat plates was one of the earlier applications of the finite element method. There are fundamental obstacles that cause the application of the finite element method to plate-bending problems to be more difficult than the problems of the preceding chapters. The governing equation for plate bending is a fourth-order differential equation and as such has a more stringent continuity requirement. An element will be discussed in this chapter that gives good results without satisfying all continuity requirements. The element was discussed by Hughs and Cohen (1978) and Hughs (1987). The discussion here relies upon the book by Cook et al. (1989).

A brief development of the theory of plates will be given. The reader who is not acquainted with the theory should refer to a more definitive treatment, for instance, Timoshenko and Woinowsky-Krieger (1959).

Relation between Force and Moment to Stress

Internal stresses in a plate produce bending moments M and shears Q as illustrated in Fig. 7-1. Moments and shears are defined as acting per unit length of plate. These internal actions are defined as follows:

$$M_x = \int_{-h/2}^{h/2} \sigma_{xx} z\, dz \qquad M_y = \int_{-h/2}^{h/2} \sigma_{yy} z\, dz \qquad M_{xy} = \int_{-h/2}^{h/2} \sigma_{xy} z\, dz$$

$$Q_x = \int_{-h/2}^{h/2} \sigma_{xz}\, dz \qquad Q_y = \int_{-h/2}^{h/2} \sigma_{yz}\, dz \tag{7.6}$$

The predominant stresses are obtained by integration of Eq. (7.6) as

$$\sigma_{xx} = \frac{M_x z}{h^3/12} \qquad \sigma_{yy} = \frac{M_y z}{h^3/12} \qquad \sigma_{xy} = \frac{M_{xy} z}{h^3/12} \tag{7.7}$$

The remaining shear stresses are also obtained by integrating Eq. (7.6):

$$\sigma_{yz} = \frac{3Q_y}{2h}\left(1 - \frac{4z^2}{h^2}\right) \qquad \sigma_{zx} = \frac{3Q_x}{2h}\left(1 - \frac{4z^2}{h^2}\right) \tag{7.8}$$

but are quite small and are maximum at the midplane of the plate, $z = 0$.

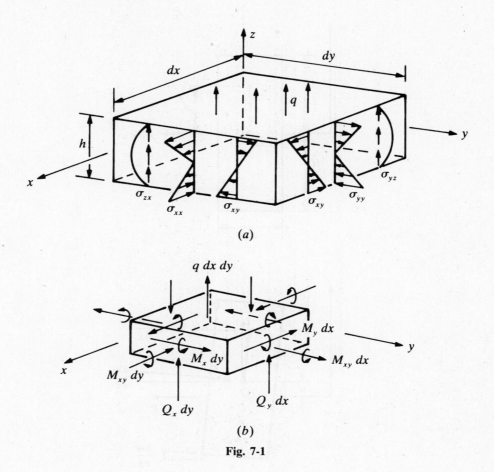

$$(a)$$

$$(b)$$

Fig. 7-1

Deformation Assumptions

Classically, thin-plate theory is somewhat similar to beam theory and is often referred to as the *Kirchhoff theory* of plates. The basic assumption is that points on the midsurface of the plate, $z = 0$, can displace only in the z direction. A straight line normal to the midsurface before bending is normal to the midsurface after bending and remains a straight line. Points on the midsurface cannot displace in the x or y direction. However, at distances z from midsurface displacements, u and v, in the x and y directions, respectively, can develop as the plate bends. These displacements are illustrated in Fig. 7-2 and are given by

$$u = -z \frac{\partial w}{\partial x} \qquad v = -z \frac{\partial w}{\partial y} \tag{7.9}$$

The corresponding strains are

$$\epsilon_{xx} = \frac{\partial u}{\partial x} = -z \frac{\partial^2 w}{\partial x^2} \qquad \epsilon_{yy} = \frac{\partial v}{\partial y} = -z \frac{\partial^2 w}{\partial y^2} \qquad \epsilon_{xy} = \frac{\partial u}{\partial y} + \frac{\partial v}{\partial x} = -2z \frac{\partial^2 w}{\partial x \, \partial y} \tag{7.10}$$

The strain-deformation relations defined by Eqs. (*7.9*) and (*7.10*) do not allow for transverse shear deformation.

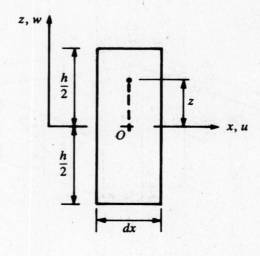

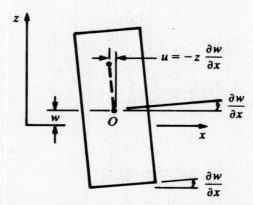

Fig. 7-2   Differential element for Kirchhoff theory.

A more complete assumption follows the same idea, except that straight lines normal to the midsurface before bending remain straight but not normal to the midsurface after bending. The deformation is shown in Fig. 7-3. The angle between the z axis and a line originally normal to the midsurface (x axis) is defined as θ_x. A similar definition holds for θ_y. The following deformation assumptions are referred to as the *Mindlin theory of plates*.

$$u = -z\theta_x \qquad v = -z\theta_y \tag{7.11}$$

The strains become

$$\epsilon_{xx} = -z\frac{\partial\theta_x}{\partial x} \qquad \epsilon_{yy} = -z\frac{\partial\theta_y}{\partial y} \qquad \epsilon_{xy} = -z\left(\frac{\partial\theta_x}{\partial y} + \frac{\partial\theta_y}{\partial x}\right)$$

$$\epsilon_{yz} = \frac{\partial w}{\partial y} - \theta_y \qquad \epsilon_{xz} = \frac{\partial w}{\partial x} - \theta_x \tag{7.12}$$

The theory accounts for transverse shear deformation and is applicable for moderately thick plates. Note that thick-plate theory reduces to thin-plate theory with the assumptions $\theta_x = \partial w/\partial x$ and $\theta_y = \partial w/\partial y$.

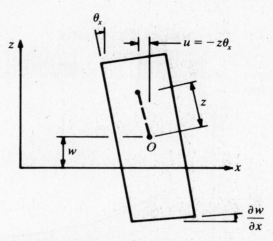

Fig. 7-3 Deformation for Mindlin theory, $\partial w/\partial x - \theta_x \neq 0$.

Stress-Strain and Moment-Curvature Relations

The stress-strain relations correspond to a linearly elastic material and following the matrix notation of previous chapters may be written

$$\{\sigma\} = [E]\{\epsilon\} \tag{7.13}$$

Homogeneous and isotropic material properties are assumed, and the strains [Eqs. (7.7) and (7.8)] are related to the strains as

$$\begin{Bmatrix} \sigma_{xx} \\ \sigma_{yy} \\ \sigma_{xy} \\ \sigma_{yz} \\ \sigma_{zx} \end{Bmatrix} = \begin{bmatrix} E/(1-\nu^2) & E\nu/(1-\nu^2) & 0 & 0 & 0 \\ E\nu/(1-\nu^2) & E/(1-\nu^2) & 0 & 0 & 0 \\ 0 & 0 & G & 0 & 0 \\ 0 & 0 & 0 & G & 0 \\ 0 & 0 & 0 & 0 & G \end{bmatrix} \begin{Bmatrix} \epsilon_{xx} \\ \epsilon_{yy} \\ \epsilon_{xy} \\ \epsilon_{yz} \\ \epsilon_{zx} \end{Bmatrix} \tag{7.14}$$

where $G = E/2(1 + \nu)$, E is Young's modulus, and ν is Poisson's ratio. Moment-curvature relations for the Mindlin theory are obtained by combining Eqs. (7.7) and (7.8) with (7.12) and substituting into Eq. (7.14).

$$\begin{Bmatrix} M_x \\ M_y \\ M_{xy} \\ Q_y \\ Q_x \end{Bmatrix} = - \begin{bmatrix} D & \nu D & 0 & 0 & 0 \\ \nu D & D & 0 & 0 & 0 \\ 0 & 0 & D(1-\nu)/2 & 0 & 0 \\ 0 & 0 & 0 & Gh & 0 \\ 0 & 0 & 0 & 0 & Gh \end{bmatrix} \begin{Bmatrix} \partial\theta_x/\partial x \\ \partial\theta_y/\partial y \\ \partial\theta_x/\partial y + \partial\theta_y/\partial x \\ \theta_y - \partial w/\partial y \\ \theta_x - \partial w/\partial x \end{Bmatrix} \tag{7.15}$$

where $D = Eh^3/[12(1 - \nu^2)]$ is the flexural rigidity.

A general plate element will be derived in Prob. 7.17 using Eq. (7.15) as a fundamental relation between moment or shear and curvature. For later reference the matrices of Eq. (7.15) will be defined as

$$\{M\} = [D_M]\{\kappa\} \tag{7.16}$$

where $[D_M]$ is the material matrix and $\{\kappa\}$ is the curvature matrix.

Solved Problems

7.1. Discuss forward difference, backward difference, and central difference finite difference approximations for the first derivative.

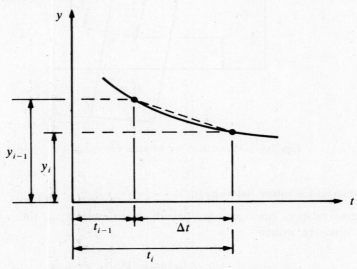

Fig. 7-4   Finite difference representation of dy/dt.

In each case the approximation is linear. Consider the finite difference analysis for the elementary equation

$$\frac{dy}{dt} + \gamma y = \delta \qquad (a)$$

with $y(0) = 2$. The solution of Eq. (a) is

$$y = \frac{\delta}{\gamma}(1 + e^{-\gamma t}) \qquad (b)$$

A curve that represents the analytical solution is sketched in Fig. 7-4. Assume that a numerical solution y_{i-1} has been computed up to point t_{i-1}. The next value of y to be computed is y_i. A forward difference model for Eq. (a) is written in terms of the previously computed value of y:

$$\left(\frac{dy}{dt}\right)_{i-1} + \gamma y_{i-1} = \delta_{i-1} \qquad (c)$$

where the finite difference approximation for the time derivative, using Fig. 7-4, is

$$\left(\frac{dy}{dt}\right)_{i-1} \simeq \frac{y_i - y_{i-1}}{\Delta t} \qquad \Delta t = t_i - t_{i-1} \qquad (d)$$

Substituting Eq. (d) into Eq. (c) gives a forward difference recurrence formula for y_i:

$$y_i = y_{i-1} + (\delta_{i-1} - \gamma y_{i-1})\,\Delta t \qquad (e)$$

A backward difference model is written in terms of the next value of y to be computed:

$$\left(\frac{dy}{dt}\right)_i + \gamma y_i = \delta_i \qquad (f)$$

with

$$\left(\frac{dy}{dt}\right)_i \simeq \frac{y_i - y_{i-1}}{\Delta t} \qquad \Delta t = t_i - t_{i-1} \qquad (g)$$

Substituting Eq. (g) into Eq. (f) gives a backward difference recurrence formula for y_i:

$$y_i(1 + \gamma \, \Delta t) = y_{i-1} + \delta_i \, \Delta t \tag{h}$$

The central difference model can be constructed assuming that y can be evaluated at the midpoint between t_i and t_{i+1}:

$$\left(\frac{dy}{dt}\right)_{i-1/2} + \gamma y_{i-1/2} = \delta_{i-1/2} \tag{i}$$

with

$$\left(\frac{dy}{dt}\right)_{i-1/2} \simeq \frac{y_i - y_{i-1}}{\Delta t} \qquad \Delta t = t_i - t_{i-1} \tag{j}$$

where $y_{i-1/2}$ is assumed as an average value:

$$y_{i-1/2} = \frac{y_i + y_{i-1}}{2} \tag{k}$$

When Eqs. (j), (k), and (i) are combined, the central difference model becomes

$$y_i\left(1 + \gamma \, \frac{\Delta t}{2}\right) = y_{i-1}\left(1 - \gamma \, \frac{\Delta t}{2}\right) + \delta_{i-1/2} \, \Delta t \tag{l}$$

A comparison of the four solutions, Eqs. (b), (e), (h), and (l), is given in Table 7.1. Let $\gamma = \delta = 1$ and $\Delta t = 0.1$. All three numerical solutions appear to be quite accurate, with the central difference solution being the most accurate for this application. However, any of the three solutions is acceptable. A smaller time increment would give a more accurate analysis but would require more computation. A larger time increment would require less computation and for the central and backward difference methods would approach the steady-state solution of $y = 1$, but it could be a poor representation of the transient behavior, which is of interest in these problems. When the time step becomes too large, the forward difference method becomes unstable and its solution totally diverges from the correct solution. A successful analysis is one that gives a good approximation for the solution and at the same time requires a minimum of computation time.

Table 7.1   Comparison of Finite Difference Solutions

Time	Exact, Eq. (b)	Forward difference, Eq. (e)	Backward difference, Eq. (h)	Central difference, Eq. (l)
0.0	2.0	2.0	2.0	2.0
0.1	1.90484	1.90000	1.90909	1.90476
0.2	1.81873	1.81000	1.82645	1.81859
0.3	1.74082	1.72900	1.75131	1.74063
0.4	1.67032	1.65610	1.68301	1.67010
0.5	1.60653	1.59049	1.62092	1.60628
0.6	1.54881	1.53144	1.56447	1.54854
0.7	1.49654	1.47830	1.51316	1.49630
0.8	1.44933	1.43047	1.46651	1.44903
0.9	1.40657	1.38742	1.42410	1.40627
1.0	1.36788	1.34868	1.38554	1.36757
1.5	1.22313	1.20589	1.23939	1.22285

7.2.   Use the Galerkin method to formulate the finite element model for transient heat conduction.

The finite element formulation is similar for one-, two-, and three-dimensional problems. In this instance, use Eq. (7.1) with $\alpha \Rightarrow \rho c$, the density times the specific heat, $\beta \Rightarrow k$, the thermal conductivity, $\gamma = 0$, $\delta \Rightarrow -Q$, and $\phi \Rightarrow T$. The assumed solution is of the form

$$T_R(x, t) = [N(x)]\{T(t)\} \tag{a}$$

The shape functions are functions of the coordinates only, and the nodal point variables $\{T\}$ are functions of t only. Following the methodology developed in Chap. 5 and substituting into the governing differential equation gives

$$[\rho c][N] \frac{\partial\{T\}}{\partial t} - [k] \frac{\partial^2[N]}{\partial x^2}\{T\} - Q = R(x; \{T\}) \tag{b}$$

Multiply by $[N]$, w_i in the Galerkin formulation, integrate over the volume, and set the result to zero:

$$\int_V \left([N]^T[\rho c][N] \frac{\partial\{T\}}{\partial t} - [N]^T[k] \frac{\partial^2[N]}{\partial x^2}\{T\} - [N]^T Q \right) dV = 0 \tag{c}$$

Application of the Green-Gauss theorem gives the final result:

$$\int_V \left([N]^T[\rho c][N] \frac{\partial\{T\}}{\partial t} + \frac{\partial[N]^T}{\partial x} [k] \frac{\partial[N]}{\partial x}\{T\} - [N]^T Q \right) dV = \int_S [N]^T[k] \frac{\partial[N]}{\partial x}\{T\} \, dS \tag{d}$$

Equation (d) can be written in the more compact form

$$[C] \frac{\partial\{T\}}{\partial t} + [K]\{T\} = \{F\} \tag{e}$$

where

$$[C] = \int_V [N]^T[\rho c][N] \, dV \tag{f}$$

The matrix $[C]$ is sometimes called the *capacitance* matrix. All applied actions have been grouped together and denoted $\{F\}$, and $[K]$ is the standard stiffness matrix. It follows that the form of Eq. (e) remains unchanged for two- or three-dimensional formulations. The only new term is $[C]$, and the finite difference in time is applied to a matrix of unknowns rather than to a single variable as in the previous problem.

7.3. Derive a general recurrence formula for the θ method and discuss its relation to the difference methods of Prob. 7.1.

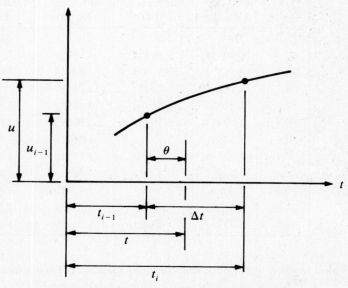

Fig. 7-5 Definition of θ for a time increment.

Assume a finite element equation similar to Eq. (e) of Prob. 7.2 with variable $\{u\}$ written with subscripts θ as defined in Fig. 7-5:

$$[C] \frac{d\{u\}_\theta}{dt} + [K]\{u\}_\theta = \{F\}_\theta \qquad (a)$$

Note that the partial derivative can be changed to a total derivative with no loss of generality. The time increment is defined as before, $\Delta t = t_i - t_{i-1}$, and θ is defined as

$$\theta = \frac{t - t_{i-1}}{\Delta t} \qquad 0 \le \theta \le 1 \qquad (b)$$

It follows that $\{u\}_\theta$ can be defined in terms of θ as

$$\{u\}_\theta \simeq (1 - \theta)\{u\}_{i-1} + \theta\{u\}_i \qquad (c)$$

or

$$\{u\}_\theta \simeq \theta(\{u\}_i - \{u\}_{i-1}) + \{u\}_i \qquad (d)$$

Also, using Eqs. (b) and (d),

$$\frac{d\{u\}_\theta}{dt} = \frac{d\{u\}_\theta}{d\theta}\frac{d\theta}{dt} = \frac{d\{u\}_\theta}{d\theta}\frac{1}{\Delta t} = \frac{\{u\}_i - \{u\}_{i-1}}{\Delta t} \qquad (e)$$

Similarly, $\{F\}_\theta$ is defined as

$$\{F\}_\theta \simeq (1 - \theta)\{F\}_{i-1} + \{F\}_i \qquad (f)$$

Substitute Eqs. (c), (e), and (f) into Eq. (a) and rearrange:

$$\frac{[C]\{u\}_i}{\Delta t_i} + \theta[K]\{u\}_i = \frac{[C]\{u\}_{i-1}}{\Delta t_i} - (1 - \theta)[K]\{u\}_{i-1} + (1 - \theta)\{F\}_{i-1} + \theta\{F\}_i \qquad (g)$$

When $\theta = 0$, Eq. (g) is the forward difference approximation. Similarly, when $\theta = 1$, the backward difference approximation can be recovered. The central difference approximation results when $\theta = \frac{1}{2}$. Equation (g) is quite general and allows the analyst to experiment with various values of θ to obtain the best numerical solution for a given problem.

7.4. The axisymmetric transient diffusion through the walls of a hollow cylinder is governed by equations similar to those of Chaps. 3 and 6 that were written in cylindrical coordinates. In fact, the governing equation is given here for reference:

$$\frac{\partial C}{\partial t} = \frac{1}{r}\frac{\partial}{\partial r}\left(rD_r\frac{\partial C}{\partial r}\right) + \frac{\partial}{\partial z}\left(D_z\frac{\partial C}{\partial z}\right) + m \qquad (a)$$

Formulate the finite element equivalent of the governing equation.

Let C be approximated by assuming $C = [N]\{C\}$ and substitute into the governing equation. Multiply by the weight function $[N]$ and integrate over the volume:

$$\int_V [N]^T[N] \frac{\partial(C)}{\partial t} dV = \int_V [N]^T\left[\frac{1}{r}\frac{\partial}{\partial r}\left(rD_r\frac{\partial[N]}{\partial r}\{C\}\right) + \frac{\partial}{\partial z}\left(D_z\frac{\partial[N]}{\partial z}\{C\}\right) + m\right] dV \qquad (b)$$

where $dV = 2\pi r \, dr \, dz$. The right-hand side of Eq. (b) can be written

$$2\pi \int_A [N]^T\left[\frac{\partial}{\partial r}\left(rD_r\frac{\partial[N]}{\partial r}\{C\}\right) + \frac{\partial}{\partial z}\left(rD_z\frac{\partial[N]}{\partial z}\{C\}\right) + mr\right] dr \, dz \qquad (c)$$

Note, because r is independent of z,

$$\frac{\partial}{\partial z}\left(D_z \frac{\partial [N]}{\partial z}\{C\}\right)r = \frac{\partial}{\partial z}\left(rD_z \frac{\partial [N]}{\partial z}\{C\}\right) \qquad (d)$$

and the first two terms of Eq. (b) have been written in a similar format. Application of the Green-Gauss theorem to these two terms will give the final form of the local finite element that defines axisymmetric transient diffusion.

$$2\pi \int_A \left([N]^T[N]\frac{\partial\{C\}}{\partial t} + \frac{\partial [N]^T}{\partial r}rD_r\frac{\partial [N]}{\partial r}\{C\} + \frac{\partial [N]^T}{\partial z}rD_z\frac{\partial [N]}{\partial z}\{C\}\right)dr\,dz$$

$$= 2\pi \int_A [N]^T mr\,dr\,dz + 2\pi \int_S \left([N]^T rD_r \frac{\partial [N]}{\partial r}n_r + [N]^T rD_z \frac{\partial [N]}{\partial z}n_z\right)dS \qquad (e)$$

The last two terms represent the flux boundary conditions. The numerical evaluation of Eq. (e) is identical to the numerical integration in cartesian coordinates with $2\pi r$ included in the computations.

7.5.   An analytical solution for the two-phase diffusion formulation of Prob. 5.25 is given by Prob. 7.21. Obtain a finite element solution for time-dependent two-phase diffusion using a strip of 10 two-dimensional linear finite elements. Assume $L = 1.0$, $\Gamma = 30$, and material parameters of $D = 0.5$, $\alpha = 10$, and $\beta = 5$. Note that the material parameters are illustrative for this analysis and do not represent a physical material.

Boundary conditions are specified for C_f in Prob. 7.21 and must be computed for C_r to obtain a finite element solution. Consider the second equation

$$\frac{dC_r(0, t)}{dt} = \beta C_f(0, t) - \alpha C_r(0, t) \qquad (a)$$

with $C_f(0, t) = \Gamma$. The equation at the boundary is

$$\frac{dC_r}{dt} + \alpha C_r = \beta \Gamma \qquad (b)$$

with solution

$$C_r = \frac{\beta \Gamma}{\alpha}(1 - e^{-\alpha t}) \qquad (c)$$

The matrix equation of Prob. 5.25 is used as the finite element model. Each finite element has two degrees of freedom per node, and the formulation is similar to that discussed in Prob. 5.16 for coupled diffusion. The diffusion term is modeled using Eq. (e) of Prob. 3.5, and all other terms are modeled using the results of Prob. 3.33 (the chemical reaction term) with K_r replaced by 1 for the time-dependent terms and α or β for the interaction terms. Visualize the local finite element stiffness matrix as

$$\begin{bmatrix} \dfrac{[N]^T[1][N]}{\Delta t} + [B]^T[D][B] + [N]^T[\beta][N] & -[N]^T[\alpha][N] \\[2ex] -[N]^T[\beta][N] & \dfrac{[N]^T[1][N]}{\Delta t} + [N]^T[\alpha][N] \end{bmatrix}\begin{Bmatrix} C_{fe} \\ C_{re} \end{Bmatrix}$$

The results were computed using an isoparametric element and are given in Table 7.2. Note that the series solution for C_r of Prob. 7.21 has very slow convergence and will require many terms to converge to an answer. The finite element (FE) solution was computed using time increments of 0.001 for $0.001 \le t \le 0.01$, then 0.01 for $0.01 \le t \le 0.1$, and then 0.1 for $0.1 \le t \le 1.5$. The steady-state solution is $C_f = 30$ and $C_r = 15$.

Table 7.2   Results for Two-phase Diffusion

Time			$x = 0$	$x = 0.1$	$x = 0.2$	$x = 0.3$	$x = 0.4$	$x = 0.5$
0.005	FE	C_f	30	7.332	0.387	0.000	0.000	0.000
		C_r	0.732	0.137	0.003	0.000	0.000	0.000
	Exact	C_f	30	4.640	0.137	0.000	0.000	0.000
		C_r	0.732	0.078	0.020	0.000	0.000	0.000
0.01	FE	C_f	30	10.991	2.014	0.092	0.001	0.000
		C_r	1.427	0.365	0.035	0.003	0.000	0.000
	Exact	C_f	30	9.282	1.316	0.078	0.002	0.000
		C_r	1.427	0.250	0.035	0.013	0.000	0.000
0.02	FE	C_f	30	14.176	4.839	1.214	0.264	0.106
		C_r	2.719	0.960	0.256	0.056	0.013	0.005
	Exact	C_f	30	13.840	4.443	0.945	0.130	0.022
		C_r	2.719	0.793	0.171	0.034	0.013	0.009
0.04	FE	C_f	30	17.505	8.689	3.701	1.473	0.877
		C_r	4.945	2.232	0.903	0.315	0.111	0.060
	Exact	C_r	30	17.575	8.702	3.577	1.259	0.641
		C_r	4.945	2.105	0.762	0.237	0.068	0.031
0.1	FE	C_f	30	21.249	14.229	9.232	6.291	5.326
		C_r	9.482	5.610	3.253	1.784	1.046	0.819
	Exact	C_f	30	21.273	14.257	9.238	6.263	5.282
		C_r	9.482	5.656	3.165	1.696	0.947	0.719
0.5	FE	C_f	30	27.226	24.782	22.881	21.682	21.271
		C_r	14.899	12.702	11.209	9.998	9.271	9.020
	Exact	C_f	30	27.603	25.412	23.779	22.706	22.338
		C_r	14.899	13.215	11.749	10.620	9.910	9.667
1.0	FE	C_f	30	29.244	28.565	28.028	27.683	27.565
		C_r	14.999	14.453	13.981	13.607	13.370	13.288
	Exact	C_f	30	29.442	28.938	28.539	28.283	28.194
		C_r	14.999	14.606	14.252	13.972	13.792	13.730

7.6.   The classical vibrating-string problem is described by Eq. (7.2). Assume free harmonic motion for the transverse motion of the string and derive the homogeneous governing differential equation.

     Assume a solution

$$v(x, t) = V(x)T(t) \qquad (a)$$

Substitute into Eq. (7.2) and separate the variables:

$$\frac{1}{V}\frac{d^2V}{dx^2} = \frac{1}{c^2T}\frac{d^2T}{dt^2} = -\left(\frac{\omega}{c}\right)^2 \qquad (b)$$

where ω is the frequency of free vibration. The solution is governed by two ordinary differential equations

$$\frac{d^2V}{dx^2} + \left(\frac{\omega}{c}\right)^2 V = 0 \qquad \frac{d^2T}{dt^2} + \omega^2 = 0 \tag{c}$$

The solution for the second equation is of the form

$$T = A \sin \omega t + B \cos \omega t$$

and it follows that Eq. (a) can be replaced with the assumption

$$v(x, t) = V(x)e^{-i\omega t} \tag{d}$$

Substituting into Eq. (7.2) gives the homogeneous form

$$\left[-\frac{d^2V}{dx^2} - \frac{\rho}{T} \omega^2 V \right] e^{-i\omega t} = 0 \tag{e}$$

or

$$\frac{d^2V}{dx^2} + \frac{\rho}{T} \omega^2 V = 0 \tag{f}$$

7.7. Use the finite element method to study the free vibration of the string described by Eq. (7.2) and Prob. 7.6. Assume boundary conditions $V(0) = V(L) = 0$. Assume (a) a three-node model with symmetry about the center of the string, (b) a three-element model without symmetry.

     The local finite element equation is derived using Eq. (f) of Prob. 7.6. Assume $V = [N]\{V\}$ and use the Galerkin method to form the weighted residual equation. Substitute V into Eq. (f), multiply by $[N]^T$, and simplify using Gauss's theorem:

$$\frac{d[N]^T}{dx} \frac{d[N]}{dx} - \lambda [N]^T [N] = 0 \tag{a}$$

where $\lambda = \rho \omega^2 / T$.

(a) The three-node model is shown in Fig. 7-6. Symmetry with respect to the centerline allows the use of one finite element. Equation (a) corresponds to Eq. (d) of Prob. 2.11 and can be written as follows (note that the length of the element is $L/2$):

$$\frac{2}{L} \begin{bmatrix} 1 & -1 \\ -1 & 1 \end{bmatrix} \begin{Bmatrix} v_1 \\ v_2 \end{Bmatrix} - \frac{\lambda L}{2} \begin{bmatrix} \frac{1}{3} & \frac{1}{6} \\ \frac{1}{6} & \frac{1}{3} \end{bmatrix} \begin{Bmatrix} v_1 \\ v_2 \end{Bmatrix} = 0$$

The boundary condition is $v_1 = 0$ and when eliminated gives $\lambda = 12/L^2$. Then, $\omega = (\lambda T/\rho)^{1/2} = (3.464/L)(T/\rho)^{1/2}$ and compares to the exact solution given by Prob. 7.24 of $\omega = (\pi/L)(T/\rho)^{1/2}$. The analyst must exercise care when using symmetry to model a vibration problem since only symmetric modes will be found. In other words, it is not good practice.

(b) The three-element model is similar to Fig. 7-6. The same model can be used except that the length is $L/3$. The global finite element equation is

$$\frac{3}{L} \begin{bmatrix} 1 & -1 & 0 & 0 \\ -1 & 2 & -1 & 0 \\ 0 & -1 & 2 & -1 \\ 0 & 0 & -1 & 1 \end{bmatrix} \begin{Bmatrix} v_1 \\ v_2 \\ v_3 \\ v_4 \end{Bmatrix} - \frac{\lambda L}{3} \begin{bmatrix} \frac{1}{3} & \frac{1}{6} & 0 & 0 \\ \frac{1}{6} & \frac{2}{3} & \frac{1}{6} & 0 \\ 0 & \frac{1}{6} & \frac{2}{3} & \frac{1}{6} \\ 0 & 0 & \frac{1}{6} & \frac{1}{3} \end{bmatrix} \begin{Bmatrix} v_1 \\ v_2 \\ v_3 \\ v_4 \end{Bmatrix} = 0$$

The boundary conditions $v_1 = v_4 = 0$ reduce the equation to the 2×2 matrix

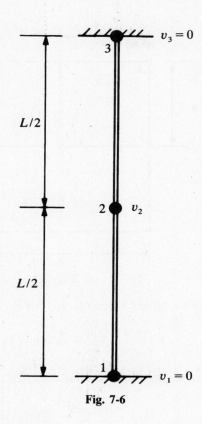

Fig. 7-6

$$\begin{bmatrix} 108 - 4\lambda L^2 & -(54 - \lambda L^2) \\ -(54 - \lambda L^2) & 108 - 4\lambda L^2 \end{bmatrix} \begin{Bmatrix} v_2 \\ v_3 \end{Bmatrix} = 0$$

or $$(108 - 4\lambda L^2)^2 - (54 - \lambda L^2)^2 = 0$$

Solving the quadratic equation gives the results

$$\omega_1 = \frac{3.2863}{L} \left(\frac{T}{\rho}\right)^{1/2} \quad \text{compared to the exact} \quad \omega_1 = \frac{3.1416}{L} \left(\frac{T}{\rho}\right)^{1/2}$$

$$\omega_2 = \frac{7.3485}{L} \left(\frac{T}{\rho}\right)^{1/2} \quad \text{compared to the exact} \quad \omega_2 = \frac{6.2823}{L} \left(\frac{T}{\rho}\right)^{1/2}$$

7.8. The rectangular vibrating membrane is the two-dimensional counterpart of the vibrating string and is described by an equation of the form

$$T \frac{\partial^2 w}{\partial x^2} + T \frac{\partial^2 w}{\partial y^2} = \rho \frac{\partial^2 w}{\partial t^2} \qquad (a)$$

where w is the deflection, T is the tension in the membrane, and ρ is the density. Formulate the finite element model for the vibrating membrane and obtain a solution using the model illustrated in Fig. 7-7.

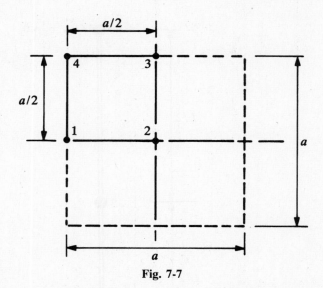

Fig. 7-7

Assuming a solution of Eq. (a) in the form $w(x, y, t) = W(x, y)e^{-i\omega t}$ will lead to an analysis of the equation

$$\frac{\partial^2 W}{\partial x^2} + \frac{\partial^2 W}{\partial y^2} + \lambda W = 0 \qquad \lambda = \frac{\rho \omega^2}{T} \tag{b}$$

Visualize that application of the Galerkin approximation for Eq. (b) will lead to the result

$$\frac{\partial [N]^T}{\partial x_i} \frac{\partial [N]}{\partial x_i} - \lambda [N]^T [N] = 0 \tag{c}$$

The membrane of Fig. 7-7 is square and can be modeled using four square elements. However, since all boundary nodes correspond to $W = 0$, the same result would be computed using symmetry with one square element. A more complete analysis would result if more elements were used and the complete membrane were modeled. As noted in Prob. 7.7, the use of symmetry in this application restricts the analysis to finding only symmetric modes. Equation (c) can be modeled using Eq. (e) of Prob. 3.5 with $k_x = k_y = 1$ and the last equation given in Prob. 3.34 with $K_r = \lambda$. The node numbering of Fig. 7-7 indicates $W_1 = W_3 = W_4 = 0$, and corresponding rows and columns can be deleted to give

$$\frac{1}{6(a/2)^2} \left[2 \frac{a^2}{4} + 2 \frac{a^2}{4} \right] - \lambda \frac{a^2}{4} \frac{1}{9} = 0$$

It follows that $\omega^2 = 24T/a^2\rho$ or $\omega = (4.899/a)(T/\rho)^{1/2}$ compares to the exact solution of $\omega = (4.443/a)(T/\rho)^{1/2}$ (see Prob. 7.26). More detail concerning the finite element analysis of the vibrating membrane is given by Bickford (1990).

7.9. The differential equation that describes the transverse motion of a beam structure is similar to Eq. (*4.1*) and can be derived following Timoshenko (1955) as

$$EI \frac{\partial^4 v}{\partial x^4} = -\rho \frac{\partial^2 v}{\partial^2 t} \tag{a}$$

where ρ is the mass density per unit length of beam. Assume free harmonic motion and derive the corresponding local beam finite element.

The beam finite element that models the left-hand side of Eq. (a) was derived and analyzed in Chap. 4. It remains to develop a model for the dynamic term on the right-hand side of the equation. The use of the Galerkin method to derive the finite element model can be of instructional value. Follow the methods of Chap. 5 and use Eqs. (g) and (h) of Prob. 4.9 as a trial solution:

$$v(x, t) = [N(x)]\{v(t)\} \tag{b}$$

where $[N(x)] = [N_1 \quad N_2 \quad N_3 \quad N_4]$ and $\{v(t)\} = [v_1 \quad \theta_1 \quad v_2 \quad \theta_2]^T$. Substitute the trial solution into the governing equation:

$$EI \frac{\partial^4 [N]}{\partial x^4} \{v\} + \rho [N] \frac{\partial^2 \{v\}}{\partial t^2} = R(x, t: \{v\}) \tag{c}$$

Multiply by the weight function and integrate over the length:

$$\int_0^L EI[N]^T \frac{\partial^4 [N]}{\partial x^4} \{v\} \, dx + \int_0^L \rho [N]^T [N] \frac{\partial^2 \{v\}}{\partial t^2} = 0 \tag{d}$$

Integrate the first term by parts twice:

$$\int_0^L \left[EI \frac{\partial^2 [N]^T}{\partial x^2} \frac{\partial^2 [N]}{\partial x^2} \{v\} + \rho [N]^T [N] \frac{\partial^2 \{v\}}{\partial t^2} \right] dx = EI \left[\frac{\partial [N]}{\partial x} \frac{\partial^2 [N]}{\partial x^2} \{v\} - [N] \frac{\partial^3 [N]}{\partial x^3} \{v\} \right]_0^L$$

The terms to the right of the equal sign represent the boundary conditions on the deflection and slope or moment and shear of Chap. 4 that were treated as joint loadings caused by applied transverse loading. The free vibration problem without transverse loading is governed by the differential equation within the volume integral. Assume a solution of the form

$$v(x, t) = V(x)e^{i\omega t}$$

and substitute into the differential equation to obtain the corresponding eigenvalue problem:

$$\int_0^L \left(EI \frac{\partial^2 [N]^T}{\partial x^2} \frac{\partial^2 [N]}{\partial x^2} \{V\} - \rho \omega^2 [N]^T [N] \{V\} \right) dx = 0 \tag{e}$$

The first term of Eq. (e) is the stiffness matrix that was derived in Chap. 4 in Eqs. (f) and (g) of Prob. 4.4. The second term of Eq. (e) defines the mass matrix:

$$[m] = \int_0^L \rho [N]^T [N] \, dx \tag{f}$$

Equations (c)–(f) of Prob. 4.9 are used to evaluate the mass matrix. The matrices inside the integral are multiplied together as follows:

$$\frac{\rho}{L^3} \begin{Bmatrix} 2x^3 - 3x^2L + L^3 \\ x^3L - 2x^2L^2 + xL^3 \\ -2x^3 + 3x^2L \\ x^3L - x^2L^2 \end{Bmatrix} \frac{1}{L^3} [2x^3 - 3x^2L + L^3 \quad x^3L - 2x^2L^2 + xL^3 \quad -2x^3 + 3x^2L \quad x^3L - x^2L^2]$$

After integrating and substituting limits, the mass matrix becomes (Logan, 1986)

$$[m] = \frac{\rho L}{420} \begin{bmatrix} 156 & 22L & 54 & -13L \\ 22L & 4L^2 & -13L & -3L^2 \\ 54 & 13L & 156 & -22L \\ -13L & -3L^2 & -22L & 4L^2 \end{bmatrix} \tag{g}$$

In summary, the eigenvalue problem for harmonic motion of a beam can be written

$$[K]\{V\} - \omega^2[m]\{V\} = 0 \tag{h}$$

7.10. Use one beam element to compute the natural frequency for free vibration of a beam that is simply supported at both ends.

Combine Eq. (f) of Prob. 4.4 and the mass matrix, Eq. (g), derived in Prob. 7.9. The boundary conditions are $v(0) = v(L) = 0$. Delete the first and third rows and columns from the finite element matrix:

$$\frac{EI}{L^3}\begin{bmatrix} 4L^2 & 2L^2 \\ 2L^2 & 4L^2 \end{bmatrix} - \frac{\omega^2 \rho L}{420}\begin{bmatrix} 4L^2 & -3L^2 \\ -3L^2 & 4L^2 \end{bmatrix} = 0 \tag{a}$$

Solving Eq. (a) gives

$$\omega_1 = \frac{10.954(EI/\rho)^{1/2}}{L^2} \quad \text{and} \quad \omega_2 = \frac{50.120(EI/\rho)^{1/2}}{L^2}$$

The exact solution can be computed as

$$\omega_n = \frac{n^2\pi^2(EI/\rho)^{1/2}}{L^2} \quad \text{or} \quad \omega_1 = \frac{9.867(EI/\rho)^{1/2}}{L^2} \quad \text{and} \quad \omega_2 = \frac{39.478(EI/\rho)^2}{L^2}$$

7.11. The free vibration problem corresponding to a more general elasticity problem can be formulated following the methods applied to the equations of elasticity in Chaps. 3, 5, and 6. A general expression can be derived and then specialized to represent a specific application. Investigate the free vibration of a finite cylinder in axisymmetric cylindrical coordinates using nine-node Lagrangian finite elements.

The equations of elasticity can be written in cartesian tensor notation:

$$\sigma_{ij,j} = \rho\,\frac{\partial^2 u_i}{\partial t^2} \qquad \sigma_{ij} = \sigma_{ji} \qquad \text{for } i \neq j \tag{a}$$

$$\sigma_{ij} = C_{ijkl}\epsilon_{kl} \tag{b}$$

$$\epsilon_{kl} = \tfrac{1}{2}(u_{k,l} + u_{l,k}) \tag{c}$$

Assume $u_i = [N]\{u\}$ as a trial solution, and it follows from Eq. (c) that

$$\{\epsilon\} = [L][N]\{u\} = [B]\{u\} \tag{d}$$

Substitute Eq. (c) into Eq. (b) and then into Eq. (a). The trial solution is then assumed, and after multiplying by the weight function and integrating, the result is

$$\int_V \left(\frac{\partial [N]^T}{\partial x_i} [C] \frac{\partial [N]}{\partial x_i} \{u\} + [N]^T \rho [N] \frac{\partial^2 \{u\}}{\partial t^2} \right) dV = \int_S [N]^T [C] \frac{\partial [N]}{\partial x_i} \{u\} \, dS \qquad (e)$$

The formulation, Eq. (e), is quite general and can represent any elasticity problem. The axisymmetric elasticity problem is defined by Eqs. (3.17) and (3.18) that were specialized in Sec. 6.7 to axisymmetric r, z coordinates. Equation (a) corresponds to Eqs. (6.6) and (6.7) with dynamic terms included. Equation (c) is a general statement that corresponds to Eq. (6.5). Equation (b) corresponds to Eq. (d) of Prob. 3.26. Note that the reader who is unfamiliar with subscript tensor notation must be aware that Eqs. (a)–(c) cannot be expanded into the equations of elasticity in any system except the cartesian system. The intent is to arrive at Eq. (e) and specialize that equation to a specific application.

Equation (e) is the general form of the free vibration problem and can be written in the alternate form, neglecting traction boundary conditions,

$$\int_V \left([B]^T [C][B]\{u\} + [N]^T \rho [N] \frac{\partial^2 \{u\}}{\partial t^2} \right) dV = 0 \qquad (f)$$

The second term of Eq. (f) is called the *consistent mass matrix* and is formulated using the shape functions rather than proportionately lumping the mass at the nodes of the element. The free vibration problem is defined using Eq. (f), and an appropriate harmonic function is assumed to represent the acceleration term. The assumed function must satisfy the displacement equations of the theory of elasticity. The governing equations, in terms of displacement, are obtained by substituting the strain-displacement equations, Eqs. (6.5), into the constitutive equations, Eqs. (3.18), and then substituting the result into the stress equilibrium equations, Eqs. (6.6) and (6.7). Functions that will satisfy the displacement equilibrium equations are

$$u(r, z, t) = U(r, z)e^{-i\omega t} \qquad v(r, z, t) = V(r, z)e^{-i\omega t} \qquad (g)$$

where ω is the circular frequency. The general form of Eq. (f) becomes

$$[K - \omega^2 M]\{U\} = 0 \qquad (h)$$

The nine-node Lagrange element is defined in Prob. 6.24. The formulation of the [B] matrix follows Eqs. (a) and (b) of Prob. 6.12 (also see Prob. 3.26) using nine-node shape functions rather than four-node shape functions, and the resulting [B] matrix is 4×18. The [C] matrix is identical to Eq. (d) of Prob. 3.26, a 4×4 matrix. The mass matrix [M] is formulated using the nine-node lagrangian shape functions according to Eq. (f). Visualize the mass matrix as

$$[M] = \int_V \begin{bmatrix} N_1 & 0 \\ 0 & N_1 \\ N_2 & 0 \\ 0 & N_2 \\ \cdot & \cdot \\ \cdot & \cdot \\ \cdot & \cdot \\ N_9 & 0 \\ 0 & N_9 \end{bmatrix}_{18 \times 2} \begin{bmatrix} \rho & 0 \\ 0 & \rho \end{bmatrix}_{2 \times 2} \begin{bmatrix} N_1 & 0 & N_2 & 0 & \cdots & N_9 & 0 \\ 0 & N_1 & 0 & N_2 & \cdots & 0 & N_9 \end{bmatrix}_{2 \times 18} dV \qquad (i)$$

where $dV = 2\pi r \, dr \, dz$. The $[\rho]$ matrix has terms in both diagonal elements since there are two degrees of freedom per node. The resulting local finite element is an 18×18 matrix.

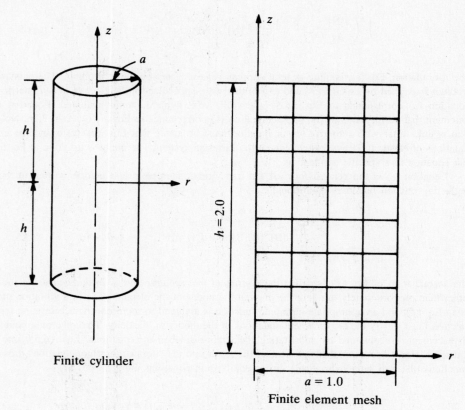

Fig. 7-8

The axisymmetric finite length cylinder of Fig. 7-8 was studied by Hutchinson (1967). The boundary conditions are $u(a, z, t) = 0$, zero displacement in the r direction along the sides of the cylinder, $w(a, z, t) = 0$, $w(r, \pm h, t) = 0$, zero displacement in the z direction on all boundaries of the cylinder, and $\sigma_{rz}(a, \pm h, t) = 0$, zero shear stress on all external surfaces of the cylinder. In this application of the finite element method, symmetry can be used; the cylinder was modeled using eight nine-node elements, resulting in 45 nodes and 90 degrees of freedom. The model is shown in Fig. 7-8 with $a = 1$, $h = 2$, and $\nu = 0.3$. Results for the first four frequencies are compared in Table 7.3.

Table 7.3 Frequencies for Finite Length Cylinder

Method	1	2	3	4
Hutchinson	2.7466	4.4597	5.5546	5.8402
Finite element	2.7485	4.4742	5.6349	5.9339

Additional results for the axisymmetric cylinder, as well as an application for the piezoelectric cylinder that was defined following the equations given by Prob. 5.19 have been given by Cheng (1988).

7.12. The differential equation that describes the buckling of a long column is similar to Eq. (*4.1*) and is derived in elementary mechanics of materials as

$$EI\frac{d^4v}{dx^4} + P\frac{d^2v}{dx^2} = 0 \qquad (a)$$

where P is a concentric axial force applied at the ends of the column. Derive the local finite element that can be used to study the buckling behavior of a column with concentric axial loading.

The derivation follows that of Prob. 7.9 where the beam finite element that models the first term of Eq. (a) was derived using the Galerkin method. This technique can be used to derive the finite element model for the second term of Eq. (a). Assume the trial solution as in Prob. 7.9:

$$v(x, t) = [N(x)]\{v(t)\} \qquad (b)$$

where $[N(x)] = [N_1 \quad N_2 \quad N_3 \quad N_4]$ and $\{v(t)\} = [v_1 \quad \theta_1 \quad v_2 \quad \theta_2]^T$. Substitute the trial solution into the governing equation. Then, multiply by the weight function and integrate over the length:

$$\int_0^L EI[N]^T \frac{d^4[N]}{dx^4}\{v\}\,dx + \int_0^L P[N]^T\frac{d^2[N]}{dx^2}\{v\}\,dx = 0 \qquad (c)$$

Integrate the first term by parts twice and the second term once. As in Prob. 7.9 there will be boundary conditions for node (joint) loadings that represent shear, moment, and axial force. Neglect the possible joint loadings and concentrate on the eigenvalue problem governing column buckling:

$$\int_0^L \left(EI\frac{d^2[N]^T}{dx^2}\frac{d^2[N]}{dx^2}\{v\} + P\frac{d[N]^T}{dx}\frac{d[N]}{dx}\{v\} \right)dx = 0 \qquad (d)$$

The first term of Eq. (d) is the stiffness matrix of Prob. 7.9. The second term must be evaluated using the derivatives of the shape functions given in Eqs. (c)–(f) of Prob. 4.9. The matrices inside the integral are multiplied together as follows:

$$\frac{\rho}{L^3}\begin{Bmatrix} 6x^2 - 6xL \\ 3x^2L - 4xL^2 + L^3 \\ -6x^2 + 6xL \\ 3x^2L - 2xL^2 \end{Bmatrix}\frac{1}{L^3}[6x^2 - 6xL \quad 3x^2L - 4xL^2 + L^3 \quad -6x^2 + 6xL \quad 3x^2L - 2xL^2]$$

After integrating and substituting limits, the matrix that represents the effect of axial force on bending stiffness becomes (Chajes, 1974)

$$[D] = \frac{P}{30L}\begin{bmatrix} 36 & 3L & -36 & 3L \\ 3L & 4L^2 & -3L & L^2 \\ -36 & -3L & 36 & -3L \\ 3L & L^2 & -3L & 4L^2 \end{bmatrix} \qquad (e)$$

In summary, the eigenvalue problem for the critical buckling load for a column can be written

$$[K]\{v\} - \lambda^2[D]\{v\} = 0 \qquad \lambda^2 = \frac{P}{EI} \qquad (f)$$

7.13. Use one finite element to compute the critical buckling load for a column pinned at both ends.

Combine Eq. (f) of Prob. 4.4 and Eq. (g) that was derived in Prob. 7.12. The boundary conditions are $v(0) = v(L) = 0$. Delete the first and third rows and columns from the finite element matrix as follows:

$$\frac{EI}{L^3}\begin{bmatrix} 4L^2 & 2L^2 \\ 2L^2 & 4L^2 \end{bmatrix} - \frac{P}{30L}\begin{bmatrix} 4L^2 & L^2 \\ L^2 & 4L^2 \end{bmatrix} = 0 \qquad (a)$$

Solving Eq. (a) gives $P_{cr} = 9.689EI/L^2$ and $P_2 = 74.311EI/L^2$. The exact solution can be computed as $P_n = n^2\pi^2EI/L^2$, or $P_{cr} = 9.867EI/L^2$ and $P_2 = 39.478EI/L^2$.

7.14. An eight-node isoparametric finite element is shown in Fig. 7-9 in ξ, η, ψ coordinates. Use the Lagrange polynomial formula to derive interpolation (shape) functions for the element.

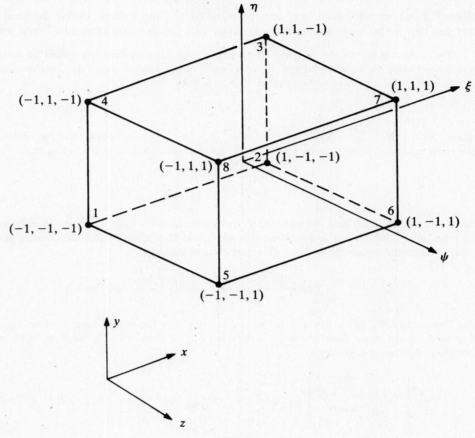

Fig. 7-9

Refer to Prob. 6.4 and node 1 of Fig. 7-9 where the plane that contains nodes 1–4 corresponds to $\psi = -1$. A linear one-dimensional interpolation formula for node 1 along the ψ axis is $L_{1\psi} = \frac{1}{2}(\psi - 1)/(-1 - 1)$. The linear one-dimensional interpolation formulas for the ξ and η directions are given in Prob. 6.4 and can be combined with $L_{1\psi}$ to give

$$N_1 = L_{1\xi}L_{1\eta}L_{1\psi} = \frac{(1 - \xi)(1 - \eta)(1 - \psi)}{8}$$

Similarly, the remaining functions are computed as

$$N_2 = L_{2\xi}L_{2\eta}L_{2\psi} = (1 + \xi)(1 - \eta)(1 - \psi)/8$$

$$N_3 = (1 + \xi)(1 + \eta)(1 - \psi)/8 \qquad N_4 = (1 - \xi)(1 + \eta)(1 - \psi)/8$$

$$N_5 = (1 - \xi)(1 - \eta)(1 + \psi)/8 \qquad N_6 = (1 + \xi)(1 - \eta)(1 + \psi)/8$$

$$N_7 = (1 + \xi)(1 + \eta)(1 + \psi)/8 \qquad N_8 = (1 + \xi)(1 + \eta)(1 - \psi)/8$$

7.15. A four-node one-dimensional isoparametric element is shown in Fig. 7-10. Derive the shape functions using the Lagrange polynomial.

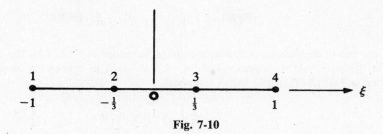

Fig. 7-10

Refer to Fig. 7-10 and substitute into Eq. (6.3):

$$L_1 = -\tfrac{9}{16}(\xi + \tfrac{1}{3})(\xi - \tfrac{1}{3})(\xi - 1) \qquad L_3 = -\tfrac{27}{16}(\xi + 1)(\xi + \tfrac{1}{3})(\xi - 1)$$

$$L_2 = \tfrac{27}{16}(\xi + 1)(\xi - \tfrac{1}{3})(\xi - 1) \qquad L_4 = \tfrac{9}{16}(\xi + 1)(\xi + \tfrac{1}{3})(\xi - \tfrac{1}{3})$$

7.16. Discuss continuity for the two-node one-dimensional linear element and its application to one-dimensional heat transfer.

Continuity for the two-node element is shown graphically in Figs. 2-4 and 2-5. The shape function is C^0 continuous at the nodes, however, its first derivative is not continuous. Therefore it can be concluded that two-node one-dimensional elements are of class C^0. There can be one exception to this result. Given the governing equation and boundary conditions

$$\frac{d^2T}{dx^2} = 0 \qquad T(0) = 0 \qquad T(L) = T_0 \qquad 0 \le x \le L$$

the solution is $T(x) = T_0 x/L$ and $dT/dx = T_0/L$. The function is continuous, and its first derivative is continuous. It follows that the function is at least of class C^1. The shape function has the same property since a linear function is the exact solution of the governing differential equation. Of course, the analysis of this problem is of little interest from a finite element viewpoint.

7.17. Derive a local stiffness matrix for plate bending using the Mindlin theory and Eq. (7.16) as the basic relationship that can be used to evaluate the strain energy that corresponds to plate bending. The use of the strain energy to derive plate elements follows the concepts developed in Chap. 4 for deriving beam finite elements.

The finite element that will be derived has been named the "heterosis" element by Hughs (1987), and the discussion here follows that of Cook et al. (1989). The development is general and applies to any element; it will be specialized to correspond to the heterosis element. The element unknowns are the plate rotations and transverse deflection and are defined in terms of nodal degrees of freedom for any element using a suitable shape function. Assume the following relation between element unknowns and nodal unknowns:

$$\begin{Bmatrix} w_e \\ \theta_{xe} \\ \theta_{ye} \end{Bmatrix} = \begin{bmatrix} N_1 & 0 & 0 & N_2 & 0 & 0 & \cdots & \text{repeat} & \cdots & N_n & 0 & 0 \\ 0 & N_1 & 0 & 0 & N_2 & 0 & \cdots & \text{for} \; n & \cdots & 0 & N_n & 0 \\ 0 & 0 & N_1 & 0 & 0 & N_2 & \cdots & \substack{\text{element} \\ \text{nodes}} & \cdots & 0 & 0 & N_n \end{bmatrix} \begin{Bmatrix} w \\ \theta_x \\ \theta_y \end{Bmatrix} \tag{a}$$

or

$$\{u\} = [N]_{3 \times 3n}\{d\} \tag{b}$$

where $\{u\}$ are the element unknowns, $[N]$ are the element shape functions, and $\{d\}$ are the nodal unknowns:

$$\{d\} = [w_1 \quad \theta_{x1} \quad \theta_{y1} \quad w_2 \quad \theta_{x2} \quad \theta_{y2} \quad \cdots \quad w_n \quad \theta_{xn} \quad \theta_{yn}]^T \tag{c}$$

In general the strain energy of deformation for the plate can be written

$$U = \frac{1}{2} \int_A \int_{-h/2}^{h/2} \{\epsilon\}^T [E]\{\epsilon\} \, dz \, dA \tag{d}$$

where A is the area of the midsurface of the plate and $\{\epsilon\}$ is defined in Eq. (7.14). The integration is through the thickness of the plate, and the equations of Sec. 7.7 are used to convert the strain energy into matrices corresponding to Eqs. (7.15) and (7.16) as follows:

$$U = \frac{1}{2} \int_A \{\kappa\}^T [D_M]\{\kappa\} \, dA \tag{e}$$

There are three degrees of freedom at each node w, θ_x, and θ_y that are defined by Eq. (a). The curvatures of Eq. (7.15) are defined in terms of nodal unknowns using an operator matrix

$$\{\kappa\} = [L][N]\{d\} = [B]\{d\} \tag{f}$$

where $[L]$ is the following 5×3 matrix:

$$[L] = \begin{bmatrix} 0 & \partial/\partial x & 0 \\ 0 & 0 & \partial/\partial y \\ 0 & \partial/\partial y & \partial/\partial x \\ -\partial/\partial y & 0 & 1 \\ -\partial/\partial x & 1 & 0 \end{bmatrix} \tag{g}$$

The strain energy can be defined as

$$U = \frac{1}{2} \int_A \{d\}^T [N]^T [L]^T [D_M][L][N]\{d\} \, dA = \frac{1}{2} \int_A \{d\}^T [B]^T [D_M][B]\{d\} \, dA \tag{h}$$

Equation (h) leads to formulation of the stiffness matrix as

$$[K] = \int_A [B]^T [D_M][B] \, dA \tag{i}$$

The derivation has been completely general, and Eq. (i) can be used to formulate a finite element using any desired shape function. The stiffness matrix will now be specialized to the heterosis element.

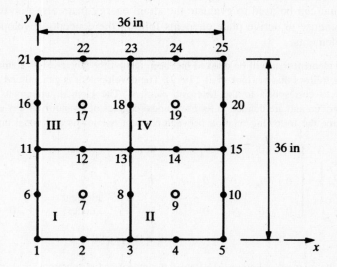

$$h = 0.25 \text{ in}, \, E = 30(10)^6 \text{ psi}, \, \nu = 0.3$$

Fig. 7-11

The heterosis element is shown in Fig. 7-11 as a nine-node quadrilateral isoparametric plate element and is in fact a combination of the nine-node Lagrangian element (see Prob. 6.24) and the standard eight-node element (see Prob. 6.25) discussed in Chap. 6. The corner and midside nodes have three degrees of freedom, θ_x, θ_y, and w, whereas the center node has only two degrees of freedom, θ_x and θ_y. The element has 26 total degrees of freedom. The Lagrange shape function is used to model the rotations θ_x and θ_y, and the standard eight-node element is used to model the transverse plate deflection w. It follows that the element stiffness matrix, Eq. (i), is split into two stiffness matrices, one to model bending and one to model transverse shear:

$$[K] = [K_b] + [K_s] \tag{j}$$

where

$$[K_b] = \int_A [B_b]^T [D_b][B_b]\, dA \tag{k}$$

$$[K_s] = \int_A [B_s]^T [D_s][B_s]\, dA \tag{l}$$

Splitting the stiffness matrix is best visualized by examining the result $[B] = [L][N]$ defined by Eq. (f). Let P denote the nine-node Lagrange shape functions (rotations) and N represent the standard eight-node shape functions (deflections). Note that it is not feasible to write a matrix equation similar to Eq. (a) to represent the combined shape functions P and N. The heterosis [B] matrix is constructed as

$$\begin{bmatrix} 0 & \partial P_1/\partial x & 0 & 0 & \partial P_2/\partial x & 0 & \cdots & \partial P_9/\partial x & 0 \\ 0 & 0 & \partial P_1/\partial y & 0 & 0 & \partial P_2/\partial y & \cdots & 0 & \partial P_9/\partial y \\ 0 & \partial P_1/\partial y & \partial P_1/\partial x & 0 & \partial P_2/\partial y & \partial P_2/\partial x & \cdots & \partial P_9/\partial y & \partial P_9/\partial x \\ -\partial N_1/\partial y & 0 & P_1 & -\partial N_2/\partial y & 0 & P_2 & \cdots & 0 & P_9 \\ -\partial N_1/\partial x & P_1 & 0 & -\partial N_2/\partial x & P_2 & 0 & \cdots & P_2 & 0 \end{bmatrix} \tag{m}$$

The matrices of Eqs. (k) and (l) can be identified as

$$\underset{[3\times 26]}{[B_b]} = \begin{bmatrix} 0 & \partial P_1/\partial x & 0 & 0 & \partial P_2/\partial x & 0 & \cdots & \partial P_9/\partial x & 0 \\ 0 & 0 & \partial P_1/\partial y & 0 & 0 & \partial P_2/\partial y & \cdots & 0 & \partial P_9/\partial y \\ 0 & \partial P_1/\partial y & \partial P_1/\partial x & 0 & \partial P_2/\partial y & \partial P_2/\partial x & \cdots & \partial P_9/\partial y & \partial P_9/\partial x \end{bmatrix} \tag{n}$$

$$\underset{[2\times 26]}{[B_s]} = \begin{bmatrix} -\partial N_1/\partial y & 0 & P_1 & -\partial N_2/\partial y & 0 & P_2 & \cdots & 0 & P_9 \\ -\partial N_1/\partial x & P_1 & 0 & -\partial N_2/\partial x & P_2 & 0 & \cdots & P_9 & 0 \end{bmatrix} \tag{o}$$

$$[D_b] = \begin{bmatrix} D & \nu D & 0 \\ \nu D & D & 0 \\ 0 & 0 & (1-\nu)D/2 \end{bmatrix} \qquad [D_s] = \begin{bmatrix} Gh & 0 \\ 0 & Gh \end{bmatrix} \tag{p}$$

7.18. The square plate of Fig. 7-11 is 36×36 in and 0.25 in thick. The plate is simply supported with a 1000-lb load applied at the center node, node 13, and is to be modeled using four heterosis elements.

(a) Assume material constants for steel and compute the deflection and rotations at each node. $E = 30(10)^6$ psi and $\nu = 0.3$.

(b) Assume material constants for aluminum and compute the deflection and rotations at each node. $E = 9.5(10)^6$ psi and $\nu = 0.29$.

The stiffness matrix is formulated according to Eq. (j) of Prob. 7.17. The load vector has only one nonzero term, and the applied load at node 13 corresponding to the first degree of freedom is 1000 lb. All deflection boundary conditions are specified as zero along the edges of the plate, and the rotations tangent to the edges of the plate are specified as zero. Results are given in Table 7.4 and can be compared with the solution given by Timoshenko and Woinowsky-Krieger (1959). Because of symmetry, results are tabulated only for nodes corresponding to element 1.

(a) Exact solution for deflection at node 13, $w_{13} = 0.35022$ in.

(b) Exact solution for deflection at node 13, $w_{13} = 1.11314$ in.

Table 7.4   Finite Element Results for Simply Supported Plate

Node	(a) Steel plate			(b) Aluminum		
	w (in)	θ_x	θ_y	w (in)	θ_x	θ_y
1	0.00000	0.00177	0.00163	0.00000	0.00564	0.00523
2	0.00000	0.00600	0.01421	0.00000	0.01918	0.45381
3	0.00000	0.00001	0.03170	0.00000	0.00004	0.10119
6	0.00000	0.01416	0.00602	0.00000	0.04520	0.01925
7	*	0.01346	0.01345	*	0.04285	0.04282
8	0.24265	0.00000	0.02765	0.77350	0.00000	0.08813
11	0.00000	0.03178	−0.00003	0.00000	0.10148	−0.00008
12	0.24263	0.02765	0.00001	0.77349	0.08812	0.00003
13	0.35008	0.00000	0.00000	1.11533	−0.00001	−0.00002

* The center node of the heterosis element does not have a deflection value.

Supplementary Problems

7.19. Assume a one-dimensional space of length L filled with a homogeneous material. The space has an initial temperature of T_i. The temperature of the surfaces of the space, defined by $x = 0$ and $x = L$, are instantaneously changed to T_F. Obtain the analytical solution for this problem.

7.20. Assume the material defined in Prob. 7.19 is 12 in thick and initially has a constant temperature distribution of 50°F. Let $\rho = 120$ psf, $c = 0.2$ Btu/(lb·°F), and $k = 0.5$ Btu/(h·ft·°F). Assume that the temperature is raised to 80°F at each surface, $x = 0$ and $x = 1$ ft. Obtain a solution using 10 equal finite elements and compare the results with the analytical solution. Use either two-node linear one-dimensional elements or a strip of four-node linear two-dimensional elements and the backward difference formulation for the time derivative.

7.21. A set of coupled equations defining two-phase diffusion was given in Prob. 5.25. Assume that the material defined by the equations occupies the space $0 \le x \le L$ and at the boundaries is suddenly subjected to a change in C_f. The boundary and initial conditions are

$$C_f(0, t) = C_f(L, t) = \Gamma \qquad t > 0 \tag{a}$$

$$C_f(x, 0) = C_r(x, 0) = 0 \qquad 0 \le x \le L \tag{b}$$

Obtain an analytical solution for this problem.

7.22. Obtain an analytical solution for the problem of axisymmetric transient diffusion defined by Eq. (a) of Prob. 7.4. Assume C is a function of r and t and reduce the problem to one dimension. Assume boundary conditions corresponding to a hollow cylinder with $C(r = 1, t) = 10$ and $\partial C(r = 2, t)/\partial r = 0$ and an initial condition of $C(r, 0) = 0$.

7.23. Obtain a four-element finite element solution for the diffusion problem defined by Probs. 7.4 and 7.22. Assume $D = 1$.

7.24. Obtain the analytical solution for the vibrating string problem defined by Eq. (7.2).

7.25. The free longitudinal vibration of a bar is governed by a differential equation similar to Eq. (2.4) and is written

$$E \frac{\partial^2 u}{\partial x^2} = \rho \frac{\partial^2 u}{\partial t^2} \qquad (a)$$

(a) Assume free-free boundary conditions, use two linear finite elements to compute a solution, and compare the results with the exact solution.

(b) Assume fixed-free boundary conditions, use two linear finite elements to compute a solution, and compare the results with the exact solution.

7.26. Obtain an analytical solution for the vibrating membrane defined in Prob. 7.8.

7.27. Use one beam element to compute the natural frequency of free vibration for a beam that is fixed at one end and free at the other end.

7.28. Obtain an analytical solution for the critical buckling load for columns with the following end conditions: (a) pinned-pinned, (b) pinned-fixed, (c) fixed-fixed, (d) fixed-free.

7.29. Use the finite element method to compute the critical buckling load for a column that is fixed at both ends. Compare the results with the exact solution.

7.30. Use the finite element method to compute the critical buckling load for a column that is fixed-free. Compare the results with the exact solution.

7.31. The quadratic three-dimensional Lagrange finite element is shown in Fig. 7-12. Use the method of Prob. 7.14 to derive the corresponding shape functions.

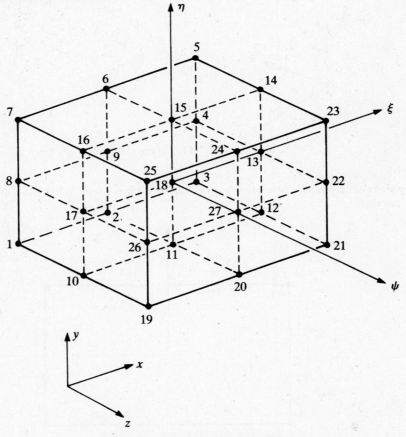

Fig. 7-12   27-node Lagrange element.

7.32. The three-dimensional isoparametric finite element shown in Fig. 7-13 is quadratic with 20 edge nodes. The element is similar to the quadratic element of Prob. 7.31 except that there are no interior nodes. Deduce the corresponding shape functions.

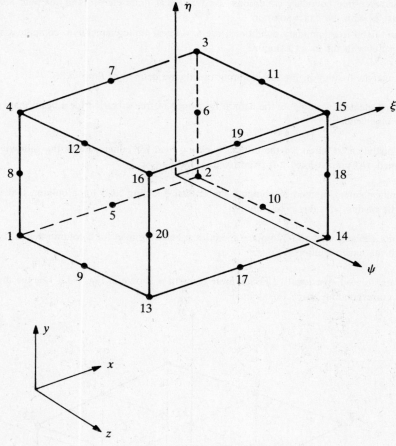

Fig. 7-13 20-node element.

7.33. A 12-node two-dimensional cubic isoparametric finite element is shown in Fig. 7-14. Deduce the corresponding shape functions.

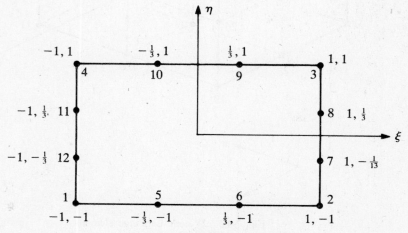

Fig. 7-14 Cubic order element.

Answers to Supplementary Problems

7.19. In elementary heat conduction problems the governing differential equation is usually written as

$$\frac{\partial^2 T}{\partial x^2} = \frac{1}{\alpha} \frac{\partial T}{\partial t} \tag{a}$$

where $\alpha = \rho c / k$ [not to be confused with the α of Eq. (7.1)]. The initial condition, $T(x, 0) = T_i - T_F = T_0$, the change in temperature. The boundary conditions can be written in terms of the temperature change and are homogeneous, $T(0, t) = T(L, t) = 0$. Equation (a) is solved using the separation of variables technique. This problem is actually elementary and can be found in most textbooks on heat conduction (Gebhart, 1961). When the initial temperature distribution is constant, the result is given as

$$T(x, t) = T_0 \frac{4}{\pi} \sum_{n=1,3}^{\infty} \frac{e^{-\lambda \alpha t}}{n} \sin \frac{n\pi x}{L} \tag{b}$$

where $\lambda = (n\pi/L)^2$.

7.20. The results were obtained using a strip of two-dimensional elements with element dimensions of 0.2×0.2 ft. The material matrix is defined as part of the capacitance matrix defined as $[N]^T [\rho c][N]$ in Eq. (d) of Prob. 7.2 and may be written as follows when the shape function matrix is defined as a 2×4 matrix:

$$[\rho c] = \begin{bmatrix} \rho c & 0 \\ 0 & 0 \end{bmatrix}$$

A similar modification is not required for [k], the thermal conductivity matrix. Results are tabulated in Table 7.5. A time increment of 0.2 h was used for 2 h and then changed to 0.5 h to complete the analysis. The results are symmetric with respect to the center of the strip and are given for only one-half of the strip.

Table 7.5   Finite Element Results for Transient Heat Transfer

x (ft)	Time (h)							
	0.4		1		5		15	
	FE	Exact	FE	Exact	FE	Exact	FE	Exact
0.0	30.00	30.00	30.00	30.00	30.00	30.00	30.00	30.00
0.1	12.97	13.16	18.56	18.73	25.68	25.77	29.40	29.46
0.2	3.96	3.64	9.73	9.82	21.79	21.97	28.86	28.97
0.3	0.98	0.61	4.41	4.27	18.70	18.95	28.43	28.57
0.4	0.23	0.05	1.89	1.60	16.72	17.01	28.15	28.34
0.5	0.09	0.01	1.18	0.86	16.04	16.34	28.06	28.25

7.21. The equations to be solved are

$$\frac{\partial C_f}{\partial t} - D \frac{\partial^2 C_f}{\partial x^2} + \beta C_f - \alpha C_r = 0 \qquad \frac{\partial C_r}{\partial t} - \beta C_f + \alpha C_r = 0$$

The solution is obtained following Gurtin and Yatomi (1979):

$$C_f(x, t) = \Gamma\left[1 - \frac{4}{\pi} \sum_{\substack{n=1 \\ n \text{ odd}}}^{\infty} \frac{(p_n - \alpha - \beta)\exp(-p_n t) - (q_n - \alpha - \beta)\exp(-q_n t)}{n(p_n - q_n)} \sin\frac{n\pi x}{L}\right]$$

$$C_r(x, t) = \frac{\beta}{\alpha}\Gamma\left[1 - \frac{4}{\pi} \sum_{\substack{n=1 \\ n \text{ odd}}}^{\infty} \frac{p_n \exp(-p_n t) - q_n \exp(-q_n t)}{n(p_n - q_n)} \sin\frac{n\pi x}{L}\right]$$

where
$$\left.\begin{array}{c} p_n \\ q_n \end{array}\right\} = \frac{1}{2}\{\alpha + \beta + \lambda_n \pm [(\alpha + \beta\lambda_n)^2 + 4\beta\lambda_n]^{1/2}\} \qquad \lambda_n = D\left(\frac{n\pi}{L}\right)^2$$

This exact analysis can be used to verify the formulation of the corresponding coupled finite element analysis.

7.22.   The governing equation is

$$D\frac{\partial^2 C}{\partial r^2} + \frac{D}{r}\frac{\partial C}{\partial r} = \frac{\partial C}{\partial t} \qquad t > 0,\ 1 \le r \le 2 \tag{a}$$

Separate the variables $C(r, t) = T(t)R(r)$ to obtain

$$\frac{d^2 R}{dr^2} + r\frac{dR}{dr} + \lambda^2 = 0 \qquad \frac{dT}{dt} + D\lambda^2 T = 0 \tag{b}$$

with solution

$$C(r, t) = T(t)R(r) = \exp\left(\frac{-D\lambda^2 t}{2}\right)[AJ_0(\lambda r) + BY_0(\lambda r)] \tag{c}$$

where J_0 and Y_0 are Bessel functions of order zero and A and B constants that are to be determined by the initial condition and boundary conditions. A general solution for Eq. (c) was given by Kardomateas (1989), and it follows that

$$C(r, t) = a + \sum_{n=1}^{\infty} \exp\left(\frac{-D\lambda_n^2 t}{2}\right)[A_n J_0(\lambda_n r) + B_n Y_0(\lambda_n r)] \tag{d}$$

where the λ_n are the real roots of

$$[\xi Y_1(2\xi)]J_0(\xi) - [\xi J_1(2\xi)]Y_0(\xi) = 0 \tag{e}$$

It follows from the boundary conditions that

$$a = 10$$

$$A_n = 10\pi[Y_0(\lambda_n)]\frac{[\lambda_n J_1(2\lambda_n)]^2}{F(\lambda_n)}$$

$$\tag{f}$$

$$B_n = -10\pi[J_0(\lambda_n)]\frac{[\lambda_n J_1(2\lambda_n)]^2}{F(\lambda_n)}$$

$$F(\lambda) = [\lambda_n J_0(\lambda_n)]^2 - [\lambda_n J_1(2\lambda_n)]^2$$

7.23.   The results, using the backward difference method, are given in Table 7.6. A time increment of 0.05 was used until t reached 0.5, was increased to 0.1 until t reached 1.0, and then was increased to 0.5 to complete the analysis.

Table 7.6 Results for Axisymmetric Diffusion

	Time							
	0.1		0.5		1.0		5.0	
r	FE	Exact	FE	Exact	FE	Exact	FE	Exact
1.00	10.000	10.000	10.000	10.000	10.000	10.000	10.000	10.000
1.25	3.830	3.854	6.665	6.654	7.892	7.904	9.904	9.948
1.50	1.041	0.935	4.314	4.286	6.390	6.411	9.836	9.912
1.75	0.239	0.136	2.952	2.902	5.504	5.529	9.796	9.890
2.00	0.097	0.023	2.523	2.458	5.222	5.244	9.783	9.883

7.24. Complete the analysis started in Prob. 7.6. The solution for the first of Eq. (c) is

$$V = C \sin\left(\frac{\omega x}{c}\right) + D \cos\left(\frac{\omega x}{c}\right)$$

with $v(0, t) = v(L, t) = 0$, and gives $D = 0$. The solution is $v(x, t) = (A \sin \omega t + B \cos \omega t) \sin(\omega x/c)$. The nontrivial result is $\sin(\omega L/c) = 0$. This result is satisfied by $\omega L/c = \pi$. Substituting c and generalizing to higher modes of vibration gives

$$\omega = \frac{n\pi}{L}\left[\frac{T}{\rho}\right]^{1/2} \qquad n = 1, 2, 3, \ldots \qquad (a)$$

7.25. The exact solution follows Probs. 7.6 and 7.24 and is given by Timoshenko (1955). Assume $u(x, t) = U(x)(A \sin \omega t + B \cos \omega t)$ and substitute into Eq. (a). It follows that $U = C \sin(\omega x/c) + D \cos(\omega x/c)$ with $c = (E/\rho)^{1/2}$.

(a) Free-free boundary conditions correspond to

$$\frac{\partial u(0)}{\partial x} = \frac{\partial u(L)}{\partial x} = 0$$

and give $\omega = (n\pi/L)(E/\rho)^{1/2}$, $n = 1, 2, 3, \ldots$. The finite element equation is similar to Eq. (a) of Prob. 7.7, and the global matrix for a two-element model is

$$\frac{2}{L}\begin{bmatrix} 1 & -1 & 0 \\ -1 & 2 & -1 \\ 0 & -1 & 1 \end{bmatrix}\begin{Bmatrix} u_1 \\ u_2 \\ u_3 \end{Bmatrix} - \frac{\lambda L}{2}\begin{bmatrix} \frac{1}{3} & \frac{1}{6} & 0 \\ \frac{1}{6} & \frac{2}{3} & \frac{1}{6} \\ 0 & \frac{1}{6} & \frac{1}{3} \end{bmatrix}\begin{Bmatrix} u_1 \\ u_2 \\ u_3 \end{Bmatrix} = 0 \qquad (a)$$

where $\lambda = \rho\omega^2/E$. Solving Eq. (a) gives $\omega = 0$ (a rigid-body mode caused by the free-free boundary conditions), $\omega_1 = (3.4641/L)(E/\rho)^{1/2}$, $\omega_2 = (6.4142/L)(E/\rho)^{1/2}$.

(b) Fixed-free boundary conditions correspond to

$$u(0) = \frac{\partial u(L)}{\partial x} = 0$$

and give $\omega = (n\pi/2L)(E/\rho)^{1/2}$, $n = 1, 3, 5, \ldots$. The global matrix is the same as Eq. (a) with $u_1 = 0$:

$$\frac{2}{L}\begin{bmatrix} 2 & -1 \\ -1 & 1 \end{bmatrix}\begin{Bmatrix} u_2 \\ u_3 \end{Bmatrix} - \frac{\lambda L}{2}\begin{bmatrix} \frac{2}{3} & \frac{1}{6} \\ \frac{1}{6} & \frac{1}{3} \end{bmatrix}\begin{Bmatrix} u_2 \\ u_3 \end{Bmatrix} = 0 \qquad (b)$$

Solving Eq. (b) gives $\omega_1 = (1.6114/L)(E/\rho)^{1/2}$, $\omega_2 = (5.6293/L)(E/\rho)^{1/2}$ that can be compared with the exact solution.

7.26. The separation of variables method of analysis is used to solve Eq. (a) of Prob. 7.8. The complete solution is given by Meirovich (1967) for a rectangular membrane with dimensions $a \times b$ as

$$\omega_{mn} = \pi \left[\left(\frac{m}{a} \right)^2 + \left(\frac{n}{b} \right)^2 \right] \left(\frac{T}{\rho} \right)^{1/2}$$

7.27. The boundary conditions are $v(0) = dv(0)/dx = 0$. The one-element beam matrix is modified by deleting the first and second rows and columns:

$$\frac{EI}{L^3} \begin{bmatrix} 12 & -6L \\ -6L & 4L^2 \end{bmatrix} - \frac{\omega^2 \rho L}{420} \begin{bmatrix} 156 & -22L \\ -22L & 4L^2 \end{bmatrix} = 0$$

Solving this equation gives

$$\omega_1 = \frac{3.534(EI/\rho)^{1/2}}{L^2} \qquad \text{and} \qquad \omega_2 = \frac{38.527(EI/\rho)^{1/2}}{L^2}$$

The exact solution can be computed following Timoshenko (1955) as

$$\omega_1 = \frac{3.515(EI/\rho)^{1/2}}{L^2} \qquad \text{and} \qquad \omega_2 = \frac{22.034(EI/\rho)^{1/2}}{L^2}$$

7.28. Solve Eq. (a) of Prob. 7.12. Refer to Chajes (1974).

$$(a) \ P_n = \frac{n^2 \pi^2 EI}{L^2} \qquad (b) \ P_{cr} = \frac{\pi^2 EI}{(0.7L)^2}$$

$$(c) \ P_{cr} = \frac{4 \pi^2 EI}{L^2} \qquad (d) \ P_{cr} = \frac{\pi^2 EI}{(2L)^2}$$

7.29. The fixed-fixed column will require a minimum of two finite elements with a length of $L/2$. Combine the two local elements and delete the rows and columns of the global matrix corresponding to $v(0) = v'(0) = v(L) = v'(L) = 0$:

$$\frac{EI}{L^3} \begin{bmatrix} 24 & 0 \\ 0 & 8L^2 \end{bmatrix} - \frac{P}{30L} \begin{bmatrix} 72 & 0 \\ 0 & 8L^2 \end{bmatrix} = 0$$

The lowest root of the eigenvalue problem is $P_{cr} = 40EI/L^2$ compared to the exact value of $P_{cr} = 4\pi^2 EI/L^2 = 39.4784EI/L^2$.

7.30. Use the results derived in Prob. 7.12 and delete the third and fourth rows and columns of the stiffness matrix corresponding to $v(L) = v'(L) = 0$:

$$\frac{EI}{L^3} \begin{bmatrix} 12 & 6L \\ 6L & 4L^2 \end{bmatrix} - \frac{P}{30L} \begin{bmatrix} 36 & 3L \\ 3L & 4L^2 \end{bmatrix} = 0$$

The lowest root of the eigenvalue problem is $P_{cr} = 2.4852EI/L^2$ compared to the exact value of $P_{cr} = 2.4674EI/L^2$.

7.31. The plane that corresponds to $\psi = -1$ contains the first nine nodes. Also, the node numbering is consecutive around the element as opposed to the numbering used in the corresponding two-dimensional element of Fig. 6-12 where corner nodes are numbered 1–4 and midside nodes are numbered 5–8. The shape functions are as follows.

Corner nodes:

$$N_1 = -(1 - \xi)(1 - \eta)(1 - \psi)\xi\eta\psi/8 \qquad N_3 = (1 + \xi)(1 - \eta)(1 - \psi)\xi\eta\psi/8$$

$$N_5 = -(1 + \xi)(1 + \eta)(1 - \psi)\xi\eta\psi/8 \qquad N_7 = (1 - \xi)(1 + \eta)(1 - \psi)\xi\eta\psi/8$$

$$N_{19} = (1 - \xi)(1 - \eta)(1 + \psi)\xi\eta\psi/8 \qquad N_{21} = -(1 + \xi)(1 - \eta)(1 + \psi)\xi\eta\psi/8$$

$$N_{23} = (1 + \xi)(1 + \eta)(1 + \psi)\xi\eta\psi/8 \qquad N_{25} = -(1 - \xi)(1 + \eta)(1 + \psi)\xi\eta\psi/8$$

Midside nodes:

$$N_2 = (1 - \xi^2)(1 - \eta)(1 - \psi)\eta\psi/4 \qquad N_4 = -(1 + \xi)(1 - \eta^2)(1 - \psi)\xi\psi/4$$

$$N_6 = -(1 - \xi^2)(1 + \eta)(1 - \psi)\eta\psi/4 \qquad N_8 = (1 - \xi)(1 - \eta^2)(1 - \psi)\xi\psi/4$$

$$N_{10} = (1 - \xi)(1 - \eta)(1 - \psi^2)\xi\eta/4 \qquad N_{12} = -(1 + \xi)(1 - \eta)(1 - \psi^2)\xi\eta/4$$

$$N_{14} = (1 + \xi)(1 + \eta)(1 - \psi^2)\xi\eta/4 \qquad N_{16} = -(1 - \xi)(1 + \eta)(1 - \psi^2)\xi\eta/4$$

$$N_{20} = -(1 - \xi^2)(1 - \eta)(1 + \psi)\eta\psi/4 \qquad N_{22} = (1 + \xi)(1 - \eta^2)(1 + \psi)\xi\psi/4$$

$$N_{24} = (1 - \xi^2)(1 + \eta)(1 + \psi)\eta\psi/4 \qquad N_{26} = -(1 - \xi)(1 - \eta^2)(1 + \psi)\xi\psi/4$$

Face center nodes:

$$N_9 = -(1 - \xi^2)(1 - \eta^2)(1 - \psi)\psi/2 \qquad N_{11} = -(1 - \xi^2)(1 - \eta)(1 - \psi^2)\eta/2$$

$$N_{13} = (1 + \xi)(1 - \eta^2)(1 - \psi^2)\xi/2 \qquad N_{15} = (1 - \xi^2)(1 + \eta)(1 - \psi^2)\eta/2$$

$$N_{17} = -(1 - \xi)(1 - \eta^2)(1 - \psi^2)\xi/2 \qquad N_{27} = (1 - \xi^2)(1 - \eta^2)(1 - \psi)\psi/2$$

Center node:

$$N_{18} = (1 - \xi^2)(1 - \eta^2)(1 - \psi)^2$$

7.32. The node numbering scheme of Fig. 7-13 shows the face that corresponds to $\psi = -1$ having nodes numbered 1–8, with corner nodes numbered 1–4 and midside nodes 5–8.

$$N_1 = (1 - \xi)(1 - \eta)(1 - \psi)(-\xi - \eta - \psi - 2)/8 \qquad N_5 = (1 + \xi^2)(1 - \eta)(1 - \psi)/4$$

$$N_2 = (1 + \xi)(1 - \eta)(1 - \psi)(\xi - \eta - \psi - 2)/8 \qquad N_6 = (1 + \xi)(1 - \eta^2)(1 - \psi)/4$$

$$N_3 = (1 + \xi)(1 + \eta)(1 - \psi)(\xi + \eta - \psi - 2)/8 \qquad N_7 = (1 - \xi^2)(1 + \eta)(1 - \psi)/4$$

$$N_4 = (1 - \xi)(1 + \eta)(1 - \psi)(-\xi + \eta - \psi - 2)/8 \qquad N_8 = (1 - \xi)(1 - \eta^2)(1 - \psi)/4$$

$$N_9 = (1 - \xi)(1 - \eta)(1 - \psi^2)/4 \qquad N_{10} = (1 + \xi)(1 - \eta)(1 - \psi^2)/4$$

$$N_{11} = (1 + \xi)(1 + \eta)(1 - \psi^2)/4 \qquad N_{12} = (1 - \xi)(1 + \eta)(1 - \psi^2)/4$$

$$N_{13} = (1 - \xi)(1 - \eta)(1 + \psi)(-\xi - \eta + \psi - 2)/8 \qquad N_{17} = (1 - \xi^2)(1 - \eta)(1 + \psi)/4$$

$$N_{14} = (1 + \xi)(1 - \eta)(1 + \psi)(\xi - \eta + \psi - 2)/8 \qquad N_{18} = (1 + \xi)(1 - \eta^2)(1 + \psi)/4$$

$$N_{15} = (1 + \xi)(1 + \eta)(1 + \psi)(\xi + \eta + \psi - 2)/8 \qquad N_{19} = (1 - \xi^2)(1 + \eta)(1 + \psi)/4$$

$$N_{16} = (1 - \xi)(1 + \eta)(1 + \psi)(-\xi + \eta + \psi - 2)/8 \qquad N_{20} = (1 - \xi)(1 - \eta^2)(1 + \psi)/4$$

7.33.

$$N_1 = \tfrac{1}{32}(1 - \xi)(1 - \eta)[-10 + 9(\xi^2 + \eta^2)]$$

$$N_2 = \tfrac{1}{32}(1 + \xi)(1 - \eta)[-10 + 9(\xi^2 + \eta^2)]$$

$$N_3 = \tfrac{1}{32}(1 + \xi)(1 + \eta)[-10 + 9(\xi^2 + \eta^2)]$$

$$N_4 = \tfrac{1}{32}(1 - \xi)(1 + \eta)[-10 + 9(\xi^2 + \eta^2)]$$

$$N_5 = \tfrac{9}{32}(1 - \xi^2)(1 - \eta)(1 - 3\xi)$$

$$N_6 = \tfrac{9}{32}(1 - \xi^2)(1 - \eta)(1 + 3\xi)$$

$$N_7 = \tfrac{9}{32}(1 + \xi)(1 - \eta^2)(1 - 3\eta)$$

$$N_8 = \tfrac{9}{32}(1 + \xi)(1 - \eta^2)(1 + 3\eta)$$

$$N_9 = \tfrac{9}{32}(1 - \xi^2)(1 + \eta)(1 + 3\xi)$$

$$N_{10} = \tfrac{9}{32}(1 - \xi^2)(1 + \eta)(1 - 3\xi)$$

$$N_{11} = \tfrac{9}{32}(1 - \xi)(1 - \eta^2)(1 + 3\eta)$$

$$N_{12} = \tfrac{9}{32}(1 - \xi)(1 - \eta^2)(1 - 3\eta)$$

Appendix

Computer Code for Coupled Steady-State Thermoelasticity

The finite element problem of coupled time-dependent thermoelasticity is derived in Chap. 5, Prob. 5.8. The steady-state two-dimensional counterpart is discussed in detail in Prob. 5.17. In this appendix the finite element code will be developed for two-dimensional steady-state thermoelasticity. The time-dependent problem is fully coupled, as can be determined from Prob. 5.8. When steady state is reached, the temperature is no longer time-dependent, and the mechanical displacements are dependent upon the stress, displacement boundary conditions, and steady-state temperature distribution within the body.

The governing differential equations are given in Prob. 5.17 and are repeated for ready reference:

$$C_{ijkl}\epsilon_{kl},_j - \beta\theta,_i = -f_i \tag{A.1}$$

$$k_{ij}\theta,_{ij} = -r \qquad k_{ij} = 0 \qquad \text{for } i \neq j \tag{A.2}$$

where ϵ_{kl} is the strain that can be defined in terms of displacement, θ is the temperature change measured from some base temperature, and f_i and r are the body force and external heat source, respectively. Material constants are C_{ijkl}, β, and k_{ij}. The corresponding finite element equations are

$$[K_{uu}]\{u\} - [K_{u\theta}]\{\theta\} = \{f\} + \{t\} \tag{A.3}$$

$$[K_{\theta\theta}]\{\theta\} = \{r\} + \{\bar{Q}\} \tag{A.4}$$

where $\{\bar{t}\}$ and $\{\bar{Q}\}$ are surface boundary conditions.

Main Program

Dimension statements for all dimensioned parameters are arranged as statement lines 100–200. Line 305 (GOSUB 10000) is a subroutine that defines the transformation matrices discussed in Prob. 5.18.

The geometry of the problem must be defined within a coordinate system, and in this application the cartesian system and an isoparametric finite element will be used. Fundamental data such as the number of nodal points, number of elements, number of degrees of freedom per node, number of different material parameters, and number of boundary conditions must be read into the computer as initial data. See lines 320–420 of the computer code.

Define the following parameters:

NP $\Rightarrow$ number of nodal points

NE $\Rightarrow$ number of elements

NODE $\Rightarrow$ number of nodes per element

NB $\Rightarrow$ number of essential boundary conditions (see Chap. 2)

NDF $\Rightarrow$ number of degrees of freedom per node

NMAT $\Rightarrow$ number of different materials

NCON $\Rightarrow$ number of material constants for each different material

The location of node points within the coordinate system and the relation between node points and elements can be defined. These were defined in Chap. 3 as the nodal coordinate array and connectivity array. Here they are defined as follows. The dimensions for the arrays are arbitrary and in this computer code are large enough to accommodate an example problem.

CORD(50,2) ⇒ nodal coordinate array (node point number, x and y locations)

NOD(20,8) ⇒ connectivity array (element number, node numbers for the element)

Geometry, material parameters, and essential boundary conditions are contained in lines 300–920 of the computer code. Natural boundary conditions are entered in lines 1000–1150.

Material properties may vary from element to element and while NMAT is the number of different materials and NCON is the number of material constants, two additional arrays are defined to store material information.

MAT(20)   ⇒ material number (each element must have a material number that indicates material for that element)

CON(5,10) ⇒ material number and corresponding magnitude for each material property

Boundary conditions are of two types, essential and natural. The essential boundary conditions are displacements u in the x direction and v in the y direction and θ the temperature change referenced to some standard. The parameter NBC is the number of separate boundary conditions, and the following arrays are required to store essential boundary condition data.

NBC(25)   ⇒ node number where the boundary condition occurs

NDOF(25) ⇒ degree of freedom, $1 \Rightarrow u$, $2 \Rightarrow v$, $3 \Rightarrow \theta$

DIS(25)   ⇒ magnitude of the boundary condition

The natural boundary conditions are surface tractions and/or heat flux and are read into the program using an IF statement to end the read process. The DATA appear as

Node number, x load, y load, heat flux normal to surface

The last node corresponding to NP must be entered as data, even if all natural boundary conditions are zero, in order to stop the read in process.

The formulation of the stiffness matrix begins with line 1200, and line 1220 defines the size of the global stiffness matrix. The isoparametric element requires numerical integration, and in this code a 2×2 gaussian quadrature is defined in lines 1230–1290 (see Prob. 6.1). A FOR-NEXT computational loop that formulates the global matrix begins at line 2070 and continues through line 2580. The subroutine that is called at line 2084 (GOSUB 4000) and returns at line 6620 (RETURN 2400) contains the computations for the local stiffness matrix for steady-state thermoelasticity. The local stiffness matrix is formulated within the subroutine, and the global stiffness matrix is assembled in lines 2410–2570.

Displacement and temperature boundary conditions are included in the global sitffness matrix in lines 2800–2990. Boundary conditions are included following the concept introduced in Prob. 2.13 of Chap. 2.

The matrix equation $[K]\{x\} = \{f\}$ is solved using a gaussian elimination subroutine that is called at line 3000 (GOSUB 9010). The subroutine is contained in lines 9000–9390 and returns to line 3300. Gaussian elimination is discussed in Chap. 1, and the method used in the computer code is illustrated in Prob. 1.16.

Displacement and temperature results for each node and each element are printed out in lines 3310–3450. Lines 3310–3330 give a quick printout of all results. Lines 3350–3450 allow for a printout at all nodes of each element.

A subroutine for computing stresses and temperature flux (GOSUB 7000) is called at line 3470. Lines 7000–8240 contain the subroutine that computes stresses and temperature flux and prints the results.

Additional discussion is necessary for the subroutines that formulate the local stiffness matrix and compute the final element results. The computer code can be adapted to any problem with any number of degrees of freedom by modifying these two subroutines and the subroutine that defines the transformation of the local stiffness matrix.

Local Stiffness Matrix

The local stiffness matrix is defined by Eqs. (*A.3*) and (*A.4*) and is contained in lines 4000–6620. The coordinates of the four-node element are defined in lines 4010–4040 using the coordinate array and the connectivity array. Note that the subroutine is called inside a FOR-NEXT loop that includes all of the elements (line 2070). Lines 4050–4090 initialize the stiffness matrix to zero, which is necessary since the same matrix S(I,J) is used for each local stiffness matrix. Lines 4200–4210 begin the FOR-NEXT loop that will be used for numerical integration. The numerical integration and geometry transformation follow the fundamental method detailed in Probs. 6.9, 6.10, and 6.11. Interpolation (shape) functions and their derivatives are defined in the subroutine of lines 9800–9930 and are derived in Prob. 6.4 for a four-node element. The jacobian matrix of Probs. 6.6 and 6.7 is constructed in lines 4270–4390 where SJ(I,J) is the jacobian and SJIN(I,J) is the inverse. A matrix of derivatives of shape functions identical to that of Eq. (b) of Prob. 6.8 is defined in lines 4410–4530. Equation (h) of Prob. 6.6 is defined in line 4550.

The stiffness matrix is formulated as a 12×12 matrix and can be written as follows using the notation of the computer code:

$$\begin{bmatrix} [K_{uu}] & - & [K_{u\theta}] \\ 0 & & [K_{\theta\theta}] \end{bmatrix} = \begin{bmatrix} [BST][D][BS] & - & [BST][BA][P] \\ [0] & & [TST][DK][TS] \end{bmatrix} \qquad (A.5)$$

Equation (A.5) follows directly from the derivation of the stiffness matrix given by Eqs. (o) and (p) of Prob. 5.17. The matrix [BS] is formulated as Eq. (a) of Prob. 6.10. The matrix [D] is the matrix [C] of Prob. 6.10, and the formulation of $[K_{uu}]$ is the same as that of Prob. 6.11. Matrices [BS], [D], and [BST] are formulated in lines 4560–5020 of the computer code.

All matrix multiplications are accomplished using the subroutine of lines 9500–9630. The subroutine will perform the computation [AM][BM] = [CM] (line 9590), and lines 5030–5260 are required to set the parameters and call the subroutine twice to compute the contribution to $[K_{uu}]$. The $[K_{uu}]$ part of the stiffness matrix is added to the local stiffness matrix in lines 5270–5310 where it is also multiplied by DA, Eq. (h) of Prob. 6.6. Note that the stiffness matrix returns from the subroutine as CM(I,J).

The $[K_{\theta\theta}]$ part of the stiffness matrix is identical to that in Prob. 6.9; it is formulated in lines 5400–5730 and is added to the stiffness matrix in lines 5740–5780. The $[K_{u\theta}]$ part of the stiffness matrix is discussed in detail in Prob. 5.30 and is formulated in lines 5900–6210. Note that in line 6030 the shape functions are used as defined in the shape function subroutine (lines 9800–9930). The $[K_{u\theta}]$ matrix is added to the stiffness matrix in lines 6220–6260. Lines 6270–6280 complete the numerical integration loop. Lines 6400–6610 contain the matrix multiplications that reorder the degrees of freedom according to Prob. 5.18 (see matrix transformation in Chap. 3).

Material constants are defined in the array CON(I,J), where I is the material number and J is the material constant. Each element has a material number that is entered as data at line 660. The material number is identified at line 2080 in the stiffness routine. Material parameters for the thermoelasticity problem are identified as follows:

CON(L1,1) = E

CON(L1,2) = ν (see Prob. 6.11)

CON(L1,3) = k_x

CON(L1,4) = k_y (see Prob. 6.9)

CON(L1,5) = β_x

CON(L1,6) = β_y (see Prob. 5.31)

Stress and Flux Computations

Element results are computed in the subroutine that is called at line 3470 and is contained in lines 7000–8240. Review Eqs. (c) and (h) of Prob. 3.10. For the thermoelasticity problem stress is computed according to Eq. (k) of Prob. 5.17. Temperature flux is computed as $\{q\} = -[k][B]\{\theta\}$ or in the notation of the computer code $\{FLX\} = -[CON][TS]\{TEMP\}$. The temperature flux is computed in lines 7960–8090. The stress is computed in two steps. The strain STA is computed in lines 7780–7860. Element stress is computed in lines 7870–7950. The location of the integration point within the finite element is computed in lines 8100–8150.

Example Problem

Input data for Prob. 5.18 are included at the end of this appendix. Results for displacement and temperature are given in the Table 5.2. The stresses and temperature flux can be compared with the analytical solution obtained by continuing the analysis of Prob. 5.18.

The stress is computed using Eq. (*e*) of Prob. 5.17:

$$\sigma_{xx} = C_{11}\epsilon_{xx} + C_{22}\epsilon_{yy} - \beta\theta$$

The one-dimensional case analyzed here becomes

$$\sigma = E\frac{du}{dx} - \beta\theta \quad \text{and} \quad q = -k\frac{dT}{dx}$$

Substituting results from Prob. 5.18 gives

$$\sigma = -\frac{\beta\theta_0}{2} \quad \text{and} \quad q = \frac{k\theta_0}{L}$$

Data

Input data for a program written in BASIC can be entered as DATA statements at the end of the program. Data for Prob. 5.18 would appear as follows. Assume a strip of material of unit length and unit width.

DATA 10,4,4,13,3,1,6	(line 330)
DATA 1,1,0	(coordinate array)
DATA 2,0,1	(lines 460–540)
DATA 3,.25,0	
DATA 4,.25,1	
DATA 5,.5,0	
DATA 6,.5,1	
DATA 7,.75,0	
DATA 8,.75,1	
DATA 9,1,0	
DATA 10,1,1	
DATA 1,1,3,4,2,1	(connectivity array and material number)
DATA 2,3,5,6,4,1	(lines 560–690)
DATA 3,5,7,8,6,1	
DATA 4,7,9,10,8,1	
DATA 1,1,0,1,1,1,0	(material data, lines 730–810)
DATA 1,1,0	(boundary conditions)
DATA 2,1,0	(lines 840–920)
DATA 1,2,0	
DATA 3,2,0	
DATA 5,2,0	
DATA 7,2,0	
DATA 9,2,0	
DATA 9,1,0	

```
DATA 10,1,0
DATA 1,3,100
DATA 2,3,100
DATA 9,3,0
DATA 10,3,0
DATA 10,0,0,0              (surface tractions and flux, lines 1010-1150)

10 REM STEADY STATE THERMOELASTICITY
20 REM STRESST.BAS __ ISOPARAMETRIC ELEMENT
30 REM FOUR NODE OR EIGHT NODE FORMULATION
100 DIM CORD(50,2),NOD(20,4),MAT(20),CON(5,10)
110 DIM NBC(25),NDOF(25),DIS(25)
120 DIM TEMP(4),UELE(8),STA(3),STR(3),FLX(2)
130 DIM SK( 50, 50),F1( 50)
140 DIM XG(3),WG(3),P(8),DEL(2,8),XJ(8),YJ(8)
150 DIM SJ(2,2),SJIN(2,2),SF(2,4),BS(3,8),BST(8,3)
160 DIM D(3,3),BA(3,2),DK(2,2),TS(2,4),TST(4,2)
170 DIM S(12,12)
180 DIM F(50),X(50),ELDIS(50)
190 DIM AM(12,12),BM(12,12),CM(12,12)
200 DIM T(12,12),TT(12,12)
300 REM
305 GOSUB 10000
306 REM RETURN FROM GOSUB
310 REM
320 REM READ INITIAL DATA
330 READ NP,NE,NODE,NB,NDF,NMAT,NCON
340 PRINT "         "
350 PRINT USING "####";NP,NE,NODE,NB,NDF,NMAT,NCON
360 PRINT "NP=",NP
370 PRINT "NE=",NE
380 PRINT "NODE=",NODE
390 PRINT "NB=",NB
400 PRINT "NDF=",NDF
410 PRINT "NMAT=",NMAT
420 PRINT "NCON=",NCON
430 REM
440 PRINT "          "
450 REM READ COORDINATE ARRAY
460 PRINT"         COORDINATE ARRAY"
470 FOR I=1 TO NP
480 READ N
490 FOR J=1 TO 2
500 READ CORD(N,J)
510 NEXT J
520 PRINT "NODE",N,CORD(N,1),CORD(N, 2)
530 INPUT "        RETURN TO CONTINUE CORD",GO$
540 NEXT I
550 REM
560 REM READ CONNECTIVITY
570 PRINT "         "
580 PRINT "         CONNECTIVITY"
```

```
590 FOR I=1 TO NE
600 READ N
610 PRINT"    ELEMENT #",N
620 FOR J=1 TO NODE
630 READ NOD(N,J)
640 PRINT "        ", NOD(N,J)
650 NEXT J
660 READ MAT(I)
670 PRINT"MATERIAL #",MAT(I)
680 INPUT "    RETURN TO CONTINUE ELEMENTS",GO$
690 NEXT I
700 REM
710 REM READ MATERIAL DATA
720 PRINT "         "
730 PRINT "       MATERIAL DATA"
740 FOR I=1 TO NMAT
750 READ L
760 PRINT "MATERIAL #",L
770 FOR J=1 TO NCON
780 READ CON(L,J)
790 PRINT L,J,CON(L,J)
800 NEXT J
810 INPUT "        RETURN TO CONTINUE MATERIALS",GO$
820 NEXT I
830 REM
840 REM READ ESSENTIAL BOUNDARY DATA
850 PRINT "         "
860 PRINT "       ESSENTIAL BOUNDARY CONDITIONS"
870 PRINT "NODE","DEGREE OF FREEDOM","VALUE"
880 FOR I=1 TO NB
890 READ NBC(I),NDOF(I),DIS(I)
900 PRINT NBC(I),NDOF(I), "        ",DIS(I)
910 NEXT I
920 INPUT "    RETURN TO CONTINUE",GO$
1000 REM
1010 REM READ NATURAL BOUNDARY CONDITIONS
1020 REM STORE IN LOAD VECTOR
1030 PRINT "       "
1040 PRINT "       NATURAL BOUNDARY CONDITIONS"
1050 READ NATNO
1060 PRINT "NODE =",NATNO
1070 FOR K=1 TO NDF
1080 READ FLUX
1090 PRINT "BOUNDARY CONDITION =",FLUX
1100 M=(NATNO-1)*NDF+K
1110 F1(M)=F1(M)+FLUX
1120 NEXT K
1130 INPUT "        RETURN TO CONTINUE B. C.",GO$
1140 IF NATNO=NP THEN 1150 ELSE 1050
1150 REM NATURAL BOUNDARY CONDITION FINISHED
1200 REM
1210 REM BEGIN TO FORM STIFFNESS MATRIX
```

```
1220 NSIZE=NP*NDF
1230 REM DEFINE XG AND WG FOR 2X2 INTEGRATION
1260 XG(1)=-.577350269#
1270 XG(2)=-XG(1)
1280 WG(1)=1
1290 WG(2)=1
2000 REM
2010 REM BEGIN TO FORM K MATRIX
2020 FOR I=1 TO NSIZE
2030 FOR J=1 TO NSIZE
2040 SK(I,J)=0
2050 NEXT J
2060 NEXT I
2070 FOR N=1 TO NE
2075 PRINT "ELEMENT",N
2080 L1=MAT(N)
2082 REM CALL GOSUB FOR LOCAL STIFFNESS
2084 GOSUB 4000
2400 REM ASSEMBLE GLOBAL STIFFNESS MATRIX
2410 I=0
2420 FOR JJ=1 TO NODE
2430 NROW=(NOD(N,JJ)-1)*NDF
2440 FOR J=1 TO NDF
2450 NROW=NROW+1
2460 I=I+1
2470 L=1
2480 FOR KK=1 TO NODE
2490 NCOL=(NOD(N,KK)-1)*NDF
2500 FOR K=1 TO NDF
2510 NCOL=NCOL+1
2520 SK(NROW,NCOL)=SK(NROW,NCOL)+S(I,L)
2530 L=L+1
2540 NEXT K
2550 NEXT KK
2560 NEXT J
2570 NEXT JJ
2580 NEXT N
2800 REM ESSENTIAL BOUNDARY CONDITION
2810 FOR N=1 TO NB
2820 I=NBC(N)
2830 NROW=(I-1)*NDF
2840 IF NDOF(N)=1 THEN 2850 ELSE 2860
2850 NROW=NROW+1
2860 IF NDOF(N)=2 THEN 2870 ELSE 2880
2870 NROW=NROW+2
2880 IF NDOF(N)=3 THEN 2890 ELSE 2900
2890 NROW=NROW+3
2900 F1(NROW)=DIS(N)
2910 SK(NROW,NROW)=1
2920 FOR J=1 TO NSIZE
2930 IF J=NROW THEN 2980 ELSE 2940
2940 ADD=SK(J,NROW)*DIS(N)
```

```
2950 F1(J)=F1(J)-ADD
2960 SK(NROW,J)=0
2970 SK(J,NROW)=0
2980 NEXT J
2990 NEXT N
3000 GOSUB 9010
3300 REM ****PRINT NODE RESULTS****
3310 FOR II=1 TO NSIZE STEP 3
3320 PRINT II,ELDIS(II),II+1,ELDIS(II+1),II+2,ELDIS(II+2)
3330 NEXT II
3340 INPUT "HIT RETURN TO CONTINUE", GO$
3350 FOR N=1 TO NE
3360 PRINT "RESULTS FOR ELEMENT", N
3380 FOR I=1 TO NODE
3390 M=NOD(N,I)
3400 MK=M*NDF
3410 PRINT " NODE   ","    U","    V","  TEMP"
3420 PRINT M,ELDIS(MK-2),ELDIS(MK-1),ELDIS(MK)
3430 NEXT I
3440 INPUT "HIT RETURN TO CONTINUE", GO$
3450 NEXT N
3460 REM COMPUTE STRESSES
3470 GOSUB 7000
3480 PRINT STRESS AND FLUX
3485 REM
3490 STOP
3500 REM END OF PROGRAM
3510 REM
4000 REM QUADRILATERAL ELEMENT
4010 FOR I=1 TO NODE
4020 XJ(I)=CORD(NOD(N,I),1)
4030 YJ(I)=CORD(NOD(N,I),2)
4040 NEXT I
4050 FOR I=1 TO NODE*NDF
4060 FOR J=1 TO NODE*NDF
4070 S(I,J)=0
4080 NEXT J
4090 NEXT I
4100 NGAU=2
4200 FOR I=1 TO NGAU
4210 FOR J=1 TO NGAU
4220 G=XG(I)
4230 H=XG(J)
4240 GOSUB 9800
4250 REM GOSUB DEFINES GEOMETRY FOR TRANSFORMATION
4260 REM FORM 2X2 JACOBIAN MATRIX
4270 SJ(1,1)=DEL(1,1)*XJ(1)+DEL(1,2)*XJ(2)
4280 SJ(1,1)=SJ(1,1)+DEL(1,3)*XJ(3)+DEL(1,4)*XJ(4)
4290 SJ(2,1)=DEL(2,1)*XJ(1)+DEL(2,2)*XJ(2)
4300 SJ(2,1)=SJ(2,1)+DEL(2,3)*XJ(3)+DEL(2,4)*XJ(4)
4310 SJ(1,2)=DEL(1,1)*YJ(1)+DEL(1,2)*YJ(2)
4320 SJ(1,2)=SJ(1,2)+DEL(1,3)*YJ(3)+DEL(1,4)*YJ(4)
```

```
4330 SJ(2,2)=DEL(2,1)*YJ(1)+DEL(2,2)*YJ(2)
4340 SJ(2,2)=SJ(2,2)+DEL(2,3)*YJ(3)+DEL(2,4)*YJ(4)
4350 DETJ=SJ(1,1)*SJ(2,2)-SJ(2,1)*SJ(1,2)
4360 SJIN(1,1)=SJ(2,2)/DETJ
4370 SJIN(1,2)=-SJ(1,2)/DETJ
4380 SJIN(2,1)=-SJ(2,1)/DETJ
4390 SJIN(2,2)=SJ(1,1)/DETJ
4400 REM COMPUTE SHAPE FUNCTIONS IN CART. COORD.
4410 REM ZERO SHAPE FUNCTION MATRIX
4420 FOR M = 1 TO 2
4430 FOR MM = 1 TO 4
4440 SF(M,MM)=0
4450 NEXT MM
4460 NEXT M
4470 FOR II=1 TO 2
4480 FOR KK=1 TO NODE
4490 FOR JJ=1 TO 2
4500 SF(II,KK)=SF(II,KK)+DEL(JJ,KK)*SJIN(II,JJ)
4510 NEXT JJ
4520 NEXT KK
4530 NEXT II
4540 REM MULT. BY DETJ AND WEIGHT FUNCTIONS
4550 DA=DETJ*WG(I)*WG(J)
4560 REM B AND B TRANSPOSE MATRICES
4570 FOR II=1 TO 3
4580 FOR JJ=1 TO 8
4590 BS(II,JJ)=0
4600 NEXT JJ
4610 NEXT II
4800 JJ=1
4810 FOR II=1 TO 8 STEP 2
4820 BS(1,II)=SF(1,JJ)
4830 BS(2,II+1)=SF(2,JJ)
4840 BS(3,II)=SF(2,JJ)
4850 BS(3,II+1)=SF(1,JJ)
4860 JJ=JJ+1
4870 NEXT II
4880 FOR II=1 TO 3
4890 FOR JJ=1 TO 8
4900 BST(JJ,II)=BS(II,JJ)
4910 NEXT JJ
4920 NEXT II
4930 C2=1-(CON(L1,2)*CON(L1,2))
4940 D(1,1)=CON(L1,1)/C2
4950 D(2,2)=D(1,1)
4960 D(1,2)=D(1,1)*CON(L1,2)
4970 D(3,3)=CON(L1,1)/(2*(1+CON(L1,2)))
4980 D(1,3)=0
4990 D(2,3)=0
5000 D(3,1)=0
5010 D(3,2)=0
5020 D(2,1)=D(1,2)
```

```
5030 FOR II=1 TO 3
5040 FOR JJ=1 TO 3
5050 AM(II,JJ)=D(II,JJ)
5060 NEXT JJ
5070 FOR KK=1 TO 8
5080 BM(II,KK)=BS(II,KK)
5090 NEXT KK
5100 NEXT II
5110 LM=3
5120 NM=8
5130 MM=3
5140 GOSUB 9500
5150 FOR II=1 TO 3
5160 FOR JJ= 1 TO 8
5170 BM(II,JJ)=CM(II,JJ)
5180 NEXT JJ
5190 FOR KK=1 TO 8
5200 AM(KK,II)=BST(KK,II)
5210 NEXT KK
5220 NEXT II
5230 LM=8
5240 NM=8
5250 MM=3
5260 GOSUB 9500
5270 FOR II=1 TO 8
5280 FOR JJ=1 TO 8
5290 S(II,JJ)=S(II,JJ)+CM(II,JJ)*DA
5300 NEXT JJ
5310 NEXT II
5400 FOR II=1 TO NODE
5410 TS(1,II)=SF(1,II)
5420 TS(2,II)=SF(2,II)
5430 TST(II,1)=TS(1,II)
5440 TST(II,2)=TS(2,II)
5450 NEXT II
5460 DK(1,1)=CON(L1,3)
5470 DK(2,2) =CON(L1,4)
5480 DK(1,2)=0
5490 DK(2,1)=0
5500 FOR II=1 TO 2
5510 FOR JJ=1 TO 2
5520 AM(II,JJ)=DK(II,JJ)
5530 NEXT JJ
5540 FOR KK=1 TO 4
5550 BM(II,KK)=TS(II,KK)
5560 NEXT KK
5570 NEXT II
5580 LM=2
5590 NM=4
5600 MM=2
5610 GOSUB 9500
5620 FOR II=1 TO 2
```

```
5630 FOR JJ=1 TO 4
5640 BM(II,JJ)=CM(II,JJ)
5650 NEXT JJ
5660 FOR KK=1 TO 4
5670 AM(KK,II)=TST(KK,II)
5680 NEXT KK
5690 NEXT II
5700 LM=4
5710 NM=4
5720 MM=2
5730 GOSUB 9500
5740 FOR II=1 TO 4
5750 FOR JJ=1 TO 4
5760 S(II+8,JJ+8)=S(II+8,JJ+8)+CM(II,JJ)*DA
5770 NEXT JJ
5780 NEXT II
5900 BA(1,1)=CON(L1,5)
5910 BA(2,2)=CON(L1,6)
5920 BA(1,2)=0
5930 BA(2,1)=0
5940 BA(3,1)=0
5950 BA(3,2)=0
5960 FOR II=1 TO 3
5970 FOR JJ=1 TO 2
5980 AM(II,JJ)=BA(II,JJ)
5990 NEXT JJ
6000 NEXT II
6010 FOR II=1 TO 2
6020 FOR KK=1 TO 4
6030 BM(II,KK)=P(KK)
6040 NEXT KK
6050 NEXT II
6060 LM=3
6070 NM=4
6080 MM=2
6090 GOSUB 9500
6100 FOR II=1 TO 3
6110 FOR JJ=1 TO 4
6120 BM(II,JJ)=CM(II,JJ)
6130 NEXT JJ
6140 FOR KK=1 TO 8
6150 AM(KK,II)=BST(KK,II)
6160 NEXT KK
6170 NEXT II
6180 LM=8
6190 NM=4
6200 MM=3
6210 GOSUB 9500
6220 FOR II=1 TO 8
6230 FOR JJ=1 TO 4
6240 S(II,JJ+8)=S(II,JJ+8)+(CM(II,JJ)*DA)
6250 NEXT JJ
```

```
6260 NEXT II
6270 NEXT J
6280 NEXT I
6400 LM=12
6410 MM=12
6420 NM=12
6430 FOR II=1 TO 12
6440 FOR JJ=1 TO 12
6450 AM(II,JJ)=T(II,JJ)
6460 BM(II,JJ)=S(II,JJ)
6470 NEXT JJ
6480 NEXT II
6490 GOSUB 9500
6500 FOR II=1 TO 12
6510 FOR JJ=1 TO 12
6520 AM(II,JJ)=CM(II,JJ)
6530 BM(II,JJ)=TT(II,JJ)
6540 NEXT JJ
6550 NEXT II
6560 GOSUB 9500
6570 FOR II=1 TO 12
6580 FOR JJ=1 TO 12
6590 S(II,JJ)=CM(II,JJ)
6600 NEXT JJ
6610 NEXT II
6620 RETURN 2400
7000 REM COMPUTE STRESSES
7010 FOR N=1 TO NE
7020 L1=MAT(N)
7030 FOR I=1 TO NODE
7040 XJ(I)=CORD(NOD(N,I),1)
7050 YJ(I)=CORD(NOD(N,I),2)
7060 NEXT I
7070 FOR I=1 TO NODE*NDF
7080 FOR J=1 TO NODE*NDF
7090 S(I,J)=0
7100 NEXT J
7110 NEXT I
7120 NGAU=2
7130 FOR IN=1 TO NGAU
7140 FOR JN=1 TO NGAU
7150 G=XG(IN)
7160 H=XG(JN)
7170 GOSUB 9800
7180 REM FORM 2X2 JACOBIAN MATRIX
7190 SJ(1,1)=DEL(1,1)*XJ(1)+DEL(1,2)*XJ(2)
7200 SJ(1,1)=SJ(1,1)+DEL(1,3)*XJ(3)+DEL(1,4)*XJ(4)
7210 SJ(2,1)=DEL(2,1)*XJ(1)+DEL(2,2)*XJ(2)
7220 SJ(2,1)=SJ(2,1)+DEL(2,3)*XJ(3)+DEL(2,4)*XJ(4)
7230 SJ(1,2)=DEL(1,1)*YJ(1)+DEL(1,2)*YJ(2)
7240 SJ(1,2)=SJ(1,2)+DEL(1,3)*YJ(3)+DEL(1,4)*YJ(4)
7250 SJ(2,2)=DEL(2,1)*YJ(1)+DEL(2,2)*YJ(2)
```

```
7260 SJ(2,2)=SJ(2,2)+DEL(2,3)*YJ(3)+DEL(2,4)*YJ(4)
7270 DETJ=SJ(1,1)*SJ(2,2)-SJ(2,1)*SJ(1,2)
7280 SJIN(1,1)=SJ(2,2)/DETJ
7290 SJIN(1,2)=-SJ(1,2)/DETJ
7300 SJIN(2,1)=-SJ(2,1)/DETJ
7310 SJIN(2,2)=SJ(1,1)/DETJ
7320 REM COMPUTE SHAPE FUNCTIONS IN CART. COORD.
7330 REM ZERO SHAPE FUNCTION MATRIX
7340 FOR M = 1 TO 2
7350 FOR MM = 1 TO 4
7360 SF(M,MM)=0
7370 NEXT MM
7380 NEXT M
7390 FOR II=1 TO 2
7400 FOR KK=1 TO NODE
7410 FOR JJ=1 TO 2
7420 SF(II,KK)=SF(II,KK)+DEL(JJ,KK)*SJIN(II,JJ)
7430 NEXT JJ
7440 NEXT KK
7450 NEXT II
7460 REM B MATRIX TO COMPUTE STRESS
7470 FOR II=1 TO 3
7480 FOR JJ=1 TO 8
7490 BS(II,JJ)=0
7500 NEXT JJ
7510 NEXT II
7520 JJ=1
7530 FOR II=1 TO 8 STEP 2
7540 BS(1,II)=SF(1,JJ)
7550 BS(2,II+1)=SF(2,JJ)
7560 BS(3,II)=SF(2,JJ)
7570 BS(3,II+1)=SF(1,JJ)
7580 JJ=JJ+1
7590 NEXT II
7600 C2=1-(CON(L1,2)*CON(L1,2))
7610 D(1,1)=CON(L1,1)/C2
7620 D(2,2)=D(1,1)
7630 D(1,2)=D(1,1)*CON(L1,2)
7640 D(3,3)=CON(L1,1)/(2*(1+CON(L1,2)))
7650 D(2,1)=D(1,2)
7660 D(1,3)=0
7670 D(2,3)=0
7680 D(3,1)=0
7690 D(3,2)=0
7700 J=1
7710 FOR I=1 TO NODE
7720 M=NOD(N,I)*NDF
7730 TEMP(I)=ELDIS(M)
7740 UELE(J)=ELDIS(M-2)
7750 UELE(J+1)=ELDIS(M-1)
7760 J=J+2
7770 NEXT I
```

```
7780 FOR I=1 TO 3
7790 STA(I)=0
7800 STR(I)=0
7810 NEXT I
7820 FOR I=1 TO 3
7830 FOR J=1 TO 8
7840 STA(I)=STA(I)+BS(I,J)*UELE(J)
7850 NEXT J
7860 NEXT I
7870 FOR I=1 TO 3
7880 FOR J=1 TO 3
7890 STR(I)=STR(I)+D(I,J)*STA(J)
7900 NEXT J
7910 NEXT I
7920 FOR I=1 TO NODE
7930 STR(1)=STR(1)+CON(L1,5)*P(I)*TEMP(I)
7940 STR(2)=STR(2)+CON(L1,6)*P(I)*TEMP(I)
7950 NEXT I
7960 FOR II=1 TO NODE
7970 TS(1,II)=SF(1,II)
7980 TS(2,II)=SF(2,II)
7990 NEXT II
8000 FOR J=1 TO NODE
8010 TS(1,J)=TS(1,J)*CON(L1,3)
8020 TS(2,J)=TS(2,J)*CON(L1,4)
8030 NEXT J
8040 FLX(1)=0
8050 FLX(2)=0
8060 FOR J=1 TO NODE
8070 FLX(1)=FLX(1)-TS(1,J)*TEMP(J)
8080 FLX(2)=FLX(2)-TS(2,J)*TEMP(J)
8090 NEXT J
8100 X=0
8110 Y=0
8120 FOR I=1 TO NODE
8130 X=X+P(I)*XJ(I)
8140 Y=Y+P(I)*YJ(I)
8150 NEXT I
8160 PRINT "ELEMENT",N
8170 PRINT "     X=",X,"     Y=",Y
8172 PRINT "     STRESS X=",STR(1)
8174 PRINT "     STRESS Y=",STR(2)
8176 PRINT "     STRESS XY=",STR(3)
8190 PRINT "     TEMP FLUX X=", FLX(1)
8191 PRINT "     TEMP FLUX Y=", FLX(2)
8200 INPUT "RETURN TO CONTINUE ",GO$
8210 NEXT JN
8220 NEXT IN
8230 NEXT N
8240 RETURN
9000 REM GAUSSIAN ELIMINATION
9010 N=NSIZE
```

```
9020 FOR I=1 TO N
9030 ELDIS(I)=0
9040 NEXT I
9050 J=1
9060 AA=SK(J,J)
9070 FOR I = J TO N
9080 IF AA = 0 THEN 9090 ELSE 9110
9090 PRINT "ZERO ON DIAGONAL, ROW",J
9100 STOP
9110 REM
9120 SK(J,I) =SK(J,I)/AA
9130 NEXT I
9140 F1(J)=F1(J)/AA
9150 IF J=N THEN 9280 ELSE 9160
9160 K=J+1
9170 C=SK(K,J)
9180 IF C=0 THEN 9230 ELSE 9190
9190 FOR I=J TO N
9200 SK(K,I)=SK(K,I)-C*SK(J,I)
9210 NEXT I
9220 F1(K)=F1(K)-C*F1(J)
9230 K=K+1
9240 IF K>N THEN 9250 ELSE 9170
9250 REM , START ANOTHER LOOP
9260 J=J+1
9270 GOTO 9060
9280 REM COMPUTE ELDIS
9290 ELDIS(N)=F1(N)
9300 J=1
9310 NN=N-J
9320 ELDIS(NN)=F1(NN)
9330 FOR I=NN TO N-1
9340 ELDIS(NN)=ELDIS(NN)-SK(NN,I+1)*ELDIS(I+1)
9350 NEXT I
9360 J=J+1
9370 IF NN=1 THEN 9380 ELSE 9310
9380 REM SUBROUTINE FINISHED
9390 RETURN 3300
9500 REM MATRIX MULTIPLICATION
9510 FOR IM=1 TO LM
9520 FOR JM=1 TO NM
9530 CM(IM,JM)=0
9540 NEXT JM
9550 NEXT IM
9560 FOR IM=1 TO LM
9570 FOR JM=1 TO NM
9580 FOR KM=1 TO MM
9590 CM(IM,JM)=CM(IM,JM)+AM(IM,KM)*BM(KM,JM)
9600 NEXT KM
9610 NEXT JM
9620 NEXT IM
9630 RETURN
```

```
9800 REM GOSUB FOR 4 NODE ELEMENT
9810 P(1)=(1-G-H+G*H)/4
9820 P(2)=(1+G-H-G*H)/4
9830 P(3)=(1+G+H+G*H)/4
9840 P(4)=(1-G+H-G*H)/4
9850 DEL(1,1)=(-1+H)/4
9860 DEL(1,2)=-DEL(1,1)
9870 DEL(1,3)=(1+H)/4
9880 DEL(1,4)=-DEL(1,3)
9890 DEL(2,1)=(-1+G)/4
9900 DEL(2,2)=(-1-G)/4
9910 DEL(2,3)=-DEL(2,2)
9920 DEL(2,4)=-DEL(2,1)
9930 RETURN
10000 REM GOSUB FOR TRANSFORMATION
10010 FOR I=1 TO 12
10020 FOR J=1 TO 12
10030 T(I,J)=0
10040 NEXT J
10050 NEXT I
10060 T(1,1)=1
10070 T(2,2)=1
10080 T(3,9)=1
10090 T(4,3)=1
10100 T(5,4)=1
10110 T(6,10)=1
10120 T(7,5)=1
10130 T(8,6)=1
10140 T(9,11)=1
10150 T(10,7)=1
10160 T(11,8)=1
10170 T(12,12)=1
10180 FOR I=1 TO 12
10190 FOR J=1 TO 12
10200 TT(I,J)=T(J,I)
10210 NEXT J
10220 NEXT I
10230 RETURN 306
12000 DATA 10,4,4,13,3,1,6
12010 DATA 1,0,0,2,0,1,3,.25,0,4,.25,1
12020 DATA 5,.5,0,6,.5,1,7,.75,0
12030 DATA 8,.75,1,9,1,0,10,1,1
12040 DATA 1,1,3,4,2,1
12050 DATA 2,3,5,6,4,1
12060 DATA 3,5,7,8,6,1
12070 DATA 4,7,9,10,8,1
12080 DATA 1,1,0,1,1,1,0
12090 DATA 1,1,0,2,1,0,1,2,0
12100 DATA 9,1,0,10,1,0
12110 DATA 1,3,100,2,3,100,9,3,0,10,3,0
12120 DATA 3,2,0,5,2,0,7,2,0,9,2,0
12130 DATA 10,0,0,0
```

Bibliography

Chapter 1

Frederick, D., and Chang, T. S. (1965). *Continuum Mechanics*, Allyn and Bacon, Boston.

Spiegel, M. R. (1959). *Schaum's Outline of Vector Analysis*, McGraw-Hill, New York.

Chapter 2

Becker, E. B., Carey, G. F., and Oden, J. T. (1981). *Finite Elements*, vol. 1, Prentice-Hall, Englewood Cliffs, N.J.

Burnett, D. S. (1987). *Finite Element Analysis*, Addison-Wesley, Reading, Mass.

Reddy, J. N. (1985). *Applied Functional Analysis and Variational Methods in Engineering*, McGraw-Hill, New York.

Segerlind, L. J. (1984). *Applied Finite Element Analysis, 2d ed.*, John Wiley, New York.

Chapter 3

Amend, J. H., Contractor, D. N., and Desai, C. S. (1976). "Oxygen Depletion and Sulfate Production in Strip Mine Spoil Dams," *Numerical Methods in Geomechanics*, vol. II, C. S. Desai (ed.), ASCE, New York, pp. 1155–1167.

Chou, P. C., and Pagano, N. J. (1967). *Elasticity*, Van Nostrand, Princeton, N.J.

Dhatt, G., and Touzot, G. (1984). *The Finite Element Method Displayed*, Wiley-Interscience, New York.

Segerlind, L. J. (1984). *Applied Finite Element Analysis, 2d ed.*, John Wiley, New York.

Stasa, F. L. (1985). *Applied Finite Element Analysis for Engineers*, Holt, Rinehart and Winston, New York.

Timoshenko, S., and Goodier, J. N. (1951). *Theory of Elasticity*, 2d ed., McGraw-Hill, New York.

Viessman, W., Jr., Lewis, G. L., and Knapp, J. W. (1989). *Introduction to Hydrology*, 3d ed., Harper and Row, New York.

Wylie, C. R., Jr. (1960). *Advanced Engineering Mathematics*, 2d ed., McGraw-Hill, New York.

Chapter 4

Bickford, W. B. (1990). *A First Course in the Finite Element Method*, Irwin, Boston.

Logan, D. L. (1986). *A First Course in the Finite Element Method*, PWS-Kent, Boston.

Scheid, F. (1988). *Schaum's Outline of Numerical Analysis*, 2d ed., McGraw-Hill, New York.

Chapter 5

Aifantis, E. C. (1979). "Continuum Basis for Diffusion in Regions with Multiple Diffusivity," *J. Appl. Phys.*, vol. 50, pp. 1334–1338.

Burnett, D. S. (1987). *Finite Element Analysis*, Addison-Wesley, Reading, Mass.

Gurtin, M. E. (1963). "Variational Principles in the Linear Theory of Viscoelasticity," *Arch. Rat. Mech. Anal.*, vol. 13, pp. 179–191.

Gurtin, M. E., and Yatomi, C. (1979). "On a Model for Two Phase Diffusion in Composite Materials," *J. Compos. Mater.*, vol. 13, pp. 126–130.

Nickell, R. E., and Sackman, J. L. (1968). "Variational Principles for Linear Coupled Thermoelasticity," *Q. J. Appl. Math.*, vol. 26, pp. 11–26.

Nye, J. F. (1957). *Physical Properties of Crystals*, Oxford University Press, London.

Oden, J. T., and Reddy, J. N. (1976). *Variational Methods in Theoretical Mechanics*, Springer-Verlag, Berlin.

Reddy, J. N. (1975). "A Note on Mixed Variational Principles for Initial-Value Problems," *Q. J. Mech. Appl. Math.*, vol. 28, pp. 123–132.

Sandhu, R. B., and Pister, K. S. (1970). "A Variational Principle for Linear, Coupled Field Problems in Continuum Mechanics," *Int. J. Eng. Sci.*, vol. 8, pp. 989–999.

Sandhu, R. B., and Pister, K. S. (1971). "Variational Principles for Boundary Value and Initial-Boundary Value Problems in Continuum Mechanics," *Int. J. Solids Struct.*, vol. 7, pp. 639–654.

Chapter 6

Burnett, D. S. (1987). *Finite Element Analysis*, Addison-Wesley, Reading, Mass.

Scheid, F. (1988). *Schaum's Outline of Numerical Analysis*, 2d ed., McGraw-Hill, New York.

Segerlind, L. J. (1984). *Applied Finite Element Analysis, 2d ed.*, John Wiley, New York.

Stasa, F. L. (1985). *Applied Finite Element Analysis for Engineers*, Holt, Rinehart and Winston, New York.

Zienkiewicz, O. C., and Taylor, R. L. (1989). *The Finite Element Method*, vol. 1, 4th ed., McGraw-Hill, New York.

Chapter 7

Bickford, W. B. (1990). *A First Course in the Finite Element Method*, Irwin, Boston.

Burnett, D. S. (1987). *Finite Element Analysis*, Addison-Wesley, Reading, Mass.

Chajes, A. (1974). *Principles of Structural Stability Theory*, Prentice-Hall, Englewood Cliffs, N.J.

Cheng, M. F. (1988). *Finite Element Analysis of Finite Axisymmetric Piezoelectric Cylinders*, Thesis, Tennessee Technological University, Cookeville, Tenn.

Cook, R. D., Malkus, D. S., and Plesha, M. E. (1989). *Concepts and Applications of Finite Element Analysis*, 3d ed., John Wiley, New York.

Gebhart, B. (1961). *Heat Transfer*, McGraw-Hill, New York.

Gurtin, M. E., and Yatomi, C. (1979). "On a Model for Two Phase Diffusion in Composite Materials," *J. Compos. Mater.*, vol. 13, pp. 126–130.

Hughes, T. J. R., and Cohen, M. (1978). "The 'Heterosis' Finite Element for Plate Bending," *Comput. Struct.*, vol. 9, pp. 445–450.

Hughes, T. J. R. (1987). *The Finite Element Method*, Prentice-Hall, Englewood Cliffs, N.J.

Hutchinson, J. R. (1967). "Axisymmetric Vibrations of a Solid Elastic Cylinder Encased in a Rigid Container," *J. Acoust. Soc. Am.*, vol. 42, pp. 398–402.

Kardomateas, G. A. (1989). "Transient Thermal Stress in Cylindrically Orthotropic Composite Tubes," *J. Appl. Mech.*, vol. 56, pp. 411–417.

Meirovitch, L. (1967). *Analytical Methods in Vibrations*, MacMillan, London.

Stasa, F. L. (1985). *Applied Finite Element Analysis for Engineers*, Holt, Rinehart and Winston, New York.

Thomson, W. T. (1965). *Vibration Theory and Applications*, Prentice-Hall, Englewood Cliffs, N.J.

Timoshenko, S.P. (1955). *Vibration Problems in Engineering*, 3d ed., Van Nostrand, Princeton, N.J.

Timoshenko, S. P., and Woinowsky-Krieger, S. (1959). *Theory of Plates and Shells*, 2d ed., McGraw-Hill, New York.

Zienkiewicz, O. C., and Taylor, R. L. (1989). *The Finite Element Method*, vol. 1, 4th ed., McGraw-Hill, New York.

Solved-Problem Index of Applications

Note: The numbers following these entries refer to problem numbers.

Index